SIMULATION-BASED LEAN SIX-SIGMA AND DESIGN FOR SIX-SIGMA

SIMULATION-BASED LEAN SIX-SIGMA AND DESIGN FOR SIX-SIGMA

BASEM EL-HAIK
RAID AL-AOMAR

WILEY-INTERSCIENCE

A JOHN WILEY & SONS, INC., PUBLICATION

Library of Congress Cataloging-in-Publication Data:

El-Haik, Basem.
    Simulation-based lean six-sigma and design for six-sigma / by Basem El-Haik,
Raid Al-Aomar.
      p. cm.
    "A Wiley-Interscience publication."
    Includes bibliographical references and index.
    ISBN-13: 978-0-471-69490-8
    ISBN-10: 0-471-69490-8
    1. Six sigma (Quality control standard) 2. Total quality management. 3. Production
engineering. I. Al-Aomar, Raid. II. Title.
    TS156.E383 2006
    658.5'62–dc22
                                         2006048112

Printed in the United States of America

10 9 8 7 6 5 4 3 2 1

To our parents, families, and friends for their continuous support

Basem and Raid
Fall 2006

CONTENTS

PREFACE

Simulation modeling within the context of six-sigma and design for six-sigma (DFSS) methods is constantly getting more attentions from black belts, green belts, and other six-sigma deployment operatives, process engineers, lean experts, and academics all over the world. This trend can easily be seen by the increasing use of simulation tools in several successful six-sigma initiatives in many Fortune 500 companies, coupled with the tremendous development in simulation software tools and applications.

For a six-sigma project, conducting experimental design and "what-if" analysis is a common key practice toward achieving significant results. Simulation models can be utilized effectively as a flexible platform for six-sigma and DFSS experimentation and analyses, which reduces the time and cost of physical experimentation and provides a visual method to validate tested scenarios. On the other hand, simulation studies often suffer from the unavailability of accurate input data and lack of a structured approach for conducting analysis. The proven and widely used six-sigma and DFSS approaches provide the simulation study with reliable simulation data as input, an accurate process map, and integrates the simulation process with a state-of-the-art set of process analyses. Hence, coupling simulation modeling with a well-structured six-sigma process compensates for such limitations and bridges the gap between modeling and process engineering. Such integration provides the synergy and infrastructure essential for successful problem solving and continuous improvement in a wide spectrum of manufacturing and service applications. To develop an appreciation for the simulation-based six-sigma methodology, the subject of this book, we first review both six-sigma and simulation approaches and then lay the background for their integration.

SIX-SIGMA DEFINED

The success of six-sigma deployment in many industries has generated enormous interest in the business world. In demonstrating such successes, six-sigma combines the power of teams and process. The power of teams implies organizational support and trained teams tackling objectives. The power of process means effective six-sigma methodology deployment, risk mitigation, project management, and an array of statistical and system thinking methods. Six-sigma focuses on the *whole quality* of a business, which includes *product* or *service quality* to external customers and *operational quality* of all internal processes, such as accounting and billing. A whole quality business with whole quality perspectives will not only provide high-quality products or services, but will also operate at lower cost and higher efficiency, because all the business processes are optimized.

In basic terms, six-sigma is a disciplined program or methodology for improving quality in all aspects of a company's products and services. In this book we adopt the DMAIC process over its five phases: define, measure, analysis, improve, and control. DMAIC represents that latest step in the evolution of the total quality management movement begun by W. Edwards Deming in the 1950s. The six-sigma initiative is credited to Mikel Harry, a statistician who is cofounder and a principal member of the Six Sigma Academy in Scottsdale, Arizona. Early corporate adopters of the program include Motorola in the 1980s, and other technology-based firms, such as General Electric, Texas Instruments, and Allied Signal.

The central theme of six-sigma is that product and process quality can be improved dramatically by understanding the relationships between the inputs to a product or process and the metrics that define the quality level of the product or process. Critical to these relationships is the concept of the voice of the customer: that quality can be defined only by the customer who will ultimately receive the outputs or benefits of a product or process. In mathematical terms, six-sigma seeks to define an array of transfer functions of the critical-to-quality characteristics (CTQs) or Y's, where $Y = f(X_1, X_2, \ldots, X_n)$ between the quality metrics of a product or process (e.g., $Y =$ the % on-time delivery for a fulfillment process) and the inputs that define and control the product or process (e.g., $X_1 =$ the number of resources available to service customers). The focus of six-sigma, then, is twofold: (1) to understand which inputs (X's) have the greatest effect on the output metrics (Y's), and (2) to control those inputs so that the outputs remain within a specified upper and/or lower specification limit.

In statistical terms, six-sigma quality means that for any given product or process quality measurement, there will be no more than 3.4 defects produced per 1 million opportunities (assuming a normal distribution). An *opportunity* is defined as any chance for nonconformance, not meeting the required specifications. The goal of six-sigma is both to center the process and to reduce the variation such that all observations of a CTQ measure are within the upper and lower limits of the specifications.

Defect-correction six-sigma methodologies are highlighted in the DMAIC process; DFSS, a proactive methodology, is characterized by its four phases: identify, characterize, optimize, and verify. The DMAIC six-sigma objective is to improve a process without redesigning the current process. DFSS puts the focus on design by doing things right the first time. The ultimate goal of DFSS is whole quality: do the right things, and do things right all the time, to achieve absolute excellence in design whether it is a service process facing a customer or an internal business process facing an employee. Superior design will deliver superior functions to generate great customer satisfaction. A DFSS entity will generate a process that delivers a service or product in the most efficient, economical, and flexible manner. Superior process design will generate a service process that exceeds customer wants and delivers these with quality and at low cost. Superior business process design will generate the most efficient, effective, economical, and flexible business process. This is what we mean by *whole quality*.

To ensure success, a six-sigma initiative must receive complete buy-in and continuous support from the highest level of a company's leadership team. In addition, a rigorous training program and dedicated staff positions will require the best and brightest minds that can be allocated to the initiative.

SIMULATION DEFINED

Simulation, in general, is a disciplined process of building a model of an existing or proposed real system and performing experiments with this model to analyze and understand the behavior of selected characteristics of a real system so as to evaluate various operational strategies to manage the real system. In abstract terms, simulation is used to describe the behavior of physical and business systems, to construct hypotheses or theories to explain behaviors, to predict the behavior of future systems, and to perform what-if scenarios for existing and proposed design alternatives.

In this book we are concerned only with stimulating processes of *transactional* nature where events can be discretely isolated and described over time. Such simulation is called *discrete event simulation* (DES). There are many benefits for simulation in a six-sigma context. Specifically, DES simulation can be used to:

- Study DMAIC/DFSS solutions before piloting or modification
- Manage risk by preventing defects and reducing costs
- Prevent or eliminate unforeseen barriers, bottlenecks, and design flaws
- Allocate necessary resources to assure stakeholder satisfaction, including customers

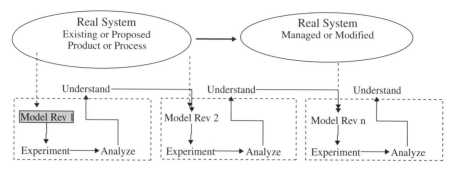

Figure P.1   Simulation high-level process.

- Build predictable, controllable, and optimal designs and solutions
- Facilitate communication among multifunctional team members

Discrete event simulation has to be designed and performed in a systematic fashion (Part II of the book). A model is an approximation of part of a real system. Model building is an important step, necessary but not sufficient. That is, a model is not simulation in itself. Simulation is a process. It is impossible to replicate all of reality in a simulation process. Only a handful of selected characteristics can be adequately modeled and analyzed to provide meaningful results. This is usually carried over several revisions involving experimentation, modification, analysis, and interpretation, as shown in Figure P.1.

Simulation can be used throughout the DFSS process, but especially in the characterize and optimize phases. In these phases, simulation is used to:

- Determine if proposed solutions meet CTQs and functional requirements
- Evaluate how high-level design (structure) and inputs will affect overall process output and performance
- Compare the performance of high-level design alternatives
- Assemble the behavior of subelements to predict overall performance
- Perform sensitivity analysis to study the importance of subelements
- Redesign the product and service as needed
- Assess the risk of failure of subelements to provide system integrity
- Compare the performance of detailed design alternatives

SIMULATION-BASED SIX-SIGMA

Over the last two decades, there has been a significant penetration of problem solving and continuous improvement using simulation modeling and six-sigma initiatives. Those approaches include primarily the design of experiment method (DOE), Taguchi methods, and simulation-based optimization methods. DFSS is a very important element of a full six-sigma deployment program, since design is the one of the important activities in which quality, cost, and cycle time can be greatly improved if the right strategy and methodologies are used. Major corporations are training many design engineers and project managers to become six-sigma green belts, black belts, and master black belts, enabling them to play a leadership role in corporate operation and process excellence. DES has become common among engineers, designers, and business planners. With the aid of DES, companies have been able to design efficient production and business systems, validate and trade off proposed design solution alternatives, troubleshoot potential problems, and improve system performance metrics. That aid usually enables companies to cut cost and meet targets while boosting sales and profit.

OBJECTIVES OF THE BOOK

The objectives of the book are:

1. To provide a good reference for six-sigma and simulation processes in the same book.
2. To provide in-depth and clear coverage of simulation-based six-sigma and our terminology for six-sigma and simulation integration (Part III).
3. To illustrate clearly complete road maps for both simulation-based lean six-sigma and simulation-based DFSS projects.
4. To present the know-how for all of the principal methods used in simulation-based six-sigma approaches and to discuss the fundamentals and background of each process clearly (Parts I and II). Case studies are provided with detailed step-by-step implementation process of each method described in Part III.
5. To help develop readers' practical skills in applying simulation-based six-sigma in transactional environments (i.e., project execution and operational and technical aspects).

BACKGROUND NEEDED

The background required to study this book is some familiarity with simple statistical methods, such as normal distribution, mean, and variance, and with simple data analysis techniques. A background in DES theory is also helpful.

SUMMARY OF CHAPTER CONTENTS

The book is organized into three parts:

- Part I, Chapters 1, 2, and 3, has six-sigma fundamentals as its theme.
- Part II, Chapters 4, 5, 6, and 7, covers simulation fundamentals.
- Part III, begins in Chapters 8, 9, and 10 by focusing on simulation-based lean six-sigma (3S-LSS) and design for six-sigma (3S-DFSS) and provides several case studies. In Chapter 11 we present elements of successful deployment of 3S methods.

Following is a summary of the chapters.

Part I

In Chapter 1 we introduce the concepts of *process* and *customer satisfaction* and highlight how customers experience a product or a service as well as its delivery process. This chapter concentrates on the service side and defines various elements of a generic process as well as tools such as process mapping. We also demystify the concept of *transaction* and event discreteness. In this chapter we explain what six-sigma is and how it has evolved over time. We explain that it is a process-based methodology and introduce the reader to process modeling with a high-level overview of process mapping, value stream mapping and value analysis as well as business process management systems. The criticality and application of measurement systems analysis is introduced. The DMAIC methodology and how it incorporates these concepts into a road map method is also explained.

Chapter 2 covers the lean six sigma (LSS) concept and discusses topics related to the integration of six-sigma and lean manufacturing. The focus in this chapter is on the details of the LSS approach, the enhancements made to six-sigma DMAIC tollgates, lean manufacturing concepts and aims, value stream mapping, and lean manufacturing techniques. The chapter also highlights the synergy and benefits of implementing an LSS system as a foundation for the proposed 3S-LSS approach.

Chapter 3 is an introduction to a high-level DFSS process. The DFSS approach introduced helps design teams frame their project with financial, cultural, and strategic implications to the business. In this chapter we form and integrate several strategic, tactical, and synergistic methodologies to enhance service DFSS capabilities and to deliver a broad set of optimized solutions. We highlights and present the service DFSS phases: identify, characterize, optimize, and verify.

Part II

In Chapter 4 we introduce the basic concepts of simulation modeling with a focus on process modeling and time-based performance measurement. We also clarify the role of simulation studies in serving the increasing needs of companies that seek continuous improvement and optimality in production and business processes. To this end, we provide an introduction to the concept, terminology, and types of models, along with a description of simulation taxonomy and a justification for utilizing simulation tools in a variety of real-world applications. Such a background is essential to establishing a basic understanding of what simulation is all about and to understanding the key simulation role in simulation-based six-sigma studies and applications.

In Chapter 5 we present the details and mechanics of DES that are essential to providing a flexible platform for use of a simulation-based six-sigma approach: elements of system modeling, events activation, random number generation, time advancement, animation, and accumulating statistics. In addition to powerful DES mechanics, we address how fast computations on today's high-speed processors, along with the growing graphics capability, contribute to the effectiveness and visualization capability of DES models. We thus provide a deeper understanding of the DES process, components, and mechanisms. Examples of manual and computer simulations are used to clarify the DES functionality.

Chapter 6 is focused on analyzing the various aspects of the simulation process and the set of techniques and steps followed when conducting a simulation study. The elements of the simulation process discussed include project scoping, conceptual modeling, data collection, model building, model analyses, and documentation. The chapter also highlights the linkage of these simulation practices to the overall scope of a six-sigma project, to the simulation software used, and to the process followed by a six-sigma team in carrying out a complete simulation study. Finally, we present the implications of simulation on six-sigma applications to system design, problem solving, and continuous improvement.

In Chapter 7 we discuss the main issues involved in analyzing simulation outputs: distinguishing between terminating and steady-state simulation, understanding the stochastic nature of simulation outcomes, determining simulation run controls (i.e., warm-up period, run length, and number of replications), and selecting the appropriate model output analysis method. We also discuss the main methods of output analysis, including statistical analysis, experimental design, and optimization.

Part III

Chapter 8 contains all the project road maps for the 3S and 3S-LSS methodologies. The DMAIC process and lean principles and concepts are integrated into a synergetic road map in a 3S-LSS environment. In Chapter 8 we also present a 3S-DFSS project road map, which highlights at a high level the identify, charcaterize, optimize, and validate phases over the seven development stages: idea creation, voice of the customer and business, concept development, preliminary design, design optimization, verification, and launch readiness. In this chapter the concept of tollgate is introduced. We also high-light the most appropriate DFSS tools and methods by the DFSS phase, indi-cating where it is most appropriate to start tool use. The road maps also recognize the concept of *tollgates*, design milestones where DFSS teams update stakeholders on development and ask that a decision be made as to whether to approve going to the next stage, recycling back to an earlier stage, or canceling the project altogether.

In Chapter 9 we discuss the details of 3S-LSS application to real-world systems using the road map presented in Chapter 8. The focus is on the prac-tical aspects of 3S-LSS use in reengineering transactional processes in both services and manufacturing. These aspects are discussed through an appli-cation of the 3S-LSS approach to a manufacturing case study. Emphasis is placed on utilizing simulation-based application of lean techniques and six-sigma analysis to improve a set of process CTQs, defined in terms of time-based performance to take advantage of simulation modeling for a system-level application of six-sigma statistical tools used in DMAIC and lean techniques.

In Chapter 10 we develop a 3S-DFSS clinic case study using the road map in Chapter 8. We show the application of several tools as they span over the project road map. The case study also highlights most appropriate DFSS tools. It indicates where it is most appropriate to start tool use such as transfer func-tions and quality function deployment. Following the DFSS road map helps accelerate the process introduction and aligns benefits for customers and stakeholders.

Chapter 11 is a practical guide to successful 3S-DFSS and 3S-LSS devel-opment, deployment, and project execution. A generic framework for suc-cessful project management and development is proposed based on the authors' experience. This framework demystifies the ambiguity involved in project selection and specifies a method for project development. We provide a practical guide for successful development of 3S projects based on the road maps discussed in Chapter 8. The guide begins by discussing the unique char-acteristics of 3S projects.

WHAT DISTINGUISHES THIS BOOK FROM OTHERS

Several six-sigma books are available on the market for both students and practitioners, as are several simulation books. We believe that none of these books integrates both methodologies to achieve the benefits outlined herein. This book constitutes an integrated problem-solving and continuous improve-ment approach based on six-sigma thinking, tools, and philosophy, together with simulation modeling flexibility and visualization. The book also includes a review of six-sigma process fundamentals (Part I), a detailed description of simulation process fundamentals (Part II), and a presentation of simulation-based six-sigma methodology (Part III).

The uniqueness of the book lies in bringing six-sigma and simulation modeling processes under the umbrella of problem solving and continuous improvement in product and service development, management, and planning. The book will not only be helpful to simulation professionals, but will also help six-sigma operatives, design engineers, project engineers, and middle-level managers to gain fundamental knowledge of six-sigma and simulation. After reading this book, readers will have a round grasp of the body of knowledge in both areas.

The book is the first book to cover completely all of "the body of knowledge of six-sigma and design for six sigma with simulation methods" outlined by the American Society for Quality including process mapping, the design of experiment method, quality function deployment, and failure mode effect analysis, among many other methods.

In the book, both simulation and contemporary six-sigma and DFSS methods are explained in detail together with practical case studies that help describe the key features of simulation-based methods. The systems approach to designing products and services, as well as problem solving, is integrated into the methods discussed. Readers are given the project life-cycle know-how for a project so that the method is understandable not only from a usability perspective (how it is used) but also from an implementation perspective (when it is used).

This book is the best way to learn about all the useful tools in six-sigma and DFSS, because it will be the only book that covers the body of knowledge used in a simulation environment. Six-sigma now dominates business decision making at all major U.S. and many international companies for both business survival and total quality improvement. The book supplements traditional techniques with emerging methods.

ACKNOWLEDGMENTS

In preparing this book we received advice and encouragement from several people. For this we thank Pearse Johnston, Mike O'Ship, Joe Smith, Onur Ulgen, and Steve Beeler. We also appreciate the help of many others, especially the faculty and students of the industrial engineering department at Jordan University of Science & Technology. We are also very grateful for the assistance of George Telecki and Rachel Witmer of John Wiley & Sons, Inc.

CONTACTING THE AUTHORS

Your comments and suggestions related to this book will be greatly appreciated and we will give them serious consideration for inclusion in future editions. We also conduct public and in-house six-sigma, DFSS, and simulation-based six-sigma workshops and provide consulting services.

Dr. Basem El-Haik can be reached by e-mail at:

basemhaik@Six-SigmaPro.com or basemhaik@gmail.com
(734) 765-5229

Dr. Raid Al-Aomar can be reached by e-mail at Jordan University of Science & Technology:

ralaomar@just.edu.jo

PART I

SIX-SIGMA FUNDAMENTALS

1

SIX-SIGMA FUNDAMENTALS

1.1 INTRODUCTION

Throughout the evolution of quality control, there has always been a preponderance of manufacturing (parts) focus. In recent years more attention has been placed on process in general; however, the application of a full suite of tools to transaction-based industry is rare and still considered risky or challenging. Only companies that have mature six-sigma[1] deployment programs see the application of design for six sigma (DFSS) to processes as an investment rather than a needless expense. Even those companies that embark on DFSS for processes seem to struggle with confusion over DFSS "process" and the process being designed.

Many business processes can benefit from DFSS, some of which are listed in Table 1.1. If measured properly, we would find that few if any of these processes perform at six sigma performance levels. The cost per transaction, timeliness, or quality (accuracy, completeness) are never where they should be and hardly world class.

---

[1] The word *sigma* refers to the lowercase Greek letter σ, a symbol used by statisticians to measure variability. As the numerical values of σ increase, the number of defects in a process falls exponentially. Six-sigma design is the ultimate goal since it means that if the same task is performed 1 million times, there will be only 3.4 defects, assuming normality.

Simulation-Based Lean Six-Sigma and Design for Six-Sigma, by Basem El-Haik and Raid Al-Aomar
Copyright © 2006 John Wiley & Sons, Inc.

TABLE 1.1   Examples of Organizational Functions

Marketing	Sales	Human Resources	Design
Brand management	Discovery	Staffing	Change control
Prospect	Account management	Training	New product
Production Control	**Sourcing**	**Information Technology**	**Finance**
Inventory control	Commodity	Help desk	Accounts payable
Scheduling	Purchasing	Training	Accounts receivable

A *transaction* is typically something that we create to serve a paying customer. Transactions are event-based and discrete in nature. (See Chapter 5 for more details.) Customers may be internal or external; if external, the term *consumer* (or *end user*) will be used for clarification purposes. Some services (e.g., dry cleaning) consist of a single process, whereas many services consist of several processes linked together. At each process, transactions occur. A transaction is the simplest process step and typically consists of an *input*, *procedures*, *resources*, and a resulting *output*. The resources can be people or machines, and the procedures can be written, learned, or even digitized in software code. It is important to understand that some services are enablers of other services, while others provide their output to the end customer. For example, transactions centered around the principal activities of an order-entry environment include transactions such as entering and delivering orders, recording payments, checking the status of orders, and monitoring the stock levels at a warehouse. Processes may involve a mix of concurrent transactions of different types and complexity either executed online or queued for deferred execution.

We experience services spanning the range from ad hoc to designed. Our experience indicates that the vast majority of services are ad hoc and have no associated metrics; many consist solely of a person with a goal and objectives. These services have a large variation in their perceived quality and are very difficult to improve. It is akin to building a house on a poor foundation.

Services affect almost every aspect of our lives. There are services such as restaurants, health care, financial, transportation, entertainment, and hospitality, and they all have the same elements in common. Services and transaction-based processes can be modeled, analyzed, and improved using discrete event simulation (DES) – hence this book.

In this chapter we cover an overview of six-sigma and its development as well as the traditional deployment for process or product improvement and its components [DMAIC (define–measure–analyze–improve–control)]. The DMAIC platform is also discussed in Chapter 2, which complements this chapter and covers topics related to the integration of six-sigma and lean manufacturing. The focus in this chapter is on the fundamental concepts of six-sigma DMAIC methodology, value stream mapping and lean manufacturing techniques, and the synergy and benefits of implementing a lean six sigma (LSS) system. LSS represents the foundation for the proposed approach of 3S-LSS (simulation-based LSS). Chapter 9 provides a detailed case study of the 3S-LSS approach. We introduce design for six sigma in Chapter 3, together with a detailed simulation-based case study in Chapter 10.

1.2 QUALITY AND SIX-SIGMA DEFINED

We all use services and interact with processes every day. When was the last time you remember feeling really good about a transaction or service? What about the last poor service you received? It is usually easier to remember painful and dissatisfying experiences than it is to remember the good ones. One of the authors recalls sending a first-class registered letter that he could not be sure if it had been received yet eight business days later, so he called the postal service provider's toll-free number and had a very professional and caring experience. It is a shame the USPS couldn't perform the same level of service in delivering the letter. Actually, the letter had been delivered but the system failed to track it. So the question is: How do we measure quality for a transaction or service?

In a traditional manufacturing environment, conformance to specifications and delivery are the common items related to quality that are measured and tracked. Often, lots are rejected because they don't have the correct supporting documentation. Quality in manufacturing, then, is conforming product, delivered on time, having all of the supporting documentation. In services, quality is measured as conformance to expectations, availability, experience of the process, and people interacting with the service delivery.

If we look at Figure 1.1, we can see that customers experience three aspects of service: (1) the specific service or product has attributes such as availability, being what was wanted, and working; (2) the process through which the service is delivered can be ease of use or value added; (3) the people (or system) should be knowledgeable and friendly. To fulfill these needs there is a service life cycle to which we apply a quality operating system.

Six-sigma is a philosophy, a measure, and a methodology that provides businesses with a perspective and tools to achieve new levels of performance in both services and products. In six-sigma, the focus is on process improvement to increase capability and reduce variation. The few vital inputs are chosen from the entire system of controllable and noise variables,[2] and the focus of improvement is on controlling these inputs.

Six-sigma as a philosophy helps companies achieve very low defects per million opportunities over long-term exposure. Six-sigma as a measure gives us a statistical scale to measure progress and benchmark other

---

[2] Noise factors are factors that cannot be controlled or can be controlled at an unaffordable cost, such as randomness and variability factors.

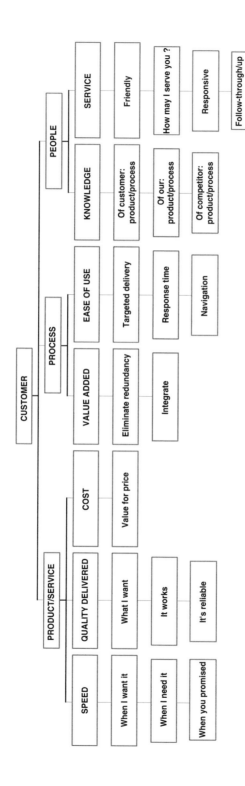

Figure 1.1 Customer experience channels.

6

companies, processes, or products. The defect-per-million opportunities (DPMO) measurement scale ranges from zero to 1 million, while the realistic sigma scale ranges from zero to six. The methodologies used in six-sigma, which we discuss in more detail in subsequent chapters, build on all of the tools that have evolved to date but put them into a data-driven framework. This framework of tools allows companies to achieve the lowest defects per million opportunities possible.

Six-sigma evolved from total quality management early efforts as discussed in El-Haik and Roy (2005). Motorola initiated the movement and it spread to Asea Brown Boveri, Texas Instruments Missile Division, and Allied Signal. It was at this juncture that Jack Welch became aware of the power of six-sigma from Larry Bossidy, and in the nature of a fast follower, committed General Electric to embracing the movement. It was GE that bridged the gap between a simple manufacturing process and product focus and what were first called *transactional processes*, and later, *commercial processes*. One of the reasons that Welch was so interested in this program was that an employee survey had just been completed and it revealed that the top-level managers of the company believed that GE had invented quality (after all, Armand Feigenbaum worked at GE); however, the vast majority of employees didn't think GE could spell "quality." Six-sigma turned out to be the methodology to accomplish Crosby's goal of zero defects. By understanding what the key process input variables are and that variation and shift can occur, we can create controls that maintain six-sigma (6σ) performance on any product or service and in any process. The Greek letter σ is used by statisticians to indicate the standard deviation (a statistical parameter) of the population of interest. Before we can clearly understand the process inputs and outputs, we need to understand process modeling.

1.3 INTRODUCTION TO PROCESS MODELING

Six-sigma is a process-focused approach to achieving new levels of performance throughout any business or organization. We need to focus on a process as a system of inputs, activities, and output(s) in order to provide a holistic approach to all the factors and the way they interact to create value or waste. When used in a productive manner, many products and services are also processes. An automated teller machine takes your account information, personal identification number, energy, and money and processes a transaction that dispenses funds or an account rebalance. A computer can take keystroke inputs, energy, and software to process bits into a word document. At the simplest level the process model can be represented by a process diagram, often called an IPO (input–process–output) diagram (Figure 1.2).

If we take the IPO concept and extend the ends to include the suppliers of the inputs and the customers of the outputs, we have the SIPOC (supplier–input–process–output–customer) (Figure 1.3). This is a very

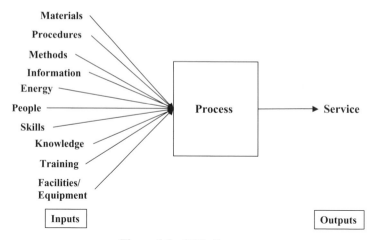

Figure 1.2   IPO diagram.

Suppliers	Inputs	Input Characteristic	Process	Outputs	Output Characteristic	Customers
			2a. What is the start of the process?			
7. Who are the suppliers of the inputs?	6. What are the inputs of the process?	8. What are the characteristics of the inputs?	1. What is the process?	3. What are the outputs of the process?	5. What are the characteristics of the outputs?	4. Who are the customers of the outputs?
			2b. What is the end of the process?			

Figure 1.3   SIPOC table.

effective tool in gathering information and modeling any process. A SIPOC tool can take the form of a table with a column for each category in the name.

1.3.1   Process Mapping

Whereas the SIPOC is a linear flow of steps, *process mapping* is a tool of displaying the relationship between process steps and allows for the display of various aspects of the process, including delays, decisions, measurements, and rework and decision loops. Process mapping builds on the SIPOC information by using standard symbols to depict varying aspects of the processes flow linked together with lines that include arrows demonstrating the direction of flow.

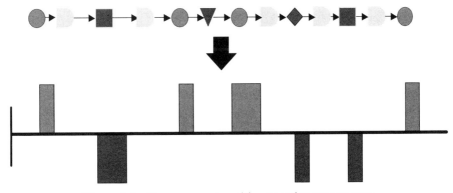

Figure 1.4 Process map transition to value stream map.

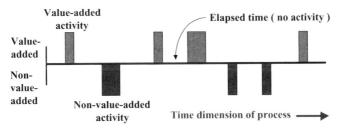

Figure 1.5 Value stream map definitions.

1.3.2 Value Stream Mapping

Process mapping can be used to develop a *value stream map* to understand how well a process is performing in terms of value and flow. Value stream maps can be performed at two levels. One can be applied directly to the process map by evaluating each step of the process map as value-added or non-value-added (Figures 1.4 and 1.5). This type of analysis has been in existence since at least the early 1980s; a good reference is the book *Hunters and the Hunted* (Swartz, 1996). This is effective if the design team is operating at a local level. However, if the design team is at more of an enterprise level and needs to be concerned about the flow of information as well as the flow of product or service, a higher-level value stream map is needed (Figure 1.6). This methodology is described best in Rother and Shook (2003).

1.4 INTRODUCTION TO BUSINESS PROCESS MANAGEMENT

Most processes are ad hoc: allow great flexibility to the people operating them. This, coupled with the lack of measurements of efficiency and effectiveness, result in the variation to which we have all become accustomed. In this case we use the term *efficiency* for the within-process step performance (often

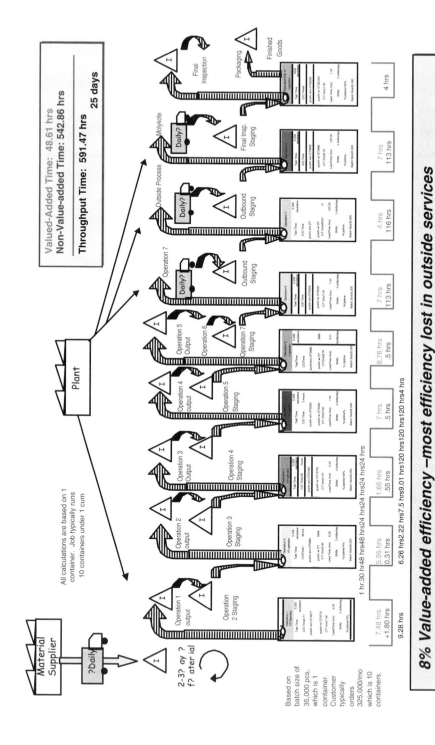

Figure 1.6 High-level value stream map example.

called the *voice of the process*). *Effectiveness* is how all of the process steps interact to perform as a system (often called the *voice of the customer*). This variation is difficult to address, due to the lack of measures that allow traceability to its cause. Transactional businesses that have embarked on six-sigma programs have learned that they have to develop process management systems and implement them in order to establish baselines from which to improve. The deployment of a business process management system (BPMS) often results in a marked improvement in performance as viewed by the customer and associates involved in the process. The benefits of implementing BPMS are magnified in cross-functional processes.

1.5 MEASUREMENT SYSTEMS ANALYSIS

Now that we have some form of documented process from among the choices – IPO, SIPOC, process map, value stream map, or BPMS – we can begin our analysis of what to fix, what to enhance, and what to design. Before we can focus on what to improve and how much to improve, we must be certain of our measurement system. Measurements can range from benchmarking to operationalization. We must determine how accurate the measurement system is in relation to a known standard. How repeatable is the measurement? How reproducible? Many service measures are the result of calculations. When performed manually, the reproducibility and repeatability can astonish you if you take the time to perform a measurement system analysis.

For example, in a supply chain we might be interested in promises kept, on-time delivery, order completeness, deflation, lead time, and acquisition cost. Many of these measures require an operational definition to provide repeatable and reproducible measures. Referring to Figure 1.7, is on-time delivery the same as on-time shipment? Many companies do not have visibility as to when a client takes delivery or processes a receipt transaction, so how do we measure these? Is it when the item arrives, when the paperwork has been completed or when the customer can actually use the item? We have seen a customer drop a supplier for a component 0.5% lower in cost only to discover that the new multiyear contract that they signed did not include transportation and cost them 3.5% more per year.

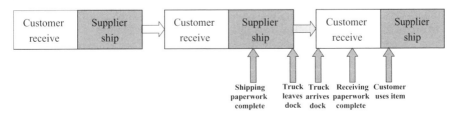

Figure 1.7   Supplier-to-customer cycle.

The majority of measures in a service or process will focus on:

- Speed
- Cost
- Quality
- Efficiency, defined as first-pass yield of a process step
- Effectiveness, as defined as the rolled throughput yield of all process steps

All of these can be made robust at a six-sigma level by creating operational definitions, defining the start and stop points, and determining sound methodologies for assessing. "If you can't measure it, you can't improve it" is a statement worth remembering, as is ensuring that adequate measurement sytems are available throughout a project life cycle.

1.6   PROCESS CAPABILITY AND SIX-SIGMA PROCESS PERFORMANCE

We determine process capability by measuring a process's performance and comparing it to the customer's needs (specifications). Process performance may not be constant and usually exhibits some form of variability. For example, we may have an accounts payable (A/P) process that has measures of accuracy and timeliness. For the first two months of a quarter, the process has few errors and is timely, but at the quarter point, the demand goes up and the A/P process exhibits more delays and errors. If the process performance is measurable in real numbers (continuous) rather than pass or fail (discrete) categories, the process variability can be modeled with a normal distribution. The normal distribution is generally used because of its robustness in modeling many real-world performance random variables. The normal distribution has two parameters: quantifying the central tendency and variation. The center is the average performance, and the degree of variation is expressed by the standard deviation. If the process cannot be measured in real numbers, we convert the pass/fail, good/bad (discrete) into a yield and then convert the yield into a sigma value. Several transformations from discrete distributions to continuous distribution can be borrowed from mathematical statistics.

If the process follows a normal probability distribution, 99.73% of the values will fall between the $\pm 3\sigma$ limits, where σ is the standard deviation and only 0.27% will be outside the $\pm 3\sigma$ limits. Since the process limits extend from -3σ to $+3\sigma$, the total spread amounts to 6σ total variation. This total spread is the process spread and is used to measure the range of process variability.

For any process performance metric there are usually some performance specification limits. These limits may be single-sided or two-sided. For the A/P process, the specification limit may be no less than 95% accuracy. For the receipt of material into a plant, it may be 2 days early and 0 days late. For a

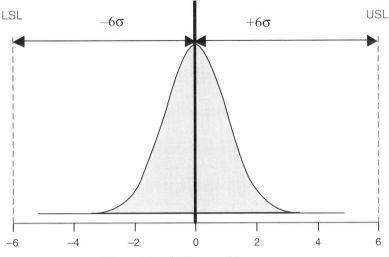

Figure 1.8 Highly capable process.

call center we may want the phone conversation to take between 2 and 4 minutes. For each of the last two double-sided specifications, they can also be stated as a target and a tolerance. The material receipt could be 1 day early ±1 day, and for the phone conversation it could be 3 minutes ± 1 minute.

If we compare the process spread with the specification spread, we can usually observe three conditions:

1. *Highly capable process* (Figure 1.8). The process spread is well within the specification spread:

$$6\sigma < USL - LSL$$
$$USL = \text{Upper Spec. Limit}; \quad LSL = \text{Lower Spec. Limit}$$

   The process is capable because it is extremely unlikely that it will yield unacceptable performance.

2. *Marginally capable process* (Figure 1.9). The process spread is approximately equal to the specification spread:

$$6\sigma = USL - LSL$$

When a process spread is nearly equal to the specification spread, the process is capable of meeting the specifications. If we remember that the process center is likely to shift from one side to the other, an amount of the output will fall outside the specification limit and will yield unacceptable performance.

3. *Incapable process* (Figure 1.10). The process spread is greater than the specification spread.

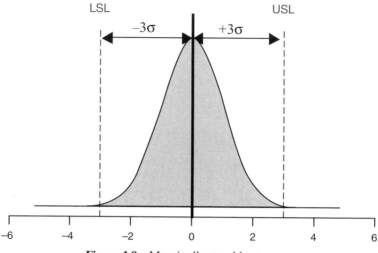

Figure 1.9   Marginally capable process.

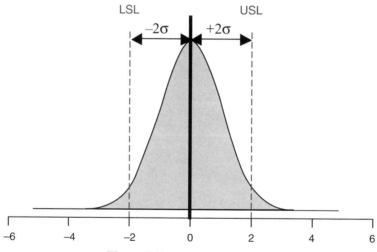

Figure 1.10   Incapable process.

$$6\sigma > USL - LSL$$

When a process spread is greater than the specification spread, the process is incapable of meeting the specifications, and a significant amount of the output will fall outside the specification limit and will yield unacceptable performance.

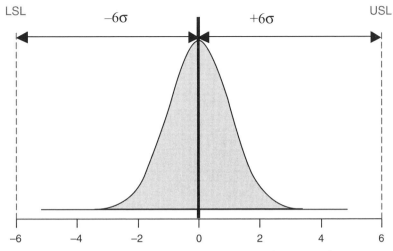

Figure 1.11   Six-sigma capable process (short-term).

1.6.1   Motorola's Six-Sigma Quality

In 1986, the Motorola Corporation won the Malcolm Baldrige National Quality Award. Motorola based its success in quality on its six-sigma program. The goal of the program was to reduce the variation in every process such that a spread of 12σ (6σ on each side of the average) fits within the process specification limits (Figure 1.11).

Motorola accounted for the process average to shift side to side over time. In this situation one side shrinks to a 4.5σ gap and the other side grows to 7.5σ (Figure 1.12). This shift accounts for 3.4 parts per million (ppm) on the small gap and a fraction of parts per billion on the large gap. So over the long term a 6σ process will generate only a 3.4-ppm defect.

To achieve six-sigma capability, it is desirable to have the process average centered within the specification window and to have the process spread over approximately one-half of the specification window. There are two approaches to accomplishing six-sigma levels of performance. When dealing with an existing process there is the process improvement method known as DMAIC, and if there is a need for a new process, it is design for six-sigma (DFSS). Both of these are discussed in the following sections.

1.7   OVERVIEW OF SIX-SIGMA IMPROVEMENT: DMAIC

Applying six-sigma methodology to improve an existing process or product follows a five-phase process:

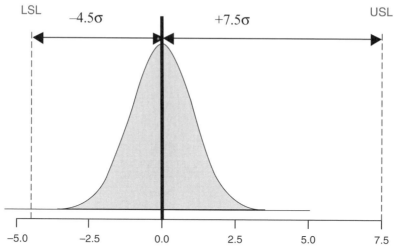

Figure 1.12   Six-sigma-capable process with long-term shift.

1. *Define.* Define the opportunity and customer requirements.
2. *Measure.* Ensure adequate measures, process stability, and initial capability.
3. *Analyze.* Analyze the data and discover the critical inputs and other factors.
4. *Improve.* Improve the process based on the new knowledge.
5. *Control.* Implement adequate controls to sustain the gain.

Each phase of the five-phase DMAIC process is described briefly below.

1.7.1   Phase 1: Define

First we create the project definition that includes the problem or opportunity statement, the objective of the project, the benefits expected, which items are in scope and which are out of scope, the team structure, and the project time line. The scope will include details such as resources, boundaries, customer segments, and timing. The next step is to determine and define customer requirements. Customers can be both external consumers and internal stakeholders. At the end of this step you should have a clear operational definition of the project metrics (called big Y's, or outputs) and their link to critical business levers as well as the goal for improving the metrics. Business levers, for example, can consist of return on invested capital, profit, customer satisfaction, and responsiveness. The last step in this phase is to define the process boundaries and high-level inputs and outputs using SIPOC as a framework and to define the data collection plan.

1.7.2 Phase 2: Measure

The first step is to make sure that we have good measures of our Y's through validation or measurement system analysis. Next, we verify that the metric is stable over time and then determine whether our baseline process capability is using the method discussed earlier. If the metric is varying wildly over time, we must address the special causes creating the instability before attempting to improve the process. The result of stabilizing the performance often provides all of the improvements desired.

Finally, in the measure phase we define all of the possible factors that affect the performance and using qualitative methods of Pareto, cause-and-effect diagrams, cause-and-effect matrices, failure modes and their effects, and detailed process mapping to narrow down to the potential influential (significant) factors (denoted as X's).

1.7.3 Phase 3: Analyze

In the analyze phase, we first use graphical analysis to search out relationships between the input factors (X's) and the outputs (Y's). In later chapters we call these output responses critical-to-quality or functional requirement outputs. Next, we follow this up with a suite of statistical analysis (Appendix A), including various forms of hypothesis testing, confidence intervals, or screening design of experiments to determine the statistical and practical significance of the factors on the project Y's. A factor may prove to be statistically significant; that is, with a certain confidence the effect is true, and there is only a small chance that it could have been by mistake. The statistically significant factor is not always practical in that it may account for only a small percentage of the effect on the Y's, in which case controlling this factor wouldn't provide much improvement. The transfer function $Y = f(X)$ for every Y measure usually represents the regression of several influential factors on the project outputs. There may be more that one project metric (output) – hence the Y's.

1.7.4 Phase 4: Improve

In the improve phase, we first identify potential solutions through team meetings and brainstorming or the use of TRIZ[3]. It is important at this point to have completed a measurement system analysis on the key factors (X's) and possibly perform some confirmation design of experiments.

The next step is to validate the solution(s) identified through a pilot run or through optimization design of experiments. Following confirmation of the improvement, a detailed project plan and cost–benefit analysis should be completed.

---

[3] TRIZ is the Russian acronym for the *theory of inventive problem solving* (TIPS). It is a systematic method developed by G. S. Altshuller to conceive creative, innovative, and predictable design solutions. See El-Haik and Roy (2005) for more details.

The last step in this phase is to implement the improvement. This is a point where change management tools can prove to be beneficial.

1.7.5   Phase 5: Control

The control phase consists of four steps. In the first step we determine the control strategy based on the new process map, failure mode and effects, and a detailed control plan. The control plan should balance between the output metric and the critical few input variables. The second step involves implementing the controls identified in the control plan. This is typically a blend of poka yokes (error-proofing) and control charts and clear roles and responsibilities and operator instructions depicted in operational method sheets. Third, we determine what the final capability of the process is with all of the improvements and controls in place.

The final step is to perform the ongoing monitoring of the process based on the frequency defined in the control plan. The DMAIC methodology has allowed businesses to achieve lasting breakthrough improvements which break the paradigm of reacting to the causes rather than the symptoms. This method allows design teams to make fact-based decisions using statistics as a compass and to implement lasting improvements that satisfy external and internal customers.

1.8   SIX-SIGMA GOES UPSTREAM: DESIGN FOR SIX-SIGMA

The DMAIC methodology is excellent when dealing with problem solving in an existing process in which reaching the entitled level of performance will provide all of the benefit required. Entitlement is the best the process or product is capable of performing with adequate control. Reviewing historical data, it is often evident as the best performance point. But what do we do if reaching entitlement is not enough or there is a need for an innovative solution never before deployed? We could continue with the typical build-it and fix-it process, or we can utilize the most powerful tools and methods available for developing an optimized, robust, de-risked design. These tools and methods can be aligned with an existing new product or process development process or used in a stand-alone manner. In this book we devote several chapters to demonstrating the DFSS tools and methodology in a transaction-based environment. Chapter 3 is the introductory chapter for DFSS, giving overviews for DFSS theory, the DFSS gated process, and DFSS application. In Chapter 10 we give a detailed description of how to deploy simulation-based DFSS.

1.9   SUMMARY

In this chapter we have explained what six-sigma is and how it has evolved over time. We explained how it is a process-based methodology and introduced the reader to process modeling with a high-level overview of IPO, process mapping, value stream mapping, and value analysis as well as a business process management system learned the criticality of understanding the measurements of the process or system and how this is accomplished with measurement systems analysis. Once we understand the goodness of our measures, we can evaluate the capability of the process to meet customer requirements and demonstrate what six-sigma capability is. Next, we moved into an explanation of the DMAIC methodology and how it incorporates these concepts into a road map method. Finally, we covered how six-sigma moves upstream to the design environment with the application of DFSS. In Chapter 3 we introduce the reader to the transaction-based design for six-sigma process. The DFSS process spans the phases: I–dentify; C–onceptualize; O–ptimize; and V–alidate; a.k.a ICOV.

2

LEAN SIX-SIGMA FUNDAMENTALS

2.1 INTRODUCTION

In this chapter we cover the lean six-sigma (LSS) concept and discuss topics related to the integration of six-sigma and lean manufacturing. The focus in this chapter is on the details of six-sigma DMAIC tollgates,[1] value stream mapping and lean manufacturing techniques, and the synergy and benefits of implementing an LSS system. LSS represents the foundation for the proposed approach of 3S-LSS (simulation-based LSS). Chapter 9 deals with the details of the 3S-LSS approach through an application case study.

The LSS method is based primarily on the combined lean manufacturing techniques and DMAIC six-sigma methodology. Six-sigma quality analysis (e.g., reducing process variability) is combined with lean manufacturing speed (e.g., reducing process lead time) within the framework of the deploying company's production or service system. This approach entails the involve ment of the majority of industrial engineering concepts and methods, such as

---

[1] In DMAIC, the transition from phase to phase is marked by milestones called *tollgate reviews*. A phase has some thickness: that is, entrance and exit criteria for the bounding tollgates. In these reviews, a decision should be made whether to proceed to the next phase of DMAIC, to recycle back for further clarification on certain decisions, or to cancel the project altogether. See Section 8.4.1.

Simulation-Based Lean Six-Sigma and Design for Six-Sigma, by Basem El-Haik and Raid Al-Aomar

modeling, work analyses, facility planning, quality control, and operations management. Special emphasis is placed on utilizing lean techniques and six-sigma analysis to improve a set of critical-to-quality measures (CTQs) or critical-to-service measures (CTSs) that are defined in terms of a set of lean time-based measures such as lead time, productivity, and inventory levels.

Main steps in the LSS approach involve developing a value stream map of the current production or service system, defining and estimating CTQs in terms of quality and lean measures, analyzing the production or service system with experimental design, improving the system with lean techniques, developing a future-state value stream map of the improved system, and defining actions to control and maintain the improvement achieved.

As discussed in Chapter 1, the six-sigma DMAIC methodology is a problem-solving approach that is based on understanding process structure and logic, defining process CTQs, measuring CTQs, applying six-sigma tools and analysis to improve process performance, and implementing methods to control the improvement achieved. This makes DMAIC a disciplined and data-driven approach that can be utilized to improve both processes and products to meet customer requirements with the least variability in the process performance measures specified.

DMAIC quality improvement of product or service attributes can be related to making process improvement (improving the quality of the manufacturing or service process and the overall production or service system). For example, the late delivery of quality products or services often affects customer satisfaction in the same manner when defective products are delivered to customers. A similar impact results from shipping orders with shortages or with the incorrect mix of products. Hence, companies often invest in more resources to increase capacity, keep large amounts of products in stock, and place enormous pressure on suppliers to meet customer demands. This is likely to increase production cost, weaken the company's competitiveness based on price, and consequently, reduce its market share. Thus, similar to focusing DMAIC on improving the quality level of products and services, DMAIC can be aimed at achieving improvement in the organization's production or service system to increase effectiveness and reduce cost. When applying DMAIC to system-level improvement, however, the focus is on time-based process performance measures such as process throughput (yield), delivery rates, and lead time.

On the other hand, lean manufacturing techniques are applied to reduce the process waste and increase process effectiveness through faster deliveries and shorter lead time. Lean techniques include the application of a pull production system, Kanbans and visual aids, Kaizen, and work analyses to reduce inventory levels, non-value-added activities, process waste, and consequently, to increase process effectiveness. This matches the DMAIC objectives for solving problems and improving the performance of current systems.

The material presented herein is intended to give the reader a high-level understanding of lean manufacturing and its role in conjunction with DMAIC methodology. Following this chapter, readers should be able to understand the basics of value stream mapping and lean techniques, to assess how LSS could be used in relation to their jobs, and to identify their needs for further learning in different aspects of LSS. This understanding lays the foundation for presenting the 3S-LSS road map in Chapter 8.

2.2 LEAN SIX-SIGMA APPROACH

Latest engineering paradigms, such as total quality management, lean and agile manufacturing, six-sigma, and Kaizen, aim collectively to help companies improve the quality of products and services, increase operations effectiveness, reduce waste and costs, and increase profit and market share. Applying these paradigms in today's dynamic production and business systems requires a special focus on analyzing time-based performance measures such as cycle time, lead time, takt time, and delivery speed. Measuring time-based performance is the essence of lean manufacturing practices. Such measures are also the focus of six-sigma projects aiming at achieving organizational excellence at both the plant and enterprise levels.

One of the latest developments in six-sigma includes the introduction and application of LSS systems. Through the synergy of lean and six-sigma in combination, LSS provides a set of methods that companies can apply to any manufacturing, transactional, or service process to reduce variability, eliminate or reduce waste (non-value-added), and cut process lead time. This is likely to result in an effective program that brings both cost-reduction results and process performance boost to achieve six-sigma quality at system-level CTQs and CTSs. Better system-level performance is achieved through LSS by adjusting quality and accuracy measures (for products or processes) to match the six-sigma level (high and less variable response). Less cost is achieved with LSS by reducing process waste, inventory, and lead time.

Numerous process gains can be obtained from applying the LSS method. Process gains are expressed primarily in terms of achieving six-sigma performance at a certain selected process CTQ and CTS with less processing cost and shorter lead time. The process lead time is a key performance metric that can be related to many system-level CTQs and CTSs, such as delivery reliability and agility. Similarly, processing cost is a key product or order attribute that can be related to the product or service price and market share. LSS therefore implies producing products and providing services with a powerful combination of less process variability and less operating cost. Figure 2.1 presents typical gains from the application of LSS to a production or business system.

The claim that six-sigma (introduced by Mikel Harry in the late 1980s) and lean manufacturing (introduced by Taiichi Ohno in the 1940s) have a complementary relationship is universally accepted. This is often explained through the following rationale: "Whereas six-sigma is focused on reducing

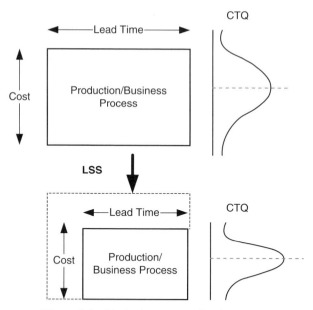

Figure 2.1   Typical process gains from LSS.

variation and improving process yield by following a problem-solving approach using statistical tools, lean is concerned primarily with eliminating or reducing waste and improving flow by following the lean principles." Hence, LSS is presented by George (2002) to combine six-sigma's focus on defects and process variation, with lean manufacturing's link to process speed, efficiency, and waste. Since then, most companies have established LSS programs, especially after the proven capability of six-sigma and lean in leading companies, including Toyota, General Electric, and Motorola.

The concept of lean manufacturing as presented by Womack (1990) has a rich and proven history that began in the automobile industry at Ford in the 1920s and at Toyota in the 1950s. Lean techniques are aimed primarily at eliminating waste in every process activity to incorporate less effort, less inventory, less time, and less space. The main lean principles include reduced inventory, pull instead of push systems, small batches, line balancing, and short process times. Other basic lean-manufacturing techniques include the 5Ss (sort, straighten, shine, standardize, and sustain), standardized work, and documentation. This is attainable by making an attempt to eliminate or reduce the seven deadly wastes of overproduction, inventory, transportation, processing, scrap, motion, and waiting or idle time.

Both the lean and the six-sigma methodologies have proven that it is possible to achieve dramatic improvements in both cost and quality by focusing on process performance. However, using either one of the two methods alone has numerous limitations. Six-sigma focuses on eliminating defects but does

TABLE 2.1   Characteristics of Six-Sigma and Lean Manufacturing

	Six-Sigma	Lean
Goal	To improve process capability and reduce variation	To reduce lead time and process waste
Focus	Process outcomes	Process flow and waste
Philosophy	Variability within specifications is cost	Time in system and overcapacity is cost
Tools	Statistical analyses	Factory physics
Application	Production and business processes	Production and business processes
Approach	DMAIC problem-solving methodology	Value stream mapping and lean techniques
Major measure	DPMO	Lead time
Major driver	CTQs/CTSs	Value-added
Project selection	Problem solving	Continuous improvement
Skills	Mainly analytical	Mainly process knowledge
Gains	Process accuracy and quality	Process effectiveness and delivery speed

not address issues related to process flow and lead time. On the other hand, lean principles are focused on reducing process waste but do not address issues related to advanced statistical tools and process capability. For example, inventory reduction in a process cannot be achieved only through a pull system of a reduced batch size. This also requires minimizing process variation by utilizing six-sigma statistical tools. Hence, most practitioners consider six-sigma and lean to be complementary.

By combining the benefits of the two methods, lean and six-sigma can be used simultaneously to target many types of process problems with the most comprehensive tool kit. Therefore, many firms view LSS as an approach that allows combining both methodologies into the integrated system of an improvement road map. Still, the differences between six-sigma and lean are profound. These differences can be addressed in terms of each method's goal, focus, philosophy, tools, application, approach, measures, drivers, skills, and gains, as shown in Table 2.1.

As shown in Table 2.1, the differences between lean and six-sigma lie primarily in the focus and techniques of both methods. In terms of goal, the elimination of waste with lean is fairly analogous to the reduction of variation in six-sigma since less waste and a better streamlined process often yield consistent production. However, the focus of lean on improving process structure, work flow, and process effectiveness is a main differentiator when compared to the focus of six-sigma in parameter design. The analyses performed in six-sigma, which are focused mostly on statistical methods, are also a differentiator when compared to those of the lean approach, which are focused primarily on analyzing factory physics and flow analyses.

Thus, developing an integrated improvement program that incorporates both lean and six-sigma tools requires more than including a few lean principles in a six-sigma study. An integrated improvement strategy has to take into consideration the differences among the two methods and how to use them effectively to the benefit of the underlying process. An integrated LSS approach to process improvement is expected to include the following major elements or components:

- Developing a current-state value stream map with information and measures that are necessary for applying six-sigma methods
- Using a current-state value stream map as a platform for the application of six-sigma and lean tools
- Applying six-sigma DMAIC methodology to process parametric adjustment
- Integrating lean techniques into DMAIC methodology to prescribe the set of process structural changes that will benefit performance
- Applying lean techniques as a continuous improvement effort to reduce process waste and increase its effectiveness
- Reflecting parametric and structural changes made to the process by developing a future-state value stream map
- Working intensively on creating a cultural change in an organization toward six-sigma accuracy and lean effectiveness in all business functions

2.3 LSS-ENHANCED DMAIC

As discussed in Chapter 1, the DMAIC process is an integral part of six-sigma and a road map for problem solving and system improvement. DMAIC tollgates provide concise definitions to process customers and their requirements and expectations. Based on such definitions, a plan is developed to collect process data and measure process performance, opportunities, and defects. A data-driven and well-structured approach is then followed to analyze process performance, solve problems, and enhance performance toward the nearly perfect six-sigma target.

DMAIC starts with a define (D) stage in which process CTQs/CTSs, performance metrics, and design parameters are identified and project objectives are clearly defined. The measure (M) step is then used to assess process performance metrics quantitatively along with relevant six-sigma measures. In the analyze (A) step, the data collected are thoroughly analyzed, gaps between current performance and goal performance are identified, and opportunities to improve are prioritized. Based on analysis, it is decided whether or not a design modification is necessary. In the improve (I) step, DMAIC strives to improve the process by adjusting process parameters, structure, and logic to eliminate defect causes and reach performance targets. This step may include

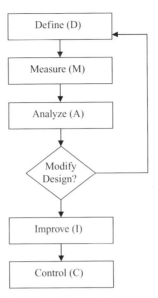

Figure 2.2   DMAIC process.

creating solutions to fix process problems and eliminate defects. Finally, in the control (C) step, DMAIC aims at controlling the improvements obtained to keep the process on the new level and prevent reverting back to low performance. This can be achieved by proper documentation and implementation of an ongoing monitoring and improvement plan along with training and incentives. An iterative approach is often followed when applying DMAIC. The five interconnected steps of DMAIC are shown in Figure 2.2.

A set of engineering tools are utilized to execute each stage of the DMAIC methodology. These tools are selected to fit the deliverables expected from each stage and to fit in the framework of the overall six-sigma project. Table 2.2 summarizes the tools and deliverables at each DMAIC stage, and Table 2.3 presents a comprehensive list of tools that can be used in six-sigma projects. It is not necessary to use all six-sigma tools at the DMAIC stages, and other engineering tools can be included. Six-sigma references, such as Breyfogle (1999), provide the details of the tools cited in Table 2.3.

In the LSS approach, the standard DMAIC methodology is refined to fit the perspective of both six-sigma and lean manufacturing. The lean curriculum is integrated into the DMAIC framework to emphasize the utilization of LSS as an approach to problem solving and continuous improvement. Within DMAIC, lean tools are focused on reducing waste and lead time, whereas six-sigma tools are aimed at reducing defects and variability.

Since the DMAIC approach consists of five stages (define, measure, analyze, improve, and control), integrating lean techniques into these five stages results in an LSS-enhanced DMAIC. Table 2.4 presents examples of lean tools that can be utilized at various stages of the LSS-enhanced DMAIC.

TABLE 2.2   Main DMAIC Tools and Deliverables

DMAIC Stage	Tools	Deliverables
Define (D)	Project charter, process map quality function deployment, benchmarking, Kano analysis, etc.	Project objectives, CTQs, design variables, resources, project plan, etc.
Measure (M)	Data collection, sampling, work measurements, sigma calculator	Measured performance, process variation (sigma value), process capability measures
Analyze (A)	Statistical analyses, charts, root-cause analysis, design of experiment, and ANOVA	Defined improvement opportunity, sources of variation, action plan
Improve (I)	Design optimization, robustness, brainstorming, validation	Selection of the best solutions, changes in deployment, adjustments to process variables
Control (C)	Error-proofing, failure made and effect analysis, statistical process control, standards, Pugh analysis	Monitoring plan, maintained performance, documentation, transfer of ownership

The enhanced LSS-DMAIC stages can be summarized as follows:

1. Define customer and process improvement requirements by prescribing the CTQs and CTSs, design parameters, and corresponding process variables. Lean measures (time-based process performance and cost measures) are expected to be principal CTQs and CTSs in the LSS approach. Process design variables include parameters that affect such time–based performance and contribute to various process waste (cost) elements. Value stream mapping (VSM) is an excellent lean practice to present a variety of time and performance elements in the process flow both graphically and analytically.

2. Measure the defined CTQs and CTSs at the current state of the process represented in the VSM. To this end, VSM of the current process state with process CTQs is developed in the LSS approach. VSM herein is a macro-level time-based representation of the process. This representation includes the stages, delays, and routings of products from material receiving to finished product shipping. Measured CTQs and CTSs may include lead time, takt time, inventory turns, delivery rates, overall equipment effectiveness, and others. Data collection of production, shipping, receiving, and maintenance records is essential in developing a close-to-reality VSM of the current state.

3. Analyze the current process state using the VSM for estimating lean measures (CTQs) at various process parameters. Many industrial engineering methods, such as flow analysis, work analyses, and scheduling production, can be used to explore opportunities of improving performance in terms of lean

TABLE 2.3   Summary of Common DMAIC Tools

• Analysis of variance	• Multivariable charts
• Attribute sampling plans	• Nonparametric tests
• Autocorrelation	• Normalized yield
• Bar charts	• NP charts
• C charts	• P charts
• Chi-square methods	• Pareto
• Continuous sampling plans	• Parts-per-million defective
• Control charts	• Performance metrics
• Correlation studies	• Pie charts
• C_p/C_{pk}	• P_p/P_{pk}
• Cross-tabulation methods	• Precontrol
• Cube plots	• Process definition
• CUMSUM charts	• Process failure mode and effect analysis
• Customer feedback surveys	• Process management
• Defect probability	• Process mapping
• Defects per million opportunities	• Product failure mode and effect analysis
• Defects per unit	• R charts
• F-tests	• Random sampling
• Fractional factorial designs	• Random strategy designs
• Full factorial designs	• Regression
• Goodness of fit	• Residual sum of square analysis
• GR&R control charts	• Response surface designs
• GR&R statistical DOE	• Rollout throughput yield
• Histograms	• Sequential sampling
• I-MR charts	• Single-factor experiments
• Indices of variability	• t-tests
• Individual charts	• Taguchi designs
• Interaction charts	• Time series analysis
• Line charts	• Tolerances
• Main effect plots	• U charts
• Mathematical models	• WEMA charts
• Mixture designs	• X-bar charts
• Monte carlo simulation	• Yield charts
• MR charts	• Z-tests

TABLE 2.4   Utilizing Lean Tools in LSS-Enhanced DMAIC

Define	Measure	Analyze	Improve	Control
Value stream mapping (VSM)	Lead time	Work analyses	SMED	Visual controls
	Takt time	Flow analysis	JIT-Kanban	Standard work
	Inventory level	Scheduling	Line balance	Kaizen

measures. It is worth mentioning that typical six-sigma analyses, such as design of experiment (DOE)–ANOVA and regression models, can also be utilized at this stage to identify critical design factors and analyze process variability. From an LSS perspective, statistical analyses will focus on parameter design, whereas lean analyses will focus on the process structural design.

4. Improve the current state by proposing a better process design (a future state). A future-state VSM is developed to quantify the improvement achieved and to represent the structure and parameters of the future state. Six-sigma's DOE can be used to set the process parameters to the levels at which process CTQs and CTSs (lean measures) are at their best (a six-sigma quality level is desirable) with the least variation (robust performance). From an LSS perspective, this is the stage where various lean manufacturing techniques are put into action to restructure the process for better performance. The focus here is on eliminating non-value-added elements from within the process by reducing lead time and excessive inventory so that process effectiveness is increased and process waste is reduced. Various lean techniques are implemented at this stage, including just-in-time (JIT)/Kanban and pull production systems batch size reduction, setup time reduction [single-minute exchange of die (SMED)], cellular manufacturing, at-the-source quality controls, and so on.

5. Control the future-state (improved) process by monitoring process variables and behavior and setting controls to maintain six-sigma and lean performance. Several lean techniques can be used to control the improved production environment and maintain the lean performance. Examples include using visual controls and displays, documenting and standardizing work procedures, and deploying a *Kaizen* (continuous improvement) plan. This will be complementary to the various six-sigma tools that are applied at the control stage of DMAIC, such as itatistical process control, failure mode and effect analysis, and error-proofing plans. The focus here is on enforcing means to sustain the LSS process gains.

2.4   LEAN MANUFACTURING

In its simplest form, *lean technology* refers to manufacturing with minimal waste. In manufacturing, waste (*muda* in Japanese) has many forms, such as material, time, unused capacity, idleness, and inventory. As a philosophy, therefore, lean manufacturing aims at eliminating waste in terms of non-value-adding operations, tasks, tools, equipment, and resources in general. Knowing that the waste in a company may represent 70% or more of its available resources emphasizes the need for seeking improvement and opens the door wide for implementing and benefiting from various lean manufacturing techniques. With this mindset, a successful lean manufacturing effort will have a significant impact on an organization's performance in terms of operating cost and delivery speed. Commonly used lean manufacturing techniques include JIT pull/Kanban systems, SMED, cellular manufacturing, the 5S system,

Kaizen, total productive maintenance, visual controls, line balancing, work-place organization, and standardized work procedures.

Although initially most of the lean efforts were focused on manufacturing operations, lean applications are increasingly being transferred from the shop floor to the office and to the service sector. Recent lean applications involve eliminating waste in the entire organization (including manufacturing and such manufacturing support functions as purchasing, scheduling, shipping, and accounting). The use of lean manufacturing is growing simply because it provides a company with the key competitive advantages of reduced costs and improved effectiveness.

Although most lean definitions include the term *elimination of waste* in every area of production, it is often more practical and more motivating to practitioners to use the term *reduction of waste* in every area of the value chain. The *value chain* is viewed as a wide spectrum of interrelated business functions, including product development, supply chain, marketing, distribution, after-sale service, customer relations, and any other business segment. The goal is to maintain high performance by establishing a highly responsive system to varying customer needs, and by producing world-class-quality products and services in the most efficient and economical manner possible. The primary results that are expected from applying lean manufacturing to a production or service process include:

- Lead-time reduction
- Inventory reduction
- Defect reduction
- Improved utilization of resource capacity
- Delivery rate improvement
- Productivity increase
- Cost per unit reduction

It is important to understand that lean application is not a matter of changing one or two process elements or moving around pieces of equipment. Clearly, lean manufacturing gains can only be attainable given a sincere effort and persistence from all stakeholders of an organization: labor, management, suppliers, and others. This is often achieved by establishing a real company-wide cultural change in the organization and a paradigm shift in management and engineering. Similar to six-sigma projects, a solid commitment from the leadership of the organization is essential for a successful lean transition. However, since there is no generic recipe for implementing lean techniques, each company is expected to customize the application of lean tools to fit the specifics of the situation. Generally speaking, each industry has its own flavor, mechanisms, and restrictions on lean techniques application.

Lean techniques application is typically focused on seven types of wastes in a production or service system, those wastes at the root of the majority of the non-value-added activities in a plant or service process. Eliminating or reducing such wastes often increases a company's profit margin without increasing prices or affecting competitiveness. Although the these wastes are often defined in the context of manufacturing environments, they can be identified similarly, with minor modifications, in service systems. This modification is related primarily to understanding the transactional nature of the process (manufacturing or service) rather than limiting the scope to production operations.

Following are the seven main types of process waste in lean manufacturing:

1. *Overproduction:* producing more than the quantity that is required and/or producing items before they are actually needed. This waste is related primarily to make-to-stock mentality and weak demand forecast.

2. *Transportation:* excessive movement of parts or material from or to storage or within the production line. Actually, all forms of material handling are often considered to be non-value-added tasks.

3. *Motion:* movements performed by workers or machines before, during or after processing. This often refers to micro-level wastes in work procedures that directly affect cycle time and effectiveness.

4. *Waiting:* parts delay in queues that are awaiting processing or idle time of operators or machines. In a similar manner, delay occurs when holding inventory for production or making customers wait for services.

5. *Processing:* non-value-added processing and use of materials, tools, and equipment. This typically represents a costly waste since it results in consuming excessive material or overusing tools and equipment.

6. *Inventory:* accumulation of raw material, work-in-process, and finished items. This is typical in push production systems functioning under fluctuations in demand and supplies.

7. *Defects:* making defective products (reworked or scrapped) as well as delivering incomplete services. This should be based on understanding the difference between the defects' definition and metrics in CTSs and CTQs.

Successful lean application depends on understanding several production concepts and methods. Making a distinction between push and pull production systems is essential to an understanding of the rationale of lean thinking. Whereas *push production systems* tend to produce items without checking the actual needs of downstream operations, *pull production systems* are driven by customer demand. As a key lean technique, therefore, a pull scheduling system emphasizes replenishment of what has been consumed by customers or producing only what is needed (driven by actual demand). A pull system typically uses some type of visual signal (Kanban), typically a card and a container. An empty container is basically a call for producing what is on a card (type and quantity). This initiates a part replenishment process upstream to warehouse material withdrawal. A pull system will control and balance production resources and is effective in reducing non-value-added activities. This is aimed at reaching a high-performance and least waste production and business environment that is often referred to as just-in-time.

Inventory reduction is another key lean objective. Parts are typically put into storage to make them instantly available at both their point of use in production and their point of delivery in shipping to customers. The goal is to reduce out-of-stock, shortages, delay, search time, and travel and material handling. This reduces delivery time and increases customer satisfaction. However, higher inventory assets often result in negative performance and financial implications (such as low retwin on investment). Thus, lean aims at reducing inventory levels by concurrently restructuring the value chain to reduce lead times, variability, and waste, so that inventory turns and delivery rates are all increased.

In lean techniques process defects are reduced by using at-the-source controls to check product quality, identify defects, and fix errors as early as possible. In-process quality control or quality at the source aims at ensuring that the products sent to the next workstation (downstream) are good-quality products. To implement quality at the source, employees must be given the means and time necessary to ensure that the parts they produce are free of defects. Examples include sampling plans, tools for measuring product dimensions, and attributes to be compared with design specifications, inspection equipment, and gauges.

In LSS, cellular manufacturing represents the opposite of departmental layout (departments of machines of similar types) and process configuration. This is achieved by grouping together machines and equipment necessary to complete the product. A cellular layout is set to better accommodate a single piece flow by eliminating the handling and queue times inherent in a traditional departmentalized layout. Finally, standardized work procedures involve documenting best practices in performing operations (e.g., best tools, best sequence, best methods) so that tasks are performed now and tomorrow and products are made in the most effective way possible.

Over the years, LSS has developed its own terminology. Thus, before exploring key lean techniques, it is essential to be familiar with lean terminology:

- *Batch size:* the number of units accumulated and moved from one workstation to the next in the product flow.
- *Bottleneck:* the slowest operation in a production flow, which limits the throughput of the entire line.
- *Cellular manufacturing:* a group of resources or pieces of equipment arranged in the most efficient manner to process a product (or a family of products). In LSS, a cell layout is typically U-shaped with a 1-unit flow.
- *Flow:* the movement of product throughout the production system.
- *Group technology:* grouping into families products of similar processing needs.

- *Just-in-time:* providing what is needed, when it is needed, in the quantity in which it is needed.
- *Kaizen (continuous improvement):* a relentless elimination of waste and seeking improvement.
- *Kanban:* a visual signal (a card and an empty container) that triggers material withdrawal or parts manufacture in a pull production system.
- *Lead time:* the time between making an order and the order delivery date. The lead time of a process is determined by dividing the WIP by the process throughput.
- *One-unit flow:* a flow of products from one workstation to the next, one piece at a time (batch size = 1).
- *Pull system:* withdrawing material and producing parts once needed.
- *Push system:* producing units based on short- and long-term forecasts and pushing them to storage, waiting for orders from customers.
- *Setup time:* the time it takes to change over the from production of one product to production of the next.
- *Takt-paced production:* setting and balancing a production line to run at the pace (takt time) required by customers.
- *Takt time:* a German word meaning musical beat or cycle; the desired interdeparture time of units from the production line. It is calculated by dividing the total daily operating time by the total daily customer demand.
- *Value:* what the customer is willing to pay for a product or service.
- *Value added:* any activity that contributes to the value of a product or service.
- *Value chain:* elements and processes needed to develop a product or service.
- *Value stream:* actions that add value to a product or process.
- *Value stream map:* a high-level graphical representation of the value chain.
- *Waste ("muda"):* tasks, requirements, and materials that do not add value to a product or service.
- *WIP inventory:* work-in-process parts in a production system.

Although lean techniques may be applied to any process improvement effort, the core lean principles are focused mainly on JIT and Kaizen philosophy. Requirements for establishing a JIT environment include implementation of the majority of lean manufacturing techniques. On the other hand, Kaizen helps to maintain JIT gains, to implement improvement plans quickly and effectively, and to open the door for creativeness and progress. Table 2.5 presents the main benefits of commonly used lean manufacturing techniques.

TABLE 2.5 Major Benefits of Lean Techniques

Lean Tool/Technque	Major Benefit
Value stream mapping	Process flow and lead-time representation
The 5S system	Workplace organization and cleanness
Identification of the 7 wastes	Analyzing opportunities for waste reduction
SMED	Setup reduction and fast changeover
Total productive maintenance	Increasing equipment availability and reducing downtime
Cellular manufacturing	Reduced flow travel and better control
Standardized work	Product consistency and high morale
Poka yoke (error-proofing)	Robust performance and fewer defects and mistakes
Cross-trained/multi skilled workforce	Team performance and shared responsibility
Just-in-time	Production stability and least waste
Single-unit flow	Takt-based production
Pull system (Kanbans)	Customer-driven production
Balanced work flow	Fewer variability and high morale
Inventory reduction	Low inventory investment
Quality at the source	Fewer defectives in production
Visual controls	Transparency and control
Kaizen	Continuous improvement

2.5 VALUE STREAM MAPPING

Lean manufacturing is focused mainly on value stream mapping and lean techniques application. Hence, lean can be viewed as VSM integration of a process value chain where VSM is utilized as a key tool for developing lean systems. In an LSS context, VSM is a critical initial step in lean conversion, where VSM is utilized as a communication tool and as a business planning tool to manage the change toward a lean environment. VSM also helps different business partners (management, suppliers, production, distributors, and customers) recognize waste and identify its causes with the process value chain. In this section we present an overview of VSM concepts, symbols, and approach.

VSM is developed based on a good understanding of lean concepts and techniques. VSM focuses on value in the context of what the customer is willing to pay for or what makes the product gain customer satisfaction. This requires the company to first hear the voice of the customer (VOC), understand customer needs, and translate them into functional requirements. This matches the theme of the define stage of the DMAIC methodology and reemphasizes the complementary relationship between lean manufacturing and six-sigma.

In addition to quality product features and functions, customer satisfaction often entails time-based lean measures such as on-time delivery, shorter lead time, and low price. These measures are the primary focus of lean manufacturing (VSM and lean techniques). VSM is, however, different from the process mapping of six-sigma since it gathers and displays a far broader range of information at a broader and higher level, from receiving raw material to delivery of finished goods, and it provides flow and performance measures.

As defined earlier, the value stream is the entire process for developing, manufacturing, and shipping a product to the customer. For current products, the value stream is typically focused on product making and delivery to customers. Once an order for a product is placed, the value stream involves all actions, activities, stages, and operations that are applied to the product in order to deliver the required amount and mix to the customer on time. The value stream starts at concept and ends at delivery to the customer. Practically, not every activity and operation in the process of product making and delivery will add direct value to the product. This will be the primary value of VSM: to specify the activities and delays in the product flow that will not add value to the product. This is what is called *waste* in lean manufacturing. The lean techniques will be then applied to eliminate or reduce such activities.

Thus, VSM is a tool that is used to graphically represent the processes or activities involved in the manufacture and delivery of a product. These activities can then be divided into value-added and non-value-added. Key process information and data and key performance measures are then added to the VSM to characterize various stages in the product flow and to quantify current-state performance. Process data include primarily cycle time, capacity, and availability. Storage and transfer activities are characterized by the distance traveled, the storage and buffer size, and the time delay. VSM includes a representation of the information flow among various flow stages. The most important performance measures in VSM are productivity and lead time.

Current-state VSM includes the flow and time dimensions of entities and information through a production or business system. From the information in the current-state map, a desired future-state map can be developed where waste is minimized and non-value-added activities are eliminated. This is achieved primarily through the application of lean techniques. Indeed, the future-state VSM helps with the testing and assessment of the impacts of various lean techniques. Its development requires significant knowledge of lean techniques and flow analysis. Designing a future state requires more art, engineering, and strategy than in present-state mapping. Improvement efforts with VSM focus mainly on:

- Reducing lead time
- Reducing scrap and rework and improving product quality
- Improving equipment and space utilization
- Reducing inventory levels
- Reducing direct and indirect material and labor costs

TABLE 2.6 Main VSM Icons

VSM Symbol	Meaning	Use
	Customer or supplier	Customer initiates orders and supplier delivers inputs
	Shipment or Supply	Supply from supplier or shipment to customer
Production Control	Production control	Planning and scheduling production
Process	Process or operation	May represent a machine, an assembly line, a department, etc.
C/T= C/O= Batch= Avail=	Process data box	Key process information: cycle time, changeover time, batch size, and availability
	Work cell	Cellular manufacturing flow
I	Inventory	Inventory (buffer) between processes
	Safety stock	Extra inventory stored to absorb demand–supply fluctuations
	Push arrow	Units produced and pushed downstream
	Pull arrow	Units requested from an upstream operation or inventory area
Monthly	Information flow	Shared information to control and schedule production activities
	Kanban card	A signal or a card containing product information used to authorize production or material transport
	Time line	Value-added and non-value-added time line

Many different formats of "maps" exist, each with its own sets of symbols and annotations. Certain symbols were proposed to standardize the VSM development process. Table 2.6 shows the commonly used VSM icons, together with their meaning and use.

The primary goal of VSM is to estimate the lead time associated with a certain product flow throughout a system. Starting from the supplier warehouse, the product flow often includes the following elements in a product's value stream:

1. Ordering materials and components from suppliers
2. Shipping materials and components from the supplier warehouse
3. Receiving and warehousing materials and components
4. Processing materials and components
5. Passing or delaying product subassemblies through buffers
6. Storing finished products prior to shipping
7. Preparing product for shipping
8. Shipping product to customer

Some of the operations in VSM will be shared or parallel, depending on the flow complexity. In general, lead time can be calculated starting with the product or service order until the delivery of products or services to the customers. In a make-to-order production environment, the total lead time consists of the following elements:

- Order-processing lead time
- Product development lead time
- Supply (purchasing) lead time
- Manufacturing (production) lead time
- Distribution (shipping) lead time

Clearly, in a make-to-stock environment, the order-processing lead time will end prior to shipping. Manufacturing lead time (MLT) is one portion of the total lead time, but it is the only component that is considered to be a manufacturing activity. The MLT can be further divided into:

- Waiting time in the buffer before processing
- Material-handling time (into the process)
- Machine or line setup time
- Run (processing) time
- Material-handling time (from the process)
- Waiting time in buffer after processing

Among these time elements, run time is the only portion that actually adds values to the products. Hence, in VSM, run time is labeled as value-added and every other step in the flow is labeled as non-value-added. Processing (run) time can be further divided into:

- Load time
- Run time
- Unload time

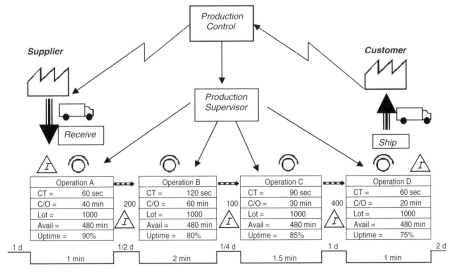

Figure 2.3   Example of a VSM.

Figure 2.3 shows an example of VSM. The time line in the VSM includes value-added (VA) time and non-value-added (NVA) time. For example, the only VA time represents the sum of the cycle times at the four operations (A, B, C, and D). All other times in the flow are NVA since they represent (in days of demand) the material-handling and inventory times. VA and NVA times add up to lead time, which is about $4\frac{3}{4}$ days plus 5.5 minutes of processing time. VA time includes processing (transformation) of a product toward a finished item (i.e., fabrication, assembly, cleaning, testing, etc.). NVA time includes delays and material handling. In addition to lead time, three other time-based measures are often used in VSM: cycle time, takt time, and available time. We obtain from VSM the percentage of VA compared to NVA time and the total lead time. Improvement is then made through the application of lean techniques to increase VA percentage and to reduce lead time.

2.6   LEAN TECHNIQUES

In this section we present an overview of the lean concept and a basic discussion of main lean techniques. Lean techniques focus on waste reduction for waste that results from any activity that adds cost without adding value, such as moving and storing. The section is not meant to be a detailed explanation of all lean techniques and methods. Instead, the focus here is on presenting the concepts and the main lean techniques that will play a major role in the integration of six-sigma and lean manufacturing in LSS. These techniques play a key role in the simulation-based LSS (3S-LSS) approach that is presented in Chapters 8 and 9.

The interesting fact about lean techniques is that they are quite interrelated and their implementation often intersects in various ways. For example, inventory reduction practices require better control over the quantity of work-in-process inventory and shorter cycle times and lead time. This typically requires a demand-driven flow control (e.g., pull/Kanban) using a smaller lot size. This, in turn, may require a better cell or line layout as well as a reduction in setup times. Capacity utilization is also increased by balancing work and resolving flow bottlenecks. Reducing production variability requires standardization of processes and operating procedure, which requires the involvement of workers (usually, operating in teams) and often leads to implementing at-the-source quality controls and the development of effective programs for error-proofing (poka yoke), continuous improvement (Kaizen), workstation order (5Ss), and cross-training.

Following is a brief discussion of the principal lean techniques, starting with the concept of pull production systems.

2.6.1 JIT/Pull Production System

In lean manufacturing, we design operations to respond to the ever-changing requirements of customers by responding to the pull, or demand, of customers. This concept contradicts the traditional batch-and-queue manufacturing approach, facilitates the planning for delivery of products to customers, and stabilizes the demand, especially when customers become confident that they can get what they want when they want it.

Thus, a pull system relies on customer demand. Similarly, a pull scheduling system emphasizes replenishment of what has been consumed or producing only what is needed by customers. A pull system typically uses some type of visual signal, such as an empty box, an open space, or a flashing light to initiate the part-replenishment process. A pull system will control and balance the resources required to produce a product and is effective in reducing non-value-added activities.

In a pull production system, *just-in-time* (JIT) is a philosophy that strives to reduce sources of waste in manufacturing by emphasizing the production of the right part in the right place at the right time. The idea is to produce or order only the parts that are needed to complete the finished products and to time their arrival at the assembly site. The basic elements of JIT were developed at Toyota in the 1950s by Taiichi Ohno and became known as the Toyota Production System. Before implementing JIT in Toyota, there were a lot of manufacturing defects in the existing system, including inventory problems, product defects, increasing costs, large lot production, and delivery delays.

Driven by the demand from customers, the subassemblies and parts required for final assembly are pulled in small batches from the supplying work centers whenever they are needed. Kanban is the key lean technique

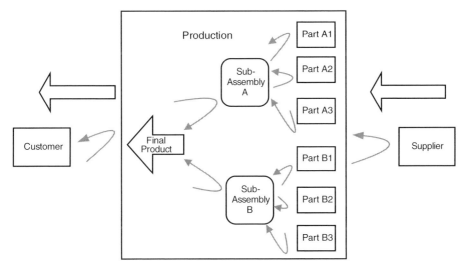

Figure 2.4 JIT pull production system.

that is used to implement JIT. This makes JIT an effective planning system that optimizes material movement and inventory levels within the production system. Figure 2.4 illustrates the JIT concept.

JIT is expected to improve profits and return on investment by reducing inventory levels and reducing production and delivery lead times. It is particularly beneficial in repetitive manufacturing processes in which the same products and components are produced over and over again. The general idea is to establish flow processes (even when the facility uses a batch process layout) by linking work centers so that there is an even, balanced flow of materials throughout the entire production process, similar to that found on an assembly line. To accomplish this, an attempt is made to reach the goals of driving all queues toward zero and achieving the ideal lot size of 1 unit (single-piece flow).

The main requirements for establishing a JIT production environment that integrates key lean techniques include the following:

- Creating a balanced workload on all workstations through uniform and stable production plan for daily production. This can be achieved by enhancing demand forecast, reducing demand fluctuations, and meeting demand fluctuations through end-item inventory rather than through fluctuations in production level.
- Reducing setup and changeover time (as discussed in Section 2.6.2).
- Reducing order and production lot size. Reducing the production lot size is related directly to reducing setup times, since it allows the production of smaller lots. Order lot size (for purchased items) can be reduced to an

economic order quantity through close cooperation with suppliers and shippers.

- Implementing better planning of production resources to meet the needs of the production process. This may require process or product redesign. This also requires a flexible workforce or multifunction workers (operators capable of operating several machines and performing maintenance tasks and quality inspections).

- Reducing production and supply lead times. Production lead times can be reduced by reconfiguring the production process so that travel and movements are reduced, receiving and shipping points are closer, and workstations are closer. This improves the coordination between successive processes. Lean techniques such as group technology and cellular manufacturing can be used to reconfigure the production process. Lead time can also be reduced by using efficient ways for material handling and by increasing machine availability through better maintenance programs. This reduces transfer and waiting times and thus, reduces the number of jobs waiting to be processed at a given machine. Facility location and cooperation with suppliers and shippers also help reduce supply and delivery lead times.

- Implementing a quality-at-source policy to reduce defective production. Since buffers are reduced in JIT, defective parts should also be reduced or eliminated from within the process and in parts received from the supplier.

- Using a Kanban (card) system to signal parts withdrawal and production. A Kanban is a card that is attached to a storage and transport container. It identifies the part number and container capacity, together with other information.

- Management support, supply chain management, and labor involvement.

- Plant layout with a cellular flow, minimal travel, and clear signaling.

- Effective production planning and inventory control with MRP.

Ultimately, JIT aims at improving response to customer needs, increasing an organization's ability to compete with others and remain competitive over the long run, and increasing the efficiency within the production process by increasing productivity and decreasing cost and wasted materials, time, and effort (in terms of overproduction, waiting, transportation, inventory, defects, etc.).

Kanbans   As one of the primary tools of JIT system, the term *Kanban* ("card signal") is sometimes used interchangeably with *JIT*. The Japanese term refers to a simple parts-movement system that depends on cards and boxes or containers to take parts from one workstation to another. *Kanban* is derived from *Kan* ("card") and *Ban* ("signal"). Kanban simply signals a cycle of replenishment for production and materials. It maintains an orderly and efficient flow of materials throughout an entire manufacturing process. It is usually a printed card or a visual aid that contains specific information, such as part name, description, and quantity.

Unlike a push MRP system, a Kanban system is a pull system, in which the Kanban (a standard empty container) is used to pull parts to the next production stage when they are needed. The unpredictable and expensive batch-and-queue method of manufacturing, coupled with the unreliable forecasting and variable lead time associated with traditional production models, is replaced by a reliable, predictable Kanban system. Kanban best fits a demand-stable, balanced, high-volume production system and may not be suitable for an entire production process.

Although there are a number of Kanban methods and variants, the concept is typically similar; following a certain demand request, a downstream operation passes a card to an upstream operation requesting parts. The card authorizes the upstream operation to begin production or to order a specified number of that component to fill the requirement.

The essence of the Kanban concept is that a supplier or a warehouse should deliver components to the production line only as and when they are needed, with no excess production or inventory. A material transport Kanban (T-Kanban) is used to signal material withdrawal from component storage. Within a manufacturing system, workstations produce the desired components only when they receive a card and an empty container, indicating that more parts will be needed in production. This limits the amount of inventory in the process. A production Kanban (P-Kanban) is used to pull or order the make of WIP. Finally, another material transport Kanban is used to order finished products from manufacturing.

Clearly, if the JIT pull system is based on a single-piece flow, only one Kanban card (signal) will be needed, with no need for containers. Usually, several Kanban containers each of a lot size are used to control production flow. Multiple cards are also needed when producing multiple products (mix of products). Figure 2.5 illustrates the Kanban concept.

The number of Kanbans required can be calculated by first determining the demand during an order lead time (in addition to a certain safety stock). Enough containers are then allocated to hold the demand in the system. Hence the number of containers is determined as follows:

Kanbans = (demand during lead time + safety stock) ÷ batch (container) size

If rounding is necessary, the number of Kanbans must be rounded up to the next-highest integer. When the machine picks up raw materials to perform an operation, it also detaches the card that was attached to the material. The card is then circulated back upstream to signal the next upstream machine to do another operation. Thus, the number of cards circulating determines the buffer size, since once all cards are attached to parts in the buffer, no more parts can be made.

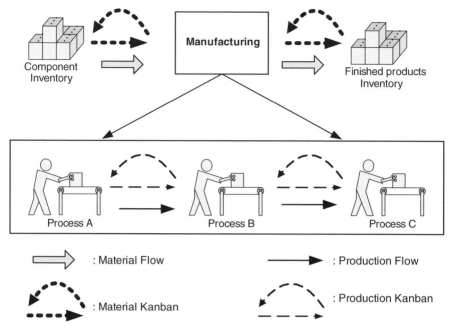

Figure 2.5 Kanban JIT production system.

Since Kanbans limit the WIP level in a production system, it is considered as one step toward eliminating buffers. To this end, the number of Kanban cards should be reduced over time and the problems that are encountered by Kanban reduction are tackled as they are exposed. The maximum stock level is determined by the number of cards (containers) that are in circulation. The plant decides on the maximum level of WIP stock to keep in the production process. This level is reduced gradually (by reducing the number of circulating Kanbans) as production problems and interruptions are minimized and processes are streamlined from supplying to shipping. The circulation of Kanbans stops if there is a production problem on the line. This acts as an alarm to initiate an effort to spot and correct the problem and prevents the accumulation of WIP inventory.

2.6.2 Single-Minute Exchange of Die

Quick changeover, or setup reduction, is a systematic means of reducing the time during which a piece of production equipment is down when changing from one product to the next. *Changeover time* is defined as the time from completion of the last good piece of one job to the first good piece of the next job. The main components of changeover time include the following:

1. Preparation for setup, which may include activities such as preparing the necessary paperwork, operator change, cleanup of the machine or work area, and preparing the required materials, dies, tools, and so on
2. Removal or installation and mounting of parts, tools, fixtures, and equipment, which may include die change and machine reconfiguration
3. Measuring, setting, and calibration of the equipment and setting machine parameters
4. Making trial pieces of the product, taking output measures, and making necessary adjustments

Setup time reduction is therefore approached by observing and analyzing current changeover procedures and practices, analyzing changeover frequency, and looking for ways to reduce each of these components and to minimize equipment downtime during the process.

Quick changeover is often referred to as *SMED* (single-minute exchange of die), which is a method of changing a process die rapidly and efficiently (if applicable) or in general for converting a process from running the current product to running the next product in the product-mix family. The term *single minute* actually means *single digit*, reducing the changeover time to less then 10 minutes. In the lean context, SMED is an approach to reduce units, time, and quality losses due to changeovers. The method, developed in Japan by Shigeo Shingo, has proven its effectiveness in many companies by reducing changeover times from hours to minutes.

Many means, actions, and tool kits can be used to help setup reductions. These may include:

- Visual controls
- Checklists
- Specially designed setup carts
- Overhang tools
- Quick fasteners and clamping tools
- Standardized dies
- Stoppers and locating pins

SMED is an important lean improvement in multiple industries, especially when the product mix includes many products and the production batch size is small. A major advantage is achieved by all manufacturers today under fierce competition in terms of product variety and delivery speed. Examples include retooling a machine to change a model, color, flavor, or size. By cutting these times down, line readiness and productivity will improve without compromising quality, safety, or profit. Consequent benefits include lower inventories, faster deliveries, and more efficiency, as well as improved morale and a high level of worker involvement.

For most manufacturers, setup times are either a source of great concern or an overlooked loss. Still, setups are key ingredients in many processes. For example, companies with high-speed machines sometimes devote more of their time to setup than they do to production itself. Such is the case with printshops and others. SMED as a lean technique focuses on changing the way in which changeovers are performed and in reducing waste in changeover materials, effort, and time. To this end, SMED takes into account all the operations required for each setup, then separates and organizes them in order to cut drastically the downtime required to perform each setup.

Like other lean techniques, SMED affects other lean techniques and helps achieve several lean objectives. The effects of SMED may include:

- Reduced lot size
- Reduced inventory
- Reduced labor cost
- Increased utilization of production resources
- Increased capacity and resolution of bottlenecks
- Reduced setup scrap and quality problems
- Improved work environment

2.6.3 Inventory Reduction

In lean manufacturing, inventory reduction is focused on eliminating excess inventory, improving inventory rates, increasing inventory turnover, and meeting on-time delivery. For example, underutilized (excess) capacity is used instead of buffer inventories to hedge against problems that may arise. In production economics, inventory ties up money that can be invested in development, training, or other revenue-generating activities. Since a lack of control often contributes to excessive inventory, analytical and computerized methods of inventory control and management often help reduce inventory levels through optimization of the inventory order quantity, reorder points, and periods between replenishments. This affects primarily the inventory of raw materials and input components.

Delivering production inputs in the right quantities and the right time frame is highly dependent on the reliability of the supply chain. Under the pressure of focusing on meeting customer demands, inventory controls are often reassessed as a reaction to material shortages. A lean inventory reduction program is therefore complementary to a company's inventory control strategy, since the lean technique focuses on the inventory within the production system as well as the finished goods inventory. Hence, inventory reduction is essential to successful implementation of modern systems of operations management such as enterprise resource planning.

JIT and Kanban are key lean techniques that help reduce inventory and improve inventory turnover. Closely related to inventory levels is the average work-in-process (WIP), the inventory within the production process (in buffers, holding areas, dispatching areas, etc.). WIP reduction is a key lean objective. There are some basic steps that any company can use to improve inventory turnover:

- For inputs inventory, we can use an inventory control system to set a realistic and economical inventory level (reorder point in a Q-system or time between orders in a P-system) for each component. The inventory level includes the quantity needed plus a certain safety stock that is determined based on the demand history. Adopt a JIT delivery system if possible.
- For WIP inventory, track the flow of each component within the production process and identify areas of excess inventory. Simulation and analytical methods can be used to estimate the required average inventory within the production process. Apply a Kanban system and visual controls to streamline the flow and avoid overproduction.
- For finished-items inventory, enhance the forecasting system and apply a pull production system. Reducing lead time can help reduce safety stocks of finished items. Identify obsolete and defective inventory and eliminate situations that cause future buildup of inventory.

2.6.4   Cellular Manufacturing

In a manufacturing context, a *cell* is a group of workstations arranged to process a certain product entirely or partially, or to perform a certain function on a group of products. An ideal cell is self-contained, with all the necessary equipment and resources. In lean manufacturing, cells are formed to facilitate production and minimize flow, travel, and material handling. To this end, equipment and workstations are arranged in a sequence that supports a smooth flow of materials and components through the process, with minimal transport or delay.

Cellular manufacturing is achieved when all of the resources required to complete the product are grouped together. Thus, a cellular layout can more easily accommodate single-piece flow by eliminating the handling and queue times inherent in a traditional departmentalized layout.

Traditionally, manufacturing cells are either functional *process-oriented* or flow *product-oriented*. Functional *departmental cells* are cells consisting of similar equipment used to perform a certain function (e.g., turning cell, milling cell). Cellular configurations can take several forms: circular, S-shaped, W-shaped, U-shaped, and so on. On the other hand, cells that are product-oriented typically run one type of high-volume product through a series of operations (i.e., straight-line flow). Figure 2.6 shows the three common types of flow layouts. In lean manufacturing, the cellular layout design is aimed at combining the flow benefits of high-volume products with the benefits of handling a variety of low-volume products. This is achieved by grouping together machines that process a certain product family.

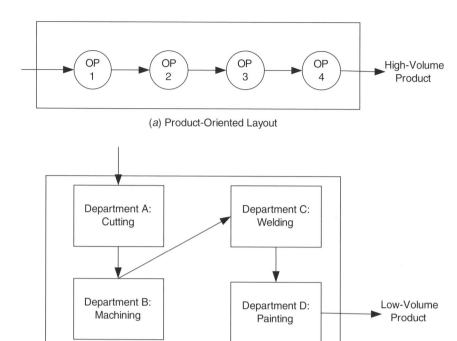

(a) Product-Oriented Layout

(b) Functional Layout

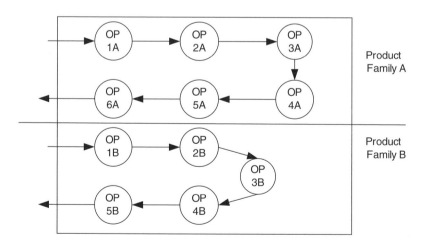

(c) Cellular Manufacturing Layout

Figure 2.6   Types of flow layouts.

Manufacturing systems consisting of functional cells are not compatible with lean manufacturing since the frequent cell-to-cell (function-to-function) flow creates several types of waste in excessive transportation, inventory, and overproduction (to better utilize the equipment). Also, as the product family grows and product flow increases in complexity, a mixed work cell (cellular manufacturing) is needed so that a series of operations for several products are performed. This type of work fits a lean manufacturing environment where high-mix, low-volume products can be smoothly manufactured. In cellular manufacturing, cells are frequently organized in a U-shape within a certain product flow so that operators can manage different machines with the least traveled distance.

Group technology, developing families of parts, is often used in conjunction with cellular manufacturing. *Group technology* refers to the process of studying a large population of different parts and then dividing them into groups of items having similar characteristics in material, geometry, processing, and so on. Group technology is often executed (groups are identified and coded) with the aid of CAD/CAM software tools, where processing information can be retrieved quickly and easily. Identifying product families implies that they can be produced by the same work cell (a group of machines, tooling, and people) with only minor changes in procedure or setup. Figure 2.7 presents an example of a cellular layout.

Cellular manufacturing provides numerous benefits to companies adopting lean technology. First, with the increasingly growing trend for higher productivity and short manufacturing lead times, cells have become popular in a variety of industries. The flexibility of manufacturing cells allows companies

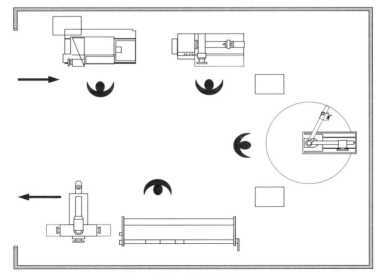

Figure 2.7 Cellular layout.

to meet customers' demands in terms of volume and variety. Such flexibility makes it possible for a company to provide customers with the right product at the right time in the right mix and in the right sequence. Second, coupled with SMED, cellular flow can help companies reach or get closer to single-piece flow, which facilitates implementation of a JIT environment and promotes Kaizen (continuous improvement) applications. Third, cellular manufacturing can help increase competitiveness by reducing costly transport and delay, shortening production lead time, reducing visibility, and saving factory space.

To implement cellular manufacturing in lean initiatives, operations are formed and linked into the most efficient combination to maximize the value-added content while minimizing waste. This entails mainly the following:

- Forming product families: evaluating which products to place in the cell (using group technology)
- Developing the most effective sequence of operations (using eliminate, combine, resequence, and simplify) with minimum transport
- Time-balancing processes (creating a balanced and continuous flow of products with minimum delay and inventory)
- Determining the cell's production capacity

2.6.5   Batch Size Reduction (Single-Unit Flow)

Small lot production (ideally, one piece) is an important component of any lean manufacturing strategy. Lot size directly affects setup time, inventory levels, and lead time. Thus, the batch size is reduced in lean manufacturing to increase agility in responding to customer needs without building up inventory.

Ordering smaller lots from suppliers helps reduce inventory holding cost but increases the ordering frequency. Hence, the *economic order quantity* (EOQ) is often determined based on a compromise between the ordering costs and the inventory holding costs:

$$EOQ = \sqrt{\frac{2(\text{annual demand})(\text{order cost})}{\text{annual holding cost per unit}}}$$

Similarly, the economic lot size (ELS) or economic production quantity (EPQ) is determined based on a balance between the costs of inventory and the costs of setup over a range of batch quantities. The lean philosophy, however, does not rely on the EOQ or the EPQ. Instead, lean promotes every operation, line, or subsystem to manufacture (order) what the downstream operation, line, or subsystem needs immediately in the smallest batch size possible, ideally 1 unit. The batch size is reduced gradually toward the 1-unit goal as more lean techniques are adopted (i.e., reduction in setup times, order lead times, and production variability). Ultimately, small batch size will increase a company's ability to make smaller lots and accept smaller orders. This also enables conversion from a make-to-stock to a make-to-order environment and facilitates use of a JIT and Kanban pull production system.

The impacts of reducing the production lot size can be clarified using a simple example. Assume that a product passes through five successive operations and it takes 1 minute on average to process the product at each operation. To process an order of 100 units using a batch size of 50, each operation in the flow will have to run 50 parts before passing them to the next operation. This results in an order lead time of 300 minutes, including a waiting time of 201 minutes before completing the first piece. With a batch size of 1-unit, the order lead time will be reduced to 104 minutes (reduced by about 66%), with a waiting time of only 5 minutes for completing the first piece. Figure 2.8 illustrates batch size reduction.

2.6.6   Balanced and Standardized Work

Work balancing and work standards are key components of lean manufacturing. They lay the groundwork for implementing a JIT production system, motivating labor, and continuous improvement. Indeed, standard operations are often sought as the best method for minimizing excess capacity, achieving production excellence, and laying the foundation for continuous improvement.

In LSS, a balanced work flow is achieved by determining the production takt time and designing workstations to operate at a cycle time equal to the takt time. Motion- and time-study techniques can be used to develop the sequence of operation, taking time measurements, and balancing work among workstations. To achieve both speed and consistency, in-process quality checks should be made part of the workload at each workstation.

Closely related to production speed and consistency, *standardized work* refers to the systematic determination and documentation of the work element sequence and process for each operation. The objective is to communicate clearly to an operator exactly how the job should be performed so that variability is minimized. As a result, each workstation will receive documentation (preferably displayed in the work area) on the takt time, work sequence, and instructions for quality checks. Other instructions for troubleshooting and maintenance can be included to train the operator on multitasking and to increase effectiveness.

Standardized work procedures mean that tasks are organized in the best sequence to ensure that products are being made in the best way every time. Standardized work procedures bring consistency (and as a result, better quality) to the workplace, and therefore they should be documented and given to anyone charged with completing the task.

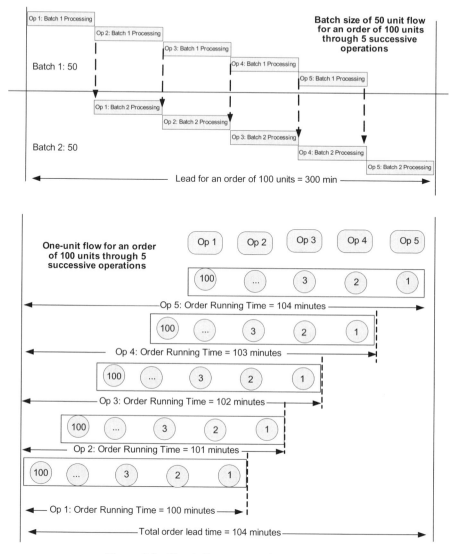

Figure 2.8   Batch flow versus single-unit flow.

2.6.7   Total Productive Maintenance

Balanced and standardized work often leads to implementing *total productive maintenance* (TPM), which aims to predict failures and to fix them before they cause machine failures and downtime. The goal of TPM is to maximize the overall equipment effectiveness by reducing equipment downtime while improving quality and capacity. This results in minimizing losses caused by breakdowns, defects, and accidents. Thus, TPM is a process to maximize the productivity of a piece of equipment for its entire life cycle.

On top of that, TPM fosters an environment for encouraging improvements in machine reliability and safety through the participation of all employees. With such an environment, TPM improves teamwork between machine operators and maintenance crew, improves safety, increases employee knowledge in machine-related processes, and creates a more involved and creative workforce.

Typically, maintenance is performed by dedicated, highly skilled technicians, not by machine operators. However, since machine operators know their machines better than anyone else, they can alert maintenance personnel and provide excellent information about machine performance. Also, machine operators often perform routine maintenance, cleaning, and lubrication. By having machine operators perform daily and routine maintenance, TPM provides the ability to better utilize maintenance technicians. This makes machine operators "process owners." Hence, TPM is based on an increased role for machine operators in maintenance efforts (preventive and corrective). TPM also addresses how to enhance equipment reliability and apply proper tools to increase the effectiveness of preventive and predictive activities.

In a lean environment, TPM attempts to minimize common equipment-related wastes, including:

- Setup and calibration
- Breakdowns and failures (short and catastrophic)
- Starving (idling) and blockage (stoppage)
- Reduced speed (low performance)
- Startup (warm-up)
- Defects (scrap, rework, rejects)

Properly implemented TPM also eliminates many machine-related bottlenecks through the following:

- Improved machine reliability
- Extended machine life
- Increased capacity without purchasing additional resources

2.6.8 Quality at the Source

In-process quality control or quality at the source means that employees are trained to make sure that the products they are sending to the next workstation are defect-free. This is essential to successful JIT/pull production systems, where only the items needed are produced, without overproduction to compensate for defective items. Defective units also contribute to increased variability in production rates and inventory levels. This undesirable result is often the focus of six-sigma projects and lean techniques. Finally, passing defective items downstream may result in shipping incomplete or nonconforming orders to customers, which will affect customer satisfaction directly and may lead to paying costly penalties or to sales cancellations.

End-of-line (predelivery) quality checks and statistical process controls are often costly and overwhelming since they tackle many sources of defects in many product types. This also makes it difficult to pinpoint the causes of quality problems. Rejected items at the end of the line are either scrapped or reworked. Scrapped units add to waste and cost in a lean context, and reworked items result in backtracking and flow complications and increased non-value-added activities such as material handling, resequencing, and rescheduling.

To successfully implement quality at the source, employees must also be given the means necessary to check product quality and to identify errors and production defects. These means and tools often include gauges, scales, special metrics and tools, and other inspection equipment. To reduce the time and effort necessary for inspection and defect identification, workers are supplied with benchmarked products or prototypes, such as sample parts, mating parts, product pictures and other visual items, charts, and illustrations.

2.6.9 5S Method

5S is a lean manufacturing technique directed at housekeeping the workplace. It focuses on developing visual order, organization, cleanliness, and standardization at the workstation. Within the lean context, 5S represents a determination to organize the workplace, to keep it neat and clean, and to maintain the discipline necessary for high-quality performance. 5S often refers to a series of activities designed to improve workplace organization and standardization so that the effectiveness of work is increased and employee morale is improved.

The 5S method is also of Japanese origin. The method is meant to create order at both the office and plant workstation as a sign of lean performance and transparency. 5S is based on the English translation of five Japanese words:

1. *Sort* through and sort out: Clean out the work area, keep in the work area what is necessary to do the work, and relocate or discard what is not actually used or needed.

2. *Set* in order and set workstation limits: Arrange needed items so that they are easy to find, easy to use, and easy to return. Principles of workstation design and motion economics are applied to arrange work items properly. The goal is to streamline production and eliminate search time and delays.

3. *Shine* and inspect through cleaning: Clean work tools and equipment and make them available at their points of use. While cleaning, inspect tools and equipment and provide the care needed to sustain performance. This may

include inspection, calibration, lubrication, and other care and preventive maintenance actions.

4. *Standardize:* Make all work areas similar so that procedures are obvious and instinctual and defects stand out. Standard signs, marks, colors, and shapes are used to standardize the workplace.

5. *Sustain:* Make the preceding four rules an integral attribute of the business or production system. Slowly but surely, an effort should be made to train employees and use 5S audits to sustain the continuity of their use. As 5S practices become natural habits, over time the total benefits of 5S will materialize and become transparent.

Other lean techniques will benefit strongly from an organized and standardized workplace. For example, Kanban use will be easier to include in a production system in which workstations are organized with clearly marked input and output stands. Prospected lean benefits of 5S include:

- Improved on-time delivery
- Improved quality and reduced defects
- Increased productivity
- Reduced lead times
- Reduced waste in materials, space, and time
- Reduced inventory and storage costs
- Reduced changeover time
- Reduced equipment downtime
- Improved safety

Ultimately with 5S, the system as a whole functions based on a workspace that allows for minimized waste. The effective approach will ensure that the time of workers at workstations will be spent doing productive (value-added) activities. An organized and clean workplace will also reduce the chance of error, rework, and injury, which are all undesirable high-cost events. Removing waste with 5S can be achieved by eliminating unnecessary tools and equipment as well as non-value-added activities such as searching, waiting, cleaning, and maintaining unnecessary tools and equipment.

2.6.10   Kaizen

Kaizen is a Japanese term that is typically referred to as *continuous improvement* or creation of a system of continuous improvement in quality, technology, processes, culture, productivity, safety, and leadership. Kaizen is considered as one of the driving forces behind Toyota's journey to excellence. This was a result of Toyota's success in establishing plans to sustain and further improve a desirable level of performance and to prevent degradation and loss

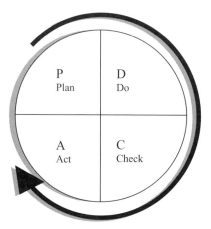

Figure 2.9   PDCA cycle of Kaizen.

of lean momentum. Many companies (automotive and nonautomotive) today are benchmarking Toyota's Kaizen model and implementing their own plans for continuous improvement.

Deploying a Kaizen effort or plan in a company implies working continuously for improvement in all facets of operation. Within a lean environment, Kaizen is based on making little changes on a regular basis (i.e., always improving productivity, safety, and effectiveness and reducing waste). These continual small improvements add up to major benefits. They result in improved productivity, improved quality, better safety, faster delivery, lower costs, and greater customer satisfaction.

Kaizen application encompasses many of the components of lean culture, such as quality circles, training, suggestion systems, preventive maintenance, and JIT philosophy. The deployment of continuous improvement effort with Kaizen can be best coordinated using a PDCA (plan, do, check, and act) cycle, shown in Figure 2.9.

The PDCA cycle was developed by Walter Shewhart, a pioneering statistician who developed statistical process control at Bell Laboratories in the United States during the 1930s. The approach was then proposed and used by William Edwards Deming in the 1950s. Coordinating continuous improvement plans with a PDCA cycle involves four stages: planning for improvement, doing or executing improvement actions, checking the implications of improvement actions, and making effective permanent actions toward improvement. The cyclical nature of PDCA captures the intended refinement and improvement essence of Kaizen. In brief, the Kaizen cycle can be viewed as a PDCA cycle having four steps:

- Establishing a plan of improvement changes
- Carrying out changes on a small scale

- Observing and evaluating the results of the changes proposed
- Implementing the approved changes on a large scale

Following is a description of the four stages of the PDCA cycle:

1. *Plan:* adopting a feasible and affordable plan to improve production and business operations. This is attained by first identifying opportunities for improvement and sources of errors and problems within the various facets of the process. Plans for improvement (fixing problems and increasing efficiency) are then specified through brainstorming sessions, benchmarking, and problem-solving techniques.

2. *Do:* making process changes to solve the problems defined and to improve system performance and efficiency. It is necessary to bear in mind that changes made toward continuous improvement should not disturb day-to-day normal production and business operations. Off-line (off-shift) experimental work can be used to test improvement changes.

3. *Check:* verifying the impact of changes (suggestions for changes) proposed in the do stage of PDCA. The focus here is to check whether or not small-scale or experimental changes are achieving the desired improvement result. Change effects on product quality and key lean practices are measured and used to approve or disapprove of the changes proposed. Some changes may lead to a new set of problems within a process. Such problems need to be identified and included in the change verification work.

4. *Act:* acting to implement the proven successful changes on a larger scale and to make these changes a permanent part of the process routine. This can be made successful through the involvement of all production or business associates of the underlying process. This includes all aspects of the business spectrum, from suppliers to customers. Monitoring measures of quality and performance can always help in ensuring the sustainability of the improvement actions made to the process. The continuous PDCA cycle takes us back to the plan stage to start new initiatives or to improve existing production or business practices.

2.7 SUMMARY

In this chapter we introduced LSS fundamentals, highlighting the integration synergy of lean manufacturing with a DMAIC approach while introducing several core principles and tools.

3

DESIGN FOR SIX-SIGMA FUNDAMENTALS

3.1 INTRODUCTION

The objective of this chapter is to introduce the transaction-based design for six-sigma (DFSS) process and theory and to lay the foundations for subsequent chapters. DFSS combines design analysis (e.g., requirements, cascading) with design synthesis (e.g., process engineering) within the framework of the deploying company's service (product) development systems. Emphasis is placed on CTQ (critical-to-quality requirement or big Y) identification, optimization, and verification using the transfer function and scorecard vehicles. A transfer function in its simplest form is a mathematical relationship between CTQ characteristics and/or their cascaded functional requirements (FRs) and critical influential factors (called X's). Scorecards help predict risks to the achievement of CTQs or FRs by monitoring and recording their mean shifts and variability performance.

DFSS is a disciplined and rigorous approach to service, process, and product design by ensuring that new designs meet customer requirements at launch. It is a design approach that ensures complete understanding of process steps, capabilities, and performance measurements by using scorecards, transfer functions, and tollgate reviews to ensure accountability of all design team

Simulation-Based Lean Six-Sigma and Design for Six-Sigma, by Basem El-Haik and Raid Al-Aomar
Copyright © 2006 John Wiley & Sons, Inc.

members, black belts, project champions, and deployment champions[1] and the rest of the organizations.

The DFSS objective is to attack the design vulnerabilities in both the conceptual and operational phases by deriving and integrating tools and methods for their elimination and reduction. Unlike the DMAIC methodology, the phases or steps of DFSS are not universally defined, as evidenced by the many customized training curricula available in the marketplace. The deployment companies will often implement the version of DFSS used by their choice of the vendor assisting in the deployment. On the other hand, a company will implement DFSS to suit its business, industry, and culture: creating its own version. However, all approaches share common themes, objectives, and tools.

DFSS is used to design or redesign a product or service. The process sigma level expected for a DFSS product or service is at least 4.5,[2] but can be 6σ or higher, depending on the entity designed. The production of such a low defect level from a product or service launch means that customer expectations and needs must be understood completely before a design can be operationalized. That is quality as defined by the customer.

The material presented herein is intended to give the reader a high-level understanding of service DFSS, and it uses and benefits. Following this chapter, readers should be able to assess how it could be used in relation to their jobs and to identify their needs for further learning. In Chapter 10 we present a simulation-based DFSS case study.

As defined in this book, DFSS has a two-track deployment and application. By *deployment* we mean the strategy adopted by the deploying company to launch the six-sigma initiative. It includes the deployment infrastructure, strategy, and plan for initiative execution (El-Haik and Roy, 2005). In what follows we are assuming that the deployment strategy is in place as a prerequisite for application and project execution. The DFSS tools are laid on top of four phases as detailed in Chapter 10 in what we will be calling the *simulation-based DFSS project road map*.

There are two distinct tracks within the six-sigma initiative as discussed in earlier chapters. The *retroactive* six-sigma DMAIC approach (see Chapter 1) takes existing process improvement as an objective, whereas *proactive* DFSS targets redesign and new service introductions in both the development and production (process) arenas.

The *service* DFSS approach can be broken down into the following phases (summarized as ICOV):

- Identify customer and design requirements. Prescribe the CTQs, design parameters, and corresponding process variables.
- Conceptualize the concepts, specifications, and technical and project risks.

---

[1] The roles and responsibilities of these six-sigma operatives and others are defined in El-Haik and Roy (2005).

[2] No more than approximately 1 defect per 1000 opportunities.

- Optimize the design transfer functions and mitigate the risks.
- Verify that the optimized design meets the intent (customer, regulatory, and deploying service function).

In this book the ICOV and DFSS acronyms are used interchangeably.

3.2 TRANSACTION-BASED DESIGN FOR SIX-SIGMA

Generally, customer-oriented design is a development process for transforming a customer's wishes into design service solutions that are useful to the customer. This process is carried over several development stages, starting at the conceptual stage. In this stage, conceiving, evaluating, and selecting good design solutions are difficult tasks with enormous consequences. It is usually the case that organizations operate in two modes: *proactive*, conceiving feasible and healthy conceptual entities, and *retroactive*, problem solving such that the design entity can live up to its committed potentials. Unfortunately, the latter mode consumes the largest portion of an organization's human and nonhuman resources. The DFSS approach highlighted in this book is designed to target both modes of operation.

DFSS is a premier approach to process design that is able to embrace and improve homegrown supportive processes (e.g., sales and marketing) within its development system. This advantage will enable the deploying company to build on current foundations while enabling it to reach unprecedented levels of achievement that exceed the set targets.

The link of the six-sigma initiative and DFSS to the company vision and annual objectives should be direct, clear, and crisp. DFSS has to be the crucial mechanism to develop and improve business performance and to drive up customer satisfaction and quality metrics. Significant improvements in all health metrics are the fundamental source of DMAIC and DFSS projects that will, in turn, transform culture one project at a time. Achieving a six-sigma culture is essential for the future well-being of the deploying company and represents the biggest return on investment beyond the obvious financial benefits. Six-sigma initiatives apply to all elements of a company's strategy, in all areas of the business, if a massive impact is really the objective.

The objective of this book is to present the service DFSS approach, concepts, and tools that eliminate or reduce both the conceptual and operational types of vulnerabilities of service entities and release such entities at six-sigma quality levels in all of their requirements.

Operational vulnerabilities take variability reduction and mean adjustment of the critical-to-quality, critical-to-cost, and critical-to-delivery requirements, the critical-to-satisfaction requirements (CTSs), as an objective and have been the subject of many knowledge fields, such as parameter design, DMAIC six-sigma, and tolerance design and tolerancing techniques. On the contrary, the conceptual vulnerabilities are usually overlooked, due to the lack of a compatible systemic approach to finding ideal solutions, ignorance of the designer, the pressure of deadlines, and budget limitations. This can be attributed, partly, to the fact that traditional quality methods can be characterized as after-the-fact practices since they use lagging information for developmental activities such as bench tests and field data. Unfortunately, this practice drives design toward endless cycles of design–test–fix–retest, creating what is broadly known as the *firefighting mode* of the design process (i.e., the creation of design-hidden factories). Companies who follow these practices usually suffer from high development costs, longer time to market, lower quality levels, and a marginal competitive edge. In addition, corrective actions to improve the conceptual vulnerabilities via operational vulnerabilities improvement means are marginally effective, if at all useful. Typically, these corrections are costly and difficult to implement as a service project progresses in the development process. Therefore, implementing DFSS in the conceptual stage is a goal that can be achieved when systematic design methods are integrated with quality concepts and methods up front. Specifically, on the technical side, we developed an approach to DFSS by borrowing from the following fundamental knowledge arenas: process engineering, quality engineering, TRIZ (Altshuller, 1964), axiomatic design[3] (Suh, 1990), and theories of probability and statistics. At the same time, there are several venues in our DFSS approach that enable transformation to a data-driven and customer-centric culture such as concurrent design teams, a deployment strategy, and a plan.

In general, most of the current design methods are empirical in nature. They represent the best thinking of the design community, which, unfortunately, lacks the design scientific base while relying on subjective judgment. When the company suffers in detrimental behavior in customer satisfaction, judgment and experience may not be sufficient to obtain an optimal six-sigma solution. We view this practice as another motivation to devise a DFSS method to address such needs.

Attention starts shifting from improving the performance during the later stages of the service design life cycle to the front-end stages, where design development take place at a higher level of abstraction (i.e., prevention versus solving). This shift is also motivated by the fact that the design decisions made during early stages of the service design life cycle have the largest impact on the total cost and quality of the system. It is often claimed that up to 80% of the total cost is committed in the concept development stage (Fredrikson, 1994). The research area of design is currently receiving increasing focus to address industry efforts to shorten lead times, cut development and manufacturing costs, lower total life-cycle cost, and improve the quality of the design entities in the form of products, services, and/or processes. It is the experience of the authors that at least 80% of design quality is also committed in the early

---

[3] A perspective design method that employs two design axioms: the independence and information axioms. See Appendix C and El-Haik (2005) for more details.

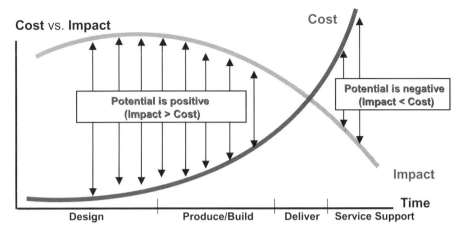

Figure 3.1   Effect of design stages on life cycle.

stages, as depicted in Figure 3.1 (see Yang and El-Haik, 2003). The "potential" in the figure is defined as the difference between the impact (influence) of the design activity at a certain design stage and the total development cost up to that stage. The *potential* is positive but decreasing as design progresses, implying reduced design freedom over time. As financial resources are committed (e.g., buying process equipment and facilities, hiring staff), the potential starts to change sign, going from positive to negative. In the consumer's hands the potential is negative, and the cost greatly exceeds the impact. At this stage, design changes for corrective actions can only be achieved at high cost, including customer dissatisfaction, warranty, marketing promotions, and in many cases under the scrutiny of the government (e.g., recall costs).

3.3   SERVICE DESIGN FOR SIX-SIGMA

Service DFSS is a structured data-driven approach to design in all aspects of service functions (e.g., human resources, marketing, sales, information technology) where deployment is launched to eliminate the defects induced by the design process and to improve customer satisfaction, sales, and revenue. To deliver on these benefits, DFSS uses such design methods as axiomatic design, creativity methods such as TRIZ, statistical techniques at all levels of design decision making in every corner of a business, identifying and optimizing the critical design factors (the X's), and validating all design decisions in the use (or surrogate) environment of the end user.

   DFSS is not an add-on but represents a cultural change within different functions and organizations where deployment is launched. It provides the means to tackle weak or new processes, driving customer and employee satisfaction. DFSS and six-sigma should be linked to the deploying company's annual objectives, vision, and mission statements. It should not be viewed as another short-lived initiative. It is a vital permanent component used to achieve leadership in design, customer satisfaction, and cultural transformation. From marketing and sales to development, operations, and finance, each business function needs to be headed by a deployment leader or a deployment champion. This local deployment team will be responsible for delivering dramatic change, thereby removing the number of customer issues and internal problems, and expediting growth. The deployment team can deliver on their objective through six-sigma operatives called *black belts* and *green belts*, who will be executing scoped projects that are in alignment with the objectives of the company. *Project champions* are responsible for scoping projects from within their realm of control and handing project charters (contracts) over to the six-sigma resource. The project champion will select projects consistent with corporate goals and remove barriers. Six-sigma resources will complete successful projects using six-sigma methodology and will train and mentor the local organization on six-sigma. The deployment leader, the highest-initiative operative, sets meaningful goals and objectives for the deployment in his or her function and drives the implementation of six-sigma publicly.

Six-sigma resources are full-time six-sigma operatives, in contrast to *green belts*, who should be completing smaller projects of their own as well as assisting black belts. They play a key role in raising the competency of the company as they drive the initiative into day-to-day operations.

Black belts are the driving force of service DFSS deployment. They are project leaders who are removed from day-to-day assignments for a period of time (usually, two years) to focus exclusively on design and improvement projects with intensive training in six-sigma tools, design techniques, problem solving, and team leadership. The black belts are trained by *master black belts*, who are hired if not homegrown.

A black belt should possess process and organization knowledge, have some basic design theory and statistical skills, and be eager to learn new tools. A black belt is a change agent in driving the initiative into his or her teams, staff function, and across the company. In doing so, their communication and leadership skills are vital. Black belts also need effective intervention skills, as they must understand why some team members may resist the six-sigma cultural transformation. Some soft training on leadership should be embedded within their training curriculum. Soft-skills training may target deployment maturity analysis, team development, business acumen, and individual leadership. In training, it is wise to share several initiative maturity indicators that are being tracked in the deployment scorecard. For example, alignment of a project to company objectives in its own scorecard (the big Y's), readiness of the project's mentoring structure, preliminary budget, team member identification, scoped project charter, and so on.

DFSS black belt training is intended to be delivered in tandem with a training project for hands-on application. The training project should be well scoped, with ample opportunity for tool application and should have cleared tollgate 0 prior to class. Usually, project presentations will be weaved into each training session.

While handling projects, the role of black belts spans several functions, such as learning, mentoring, teaching, and coaching. As a mentor, a black belt cultivates a network of experts in the project on hand, working with process operators, process owners, and all levels of management. To become self-sustained, the deployment team may need to task their black belts with providing formal training to green belts and team members.

Service DFSS is a disciplined methodology that applies the transfer function [CTSs = $f(X)$] to ensure that customer expectations are met, embeds customer expectations into the design, predicts design performance prior to setting up the pilot run builds performance measurement systems (scorecards) into the design to ensure effective ongoing process management, and leverages a common language for design within a design tollgate process.

DFSS projects can be categorized as design or redesign of an entity: whether product, process, or service. *Creative design* is the term that we will use to indicate new design, design from scratch, and incremental design to indicate the redesign case or design from a datum. In the latter case, some data can be used to establish a baseline for current performance. The degree of deviation of the redesign from the datum is the key factor in deciding on the usefulness of relative existing data. Service DFSS projects can arise from historical sources (e.g., service redesign due to customer issues) or from proactive sources such as growth and innovation (new service introduction). In either case, the service DFSS project requires greater emphasis on:

- A voice-of-the-customer collection scheme
- Addressing all (multiple) CTQs/CTSs as cascaded by the customer
- Assessing and mitigating technical failure modes and project risks in the project's own environment as they linked to the tollgate process reviews
- Project management, with some plan for communication to all affected parties and budget management
- A detailed project change management process

3.4   SERVICE DFSS: THE ICOV PROCESS

As mentioned in Section 3.1, DFSS has four phases – identify, characterize, optimize, and verify – over seven development stages. The acronym ICOV is used to denote these four phases. The service life cycle is depicted in Figure 3.2. Notice the position of the service ICOV phases of a design project.

Naturally, the process of service design begins when there is a need, an impetus. People create the need, whether it is a problem to be solved (if the waiting time of a call center becomes very long, the customer satisfaction process need to be redesigned) or a new invention. Design objective and scope

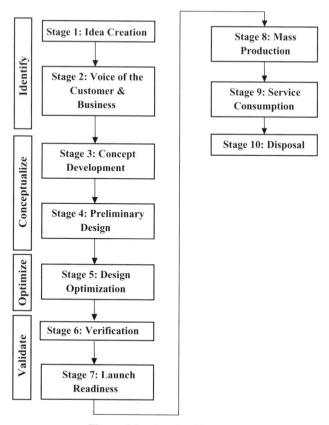

Figure 3.2 Service life cycle.

are critical in the impetus stage. A design project charter should describe simply and clearly what is to be designed. It cannot be vague. Writing a clearly stated design charter is just one step. In stage 2 the design team must write down all the information it may need, in particular the voice of the customer (VOC) and the voice of the business (VOB). With the help of the quality function deployment process, such consideration will lead the definition of the service design functional requirements to be grouped later into processes (systems and subsystems) and subprocesses (components). A functional requirement must contribute to an innovation or to a solution of the objective described in the design charter. Another question that should be on the minds of the team members relates to how the end result will look. The simplicity, comprehensiveness, and interfaces should make the service attractive. What options are available to the team? And at what cost? Do they have the right physical properties, such as bandwidth, completeness, language, and reliability? Will it be difficult to operate and maintain? Consider what methods you will need to process, store, and deliver the service.

In stage 3 the design team should produce a number of solutions. It is very important that they write or draw every idea on paper as it occurs to them. This will help them remember and describe the ideas more clearly. It is also easier to discuss them with other people if drawings are available. These first drawings do not have to be very detailed or accurate. Sketches will suffice and should be made quickly. The important thing is to record all ideas and develop solutions in the preliminary design stage (stage 4). The design team may find that they like several of the solutions. Eventually, the design team must choose one. Usually, careful comparison with the original design charter will help them to select the best solution subject to the constraints of cost, technology, and skills available. Deciding among several solutions possible is not always easy. It helps to summarize the design requirements and solutions and put the summary in a *morphological matrix*, which is a way to show all functions and corresponding possible design parameters (solutions). An overall design alternative set is synthesized from this matrix that are conceptually high potential and feasible solutions. Which solution would they choose? The *Pugh matrix*, a concept selection tool named after Stuart Pugh, can be used. The solution selected will be subjected to a thorough design optimization stage (stage 5). This optimization could be deterministic and/or statistical in nature. On the statistical front, the design solution will be made insensitive to uncontrollable factors (called *noise factors*) that may affect its performance. Factors such as customer usage profile, environment, and production case-to-case variation should be considered as noise. To assist on this noise insensitivity task, we rely on the transfer function as an appropriate vehicle. In stage 5 the team needs to prepare detailed documentation of the optimized solution, which must include all of the information needed to produce the service. Consideration for design documentation, process maps, operational instructions, software code, communication, marketing, and so on, should be put in place. In stage 6 the team can make a model assuming the availability of the transfer functions and later a prototype, or they can go directly to making a prototype or pilot. A model is a full-size or small-scale simulation of a process or product. Architects, engineers, and most designers use models. Models are one more step in communicating the functionality of the solution. For most people it is easier to understand the project end result when it is shown in three-dimensional form. A scale model is used when the design scope is very large. A prototype is the first working version of the team's solution. It is generally full size and often uses homegrown expertise. Design verification (stage 6), which includes testing and evaluation, is basically an effort to answer very basic questions: Does it work? (That is, does it meet the design charter?) If failures are discovered, will modifications improve the solution? These questions have to be answered. Once satisfactory answers are in hand, the team can move to the next development and design stage.

In stage 7 the team needs to prepare the production facilities where the service will be produced for launch. At this stage they should assure that the service is marketable and that no competitors will beat them to the market.

The team, together with the project stakeholders, must decide how many units to make. Similar to products, service may be mass-produced in low or high volume. The task of making the service is divided into jobs. Each worker trains to do his or her assigned job. As workers complete their special jobs, the service or product takes shape. Following stage 7, mass production saves time and other resources. Since workers train to do a certain job, each becomes skilled in that job.

3.5 SERVICE DFSS: THE ICOV PROCESS IN SERVICE DEVELOPMENT

Due to the fact that service DFSS integrates well with a service life-cycle system, it is an event-driven process: in particular in the development (design) stage. In this stage, milestones occur when the entrance criteria (inputs) are satisfied. At these milestones, the stakeholders, including the project champion, process or design owner, and deployment champion (if necessary) conduct *tollgate reviews*. A development stage has some thickness, that is, entrance and exit criteria for the bounding tollgates. The ICOV DFSS phases as well as the seven stages of the development process are depicted in Figure 3.3. In these reviews a decision should be made as to whether to proceed to the next phase of development, recycle back for further clarification on certain decisions, or cancel the project altogether. Cancellation of problematic projects as early as possible is a good thing. It stops nonconforming projects from progressing further while consuming resources and frustrating people. In any case, a black belt should quantify the size of the benefits of the design project in language that will have an impact on upper management, identify major opportunities for improving customer dissatisfaction and associated threats to salability, and stimulate improvements through publication of the DFSS approach.

In tollgate reviews, work proceeds when the exit criteria (required decisions) are made. As a DFSS deployment side bonus, a standard measure of development progress across the deploying company using a common development terminology is achieved. Consistent exit criteria from each tollgate include both the service DFSS's own deliverables, due to the application of the approach itself, and the business unit- or function-specific deliverables.

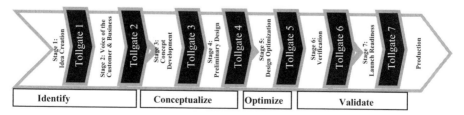

Figure 3.3 ICOV DFSS process.

3.6 SUMMARY

Service DFSS offers a robust set of tools and processes that address many of today's complex business design problems. The DFSS approach helps design teams frame their project based on a process with financial, cultural, and strategic implications to the business. The service DFSS comprehensive tools and methods described in this book allow teams to assess service issues quickly and identify financial and operational improvements that reduce costs, optimize investments, and maximize returns. Service DFSS leverages a flexible and nimble organization and maintains low development costs, allowing deploying companies to pass these benefits on to their customers. Service DFSS employs a unique gated process that allows teams to build tailormade approaches (i.e., not all the tools need to be used in each project). Therefore, it can be designed to accommodate the specific needs of the project charter. Project by project, the competency level of the design teams will be enhanced, leading to deeper knowledge and broader experience.

In this chapter we formed and integrated several strategic and tactical methodologies that produce synergies to enhance service DFSS capabilities to deliver a broad set of optimized solutions. The methods presented in this book have a wide spread of applications to help design teams and others in different project portfolios (e.g., staffing and other human resources functions, finance, operations and supply chain functions, organizational development, financial services, training, technology, software tools and methods).

Service DFSS provides a unique commitment to project customers by guaranteeing agreed-upon financial and other results. Each project must have measurable outcomes, and the design team is responsible for defining and achieving those outcomes. A service DFSS approach ensures these outcomes through risk identification and mitigation plans, variable (DFSS tools that are used over many stages)- and fixed (a DFSS tool that is used once)-tool structures, and advanced conceptual tools. The DFSS principles and structure should motivate design teams to provide business and customers with a substantial return on their design investment.

PART II

SIMULATION FUNDAMENTALS

4

BASIC SIMULATION CONCEPTS

4.1 INTRODUCTION

Production and business systems are key building blocks in the structure of modern industrial societies. Companies and industrial firms, through which production and business operations are usually performed, represent the major sector of today's global economy. Therefore, in the last decade, companies have made continuous improvement in their production and business systems a milestone in their strategic planning for the new millennium.

It is usually asserted that production and business operations have the potential to strengthen or weaken a company's competitive ability. To remain competitive, companies have to maintain a high level of performance by maintaining high quality, low cost, short manufacturing lead times, and a high level of customer satisfaction. As a result of fierce competition and decreasing business safety margins, efficient and robust production and business operations have become a necessity for survival in the marketplace.

Many industrial engineering subjects, such as operations research, quality control, and simulation, offer robust and efficient design and problem-solving tools with the ultimate aim of performance enhancement. Examples of this performance include the throughput of a factory, the quality of a product, or the profit of an organization. Simulation modeling as an industrial engineering tool for system design and improvement has undergone tremendous development in the past decade. This can be pictured through the growing

Simulation-Based Lean Six-Sigma and Design for Six-Sigma, by Basem El-Haik and Raid Al-Aomar
Copyright © 2006 John Wiley & Sons, Inc.

capabilities of simulation software tools and the application of simulation solutions to a variety of real-world problems in different business arenas.

With the aid of simulation, companies have been able to design efficient production and business systems, validate and trade off proposed design solution alternatives, troubleshoot potential problems, improve systems performance metrics, and consequently, cut cost, meet targets, and boost sales and profits. In the last decade, simulation modeling has been playing a major role in designing, analyzing, and optimizing engineering systems in a wide range of industrial and business applications.

Recently, six-sigma practitioners started to recognize the essential simulation role in system design, problem solving, and continuous improvement. System-level simulation in particular has become a key six-sigma tool for representing and measuring the time-based performance of real-world stochastic production and business systems. Examples include six-sigma studies that are focused on enhancing the productivity, quality, inventory, flow, efficiency, and lead time in production and business systems. In these studies, simulation is particularly essential for system representation, performance evaluation, experimental design, what-if analysis, and optimization. With such capability, simulation modeling can be utilized as an important tool in six-sigma applications for both DFSS and DMAIC.

Before presenting the details of simulation utilization in the 3S-LSS and 3S-DFSS approaches, we provide an introductory to the basic concepts of simulation modeling, with a focus on process modeling and time-based performance measurement. We also clarify the role of simulation studies in serving the increasing needs of companies that seek continuous improvement and optimality to production and business processes. To this end, we provide an introduction to the concept, terminology, and types of models, together with a description of simulation taxonomy and a justification for utilizing simulation tools in a variety of real-world applications. Such a background is essential to establishing a basic understanding of what simulation is all about and to understanding the key simulation role in simulation-based six-sigma studies and applications.

4.2 SYSTEM MODELING

System modeling as a term includes two important commonly used concepts: *system* and *modeling*. It is imperative to clarify such concepts before attempting to focus on their relevance to the "simulation" topic. Hence, in this section we introduce these two concepts and provide a generic classification of the various types of system models.

4.2.1 System Concept

The word *system* is commonly used in its broad meaning in a variety of engineering and nonengineering fields. In simple words, a system is often referred

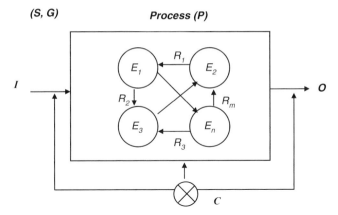

Figure 4.1   Definition of the system concept.

to as a set of elements or operations that are organized and logically related toward the attainment of a certain goal or objective. To attain the intended goal or to serve the desired function, it is necessary for the system to receive a set of inputs, to process them correctly, and to produce the required outcomes. To sustain such flow, some control is required to govern system behavior. Given such a definition, we can analyze any system (S) based on the architecture shown in Figure 4.1.

As shown in the figure, each system can be defined primarily in terms of a set of inputs (I) and a process (P) that transforms and changes the characteristics of system inputs to produce a specific set of outputs (O). The process is performed through a sets of system elements (E) governed by a set of relationships (R). An overall goal (G) is often defined to represent the purpose and objective of the system. To sustain a flawless flow and correct functionality of I–P–O, some type of control (C) is essentially applied to system inputs, process, and outputs. Thus, building a system requires primarily six basic elements:

1. Specifying the set of inputs (I) and the specifications required to produce specified outcomes (O)
2. Listing system elements $S = (E_1, E_2, E_3, \ldots, E_n)$ and defining the characteristics and the individual role of each element
3. Analyzing and understanding the logical relationships ($R_1, R_2, R_3, \ldots, R_m$) among the set of system elements defined
4. Specifying the set of outcomes (O) that should be produced and their specifications in order to reach the goal (G) and objective specified
5. Specifying the system controls (C) and their rules in monitoring and adjusting I–P–O flow to meet inputs (I), process (P), and output (O) specifications

6. Defining the goal (G) or the overall system objective and relating system structure (inputs, elements, relationships, controls, and outputs) to goal attainment

Based on this understanding of the system concept, it becomes essential that any arrangement of things or objects to be called a system have a defined set of elements. Such elements should have some logical relationships among them, and there should be some type of goal, objective, or useful outcome from the interaction of the system elements defined. Transforming system inputs into desired outputs is often performed through the system resources. Correct processing is often supported by controls and inventory systems to assure quality and maintain steady performance. Understanding the system concept is our gateway to the largest and widest subject of system analysis and design, which represents a key skill in simulation modeling.

Common examples of system concept include a classroom, a computer system, and a plant. We now apply the definition of system to a classroom. This leads to the following analyses:

1. Students of certain qualifications of age, academic level, major, and so on, are specified as the set of inputs (I). Upon graduation, students are also defined as system outcome (O).
2. The overall goal (G) is set to provide a good-quality education to students attending classes.
3. To achieve G, students are subject to various elements of the educational process (P) in the classroom, which involves attendance, participation in class activities, submitting assignments, passing exams, and so on, in order to complete the class with certain qualifications and skills.
4. The set of system elements is defined as follows:

$$S = [\text{tables, chairs, students, instructor, books, whiteboard}]$$

5. The elements defined in S are logically related through a set of relationships (R). For example, chairs are located around the tables, tables are sit to face the instructor, students sit on chairs, books are put on tables in front of students, the instructor stands in front of students and uses the whiteboard to explain the concepts, and so on.
6. Interaction of elements as explained in item 5 leads to the attainment of the goal (G) of learning or educating students on a certain subject.
7. Finally, the instructor applies regulations and rules of conduct in the classroom, attendance and grading policies, and so on, as a sort of control (C) on the education process.

Finally, it is worth mentioning that the term *system* covers products and processes. A product system could be an automobile, a cellular phone, a computer, or a calculator. Any of these products involves the defined components of the system in terms of inputs, outputs, elements, relationships, controls, and goal. Try to analyze from this perspective all the examples described. On the other hand, a process system can be a manufacturing process, an assembly line, a power plant, a business process such as banking operations, a logistic system, or an educational system. Similarly, any of these processes involves the system components defined in terms of inputs, outputs, elements, relationships, controls, and goal. Try to analyze all the examples from this perspective, too.

4.2.2 Modeling Concept

The word *modeling* refers to the process of representing a system (a product or a process) with a model that is easier to understand than the actual model and less expensive to build. The system representation in the model implies taking into account the components of the system, as discussed in Section 4.2.1. This includes representing system elements, relationships, goal, inputs, controls, and outputs. Modeling a system therefore has two prerequisites:

1. Understanding the structure of the actual (real-world) system and the functionality and characteristics of each system component and relationship. It is imperative to be familiar with a system before attempting to model it and to understand its purpose and functionality before attempting to establish a useful representation of its behavior. For example, the modeler needs to be familiar with the production system of building vehicles before attempting to model a vehicle body shop or a vehicle assembly operation. Similarly, the modeler needs to be familiar with various types of bank transactions before attempting to model banking operations.

2. Being familiar with different modeling and system representation techniques and methods. This skill is essential to choose the appropriate modeling technique for representing the underlying real-world system. Due to budgetary and time constraints, the model selected should be practically and economically feasible as well as beneficial in meeting the ultimate goal of modeling. As we discuss in Section 4.2.3, several model types can be used to create a system model, and selection of the most feasible modeling method is a decision based on economy, attainability, and usefulness.

The key question to be answered by the modeler is how to model a system of interest. The answer is a combination of art and science that is used for abstracting a real-world system into an educational model. The model should be clear, comprehensive, and accurate so that we can rely on its representation in understanding system functionality, analyzing its various postures, and predicting its future behavior. From this perspective, and as shown in Figure 4.2, *system modeling* is the process of transferring the actual system into a model that can be used to replace the actual one for system analysis and system improvement. The objects of a real-world system are replaced by

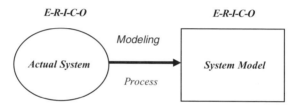

Figure 4.2   Process of system modeling.

objects of representation and symbols of indication, including the set of system elements (*E*), element relationships (*R*), system inputs (*I*), controls (*C*), and system outputs (*O*). The actual system E–R–I–C–O is mimicked thoroughly in the system model, leading to a representation that captures the characteristics of the real-world process. Of course, we should always keep in mind that the system model is just an approximation of the actual system. However, this approximation should not overlook the key system characteristics.

The science of system modeling stems from learning modeling methodologies and having fluency in analytical skills and logical thinking. The art aspect of modeling involves representation, graphical and abstraction skills of objects, relationships, and structure. It is simply being able to imagine system functionality, operation, and element interaction and possessing the capability of mapping relationships and representing behavior. For example, take a view of an assembly line, understand its functionality and structure, and try to represent it in a process diagram. Does the diagram explain everything you need for the sake of analyzing the assembly line? What else should be quantified and measured?

4.2.3   Types of Models

As mentioned in Section 4.2.2, several modeling methods can be used to develop a system model. The analyst decides on the modeling approach by choosing a certain type of model to represent the actual system. The analyst choice will be based on several criteria, such as system complexity, the objectives of system modeling, and the cost of modeling. As shown in Figure 4.3, we can classify the various types of models into four major categories: physical models, graphical models, mathematical models, and computer models. Following is a summary of these types of models.

Physical Models   Physical models are tangible prototypes of actual products or processes. Prototypes can use a 1:1 scale or any other feasible scale of choice. Such models provide a close-to-reality direct representation of the actual system and can be used to demonstrate the system's structure, the role of each system element, and the actual functionality of the system of interest in a physical manner. They help designers achieve a deeper understanding of system structure and details and to try out various configurations of design

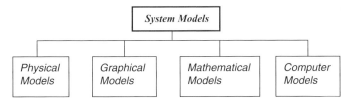

Figure 4.3 Types of system models.

Figure 4.4 Examples of product prototypes. (From http://www.centerpointdrafting. com.)

elements before the product is actually built and the process deployed. They are used primarily in engineering of large-scale projects to examine a limited set of behavioral characteristics in the system. For example, a stream and dam model is used to simulate the rate of outflow from the dam under a variety of circumstances. Prototypes built for vehicles in new car programs or for bridges, buildings, and other architectural designs are additional examples of physical models. Physical models such as car prototypes are built from clay or wood. Three-dimensional prototype printing machines are also available to develop prototypes using different materials. Various techniques of rapid prototyping and reverse engineering are also used to develop product and process prototypes. Figure 4.4 shows examples of product prototypes. Physical models can also be operational models, such as flight simulators and real-time simulators of chemical operations. Another form of physical model can be found in Lego-type machine and conveyor structures, and plants or reactor models. The benefit of physical models is the direct and easy-to-understand tangible representation of the underlying system. However, there are several limitations to physical models. These limitations include the cost of physical modeling, which could be enormous in some cases. Not all systems can be modeled with physical prototypes. Some systems are too complex to be prototyped. Other physical models might be time consuming and require superior crafting and development skills to be built. For example, think of building a physical model for a car's internal combustion engine or an assembly line of personal computers. What types of cost, time, complexity, and skill would be involved in modeling such systems?

Graphical Models Graphical models are abstractions of actual products or processes using graphical tools. This type of modeling starts with paper and pencil sketches, progresses to engineering drawings, and ends with pictures and movies of the system of interest. Common system graphical representations include a system layout, flow diagrams, block diagrams, network diagrams, process maps, and operations charts. Since graphical representations are static models, three-dimensional animations and clip videos are often used to illustrate the operation of a dynamic system or the assembly process of a product. This adds to the difficulty of abstracting a complex system graphically, especially when the underlying product and process are still in the design phase. Such graphical tools often oversimplify the reality of the system in blocks and arrows and do not provide technical and functionality details of the process. The absence of dynamic functionality in graphical representations makes it difficult to try out what-if scenarios and to explain how the system responds to various changes in model parameters and operating conditions. Such limitations have made most graphical tools part of or a prerequisite to other modeling techniques, such as physical, mathematical, and computer models. Figure 4.5 presents an example of a graphical model (the operations chart for a can opener).

Mathematical Models Mathematical modeling is the process of representing system behavior with formulas or mathematical equations. Such models are symbolic representations of systems functionality, decision (control) variables, response, and constraints. They assist in formulating the system design problem in a form that is solvable using graphical and calculus-based methods. Mathematical models use mathematical equations, probabilistic models, and statistical methods to represent the fundamental relationships among system components. The equations can be derived in a number of ways. Many of them come from extensive scientific studies that have formulated a mathematical relationship and then tested it against real data. Design formulas for stress–strain analyses and mathematical programming models such as linear and goal programming are examples of mathematical models. Figure 4.6 shows a mathematical model built using the MATLAB software tool. Some mathematical models are based on empirical models such as regression models and transfer functions. Mathematical models are used extensively nowadays to analyze weather events, earthquake risks, and population dynamics. Typically, a mathematical formula is a closed-form relationship between a dependent variable (Y) and one or more independent variables (X) in the form $Y = f(X)$. Such a formula can be linear or nonlinear with a certain parabolic order as an

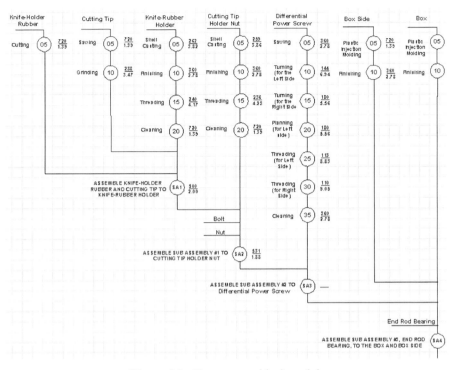

Figure 4.5 Process graphical model.

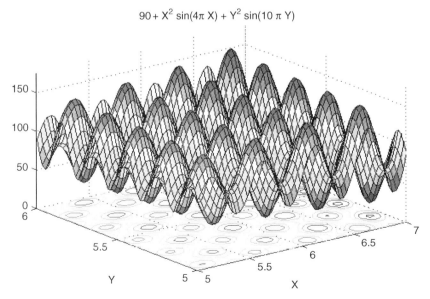

Figure 4.6 Two-dimensional mathematical model.

approximation. The dependent variable is often selected to measure a key characteristic of the system, such as the speed of a vehicle, the yield of a process, the profit of an organization, the demand of a product, and so on. Independent variables of the formula represent the key or critical parameters on which system response (the key characteristic) depends. Examples include the number of operators in a plant, sales prices of products, operating expenses, machine cycle times, the number of drive-through windows in a fast-food restaurant, and so on. Expressing the system response as a function of system parameters also provides an objective function for system experimental design, improvement studies, and optimization. A typical example is the use of linear programming models for applications of capital budgeting, production planning, resources allocation, and facility location. Other examples include queuing models, Markov chains, linear and nonlinear regression models, break-even analysis, forecasting models, and economic order quantity models. Unfortunately, not all system responses can be modeled using mathematical formulas. The complexity of most real-world systems precludes the application of such models. Hence, a set of simplification assumptions must often accompany the application of mathematical models for the formulas to hold. Such assumptions often lead to impractical results that limit the chance of implementing or even, in some cases, considering such results. For example, think of developing a formula that computes a production system throughput given parameters such as machine cycle times, speeds of conveyance systems, number of assembly operators, sizes of system buffers, and plant operating pattern. What type of mathematical model would you use to approximate such a response? How representative will the mathematical model be? Can you use the throughput numbers to plan schedule deliveries to customers?

Computer Models Computer models are numerical, graphical, and logical representation of a system (a product or a process) that utilizes the capability of a computer in fast computations, large capacity, consistency, animation, and accuracy. Computer simulation models, which represent the middleware of modeling, are virtual representations of real-world products and processes on the computer. Computer simulations of products and processes are developed using different application programs and software tools. For example, a computer program can be used to perform finite element analysis to analyze stresses and strains for a certain product design, as shown in Figure 4.7. Similarly, several mathematical models that represent complex mathematical operations, control systems, fluid mechanics, computer algorithms, and others can be built, animated, and analyzed with computer tools. Software tools are also available to develop static and dynamic animations of many industrial processes.

Accurate and well-built computer models compensate for the limitations of the other types of models. They are built using software tools, which is easier, faster, and cheaper than building physical models. In addition, the flexibility of computer models allow for quick changes, easy testing of what-ifs,

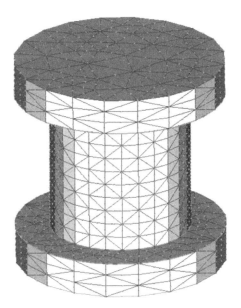

Figure 4.7 Finite element analysis computer model.

and quick evaluation of system performance for experimental design and optimization studies. Computer models also provide the benefits of graphical models with modern animation and logical presentation tools. Indeed, most computer models are built using graphical modeling tools. Compared to mathematical models, computer models are much more realistic and efficient. They utilize computer capabilities for more accurate approximations, they run complicated computations in little time, and they can measure system performance without the need for a closed-form definition of the system objective function. Such capabilities made computer models the most common modeling techniques in today's production and business applications. Limitations of computer models include computer limitations, application software limitations, the limitations of system complexity, and limitations of analysts.

Discrete event simulation (DES) (discussed in detail in Chapter 5) is computer simulation that mimics the operation of real-world processes as they evolve over time. This type of computer model is the major focus of this book, where simulation is used as a template for applying lean six-sigma and design for six-sigma methods. DES computer modeling assists in capturing the dynamics and logics of system processes and estimating the system's long-term performance under stochastic conditions. Moreover, DES models allow the user to ask what-if questions about a system, present changes that are made in the physical conditions, and run the system (often for long periods) to simulate the impact of such changes. The model results are then compared to gain insight into the behavior of the system. For example, a DES plant model can be used to estimate the assembly line throughput by running the model

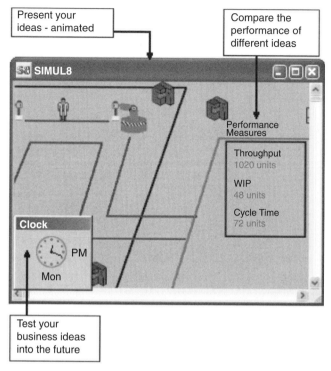

Figure 4.8   DES environment in the Simul8 software package.

dynamically and tracking its throughput hour by hour or shift by shift. The model can be also used to assess multiple production scenarios based on a long-term throughput average. As shown in Figure 4.8, simulation software tools provide a flexible environment of modeling and analyses that makes DES incomparable to mathematical and physical models.

4.3   SIMULATION MODELING

As shown earlier, *modeling* is the art and science of capturing the functionality and the relevant characteristics of real-world systems and presenting such systems in a form that facilitates system knowledge, analyses, improvement, and optimization. Physical, graphical, mathematical, and computer models are the major types of models developed in different engineering applications. This section focuses on defining the *simulation* concept, developing the taxonomy of various types of simulation models, and explaining the role of simulation in planning, designing, and improving the performance of modern business and production operations.

4.3.1 Simulation Defined

Simulation is widely used term in reference to computer simulation models that represent a product or a process. Such models are built based on both mathematical and logical relationships impeded within the system structure. For example, finite element analysis is the mathematical basis for a camshaft product simulation model, and the operation of a factory is the logical design basis for a plant process model.

System simulation or *simulation modeling*, in general, is the mimicking activity of the operation of a real system, in a computer, with a focus on process flow, logic, and dynamics. Therefore, simulation means making a simplified representation of an original. Just as a model aircraft captures the many of the important physical features of a real aircraft, so a simulation model captures the important operational features of a real system. Examples of system simulation are the day-to-day operation of a bank, the value of a stock portfolio over a time period, the running of an assembly line in a factory, and the staff assignment of a hospital.

Therefore, instead of attempting to build extensive mathematical models by experts, simulation software tools has made it possible to model and analyze the operation of a real system by engineers and managers, not only programmers. This allows engineers to collect pertinent information about the behavior of the system by executing a computerized model instead of observing the real one.

In conclusion, the primary requirements for simulation are a simulation analyst (e.g., a six-sigma core team member or the team itself) a computer, and a simulation language or software tool. The analyst is the key player in conducting the simulation study. He or she is responsible for understanding the real-world system and analyzing its elements, logic, inputs, outputs, and goals, in the manner presented in Section 4.2.1. The analyst then operates the computer system and uses the software tool to build, validate, and verify the simulation model. Finally, the analyst analyzes the results obtained from running the simulation model and conducts experimental design with the model in order to draw conclusions on model behavior and determine the best parameter settings.

The computer system provides the hardware and software tools required to operate and run the simulation. The simulation software or language provides the platform and environment that facilitates model building, testing, debugging, and running. The simulation analyst utilizes the simulation software on a capable computer system to develop a system simulation model that can be used as a practical (close-to-reality) representation of the actual system. Figure 4.9 presents the simulation concept.

4.3.2 Simulation Taxonomy

Based on the internal representation scheme selected, simulation models can be *discrete*, *continuous*, or *combined*. Discrete event simulation models, which

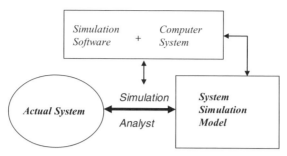

Figure 4.9   Simulation concept.

are the focus of this book, are the most common among simulation types. DES models are based on a discrete internal representation of model variables (variables that change their state at discrete points in time). DES mechanics are discussed in detail in Chapter 5.

In general, discrete simulation models focus on modeling discrete variables that are presented by random or probabilistic models, where the state of the system changes in discrete points in time. A discrete variable can be the number of customers in a bank, products and components in an assembly process, or cars in a drive-through restaurant.

Continuous simulation models, on the other hand, focus on continuous variables, random or probabilistic, where the state of the system changes continuously. Examples of continuous variables include the level of water behind a dam, chemical processes in a petroleum refinery, and the flow of fluids in distribution pipes.

Combined simulation models include both discrete and continuous elements. For example, separate (discrete) fluid containers arrive to a chemical process where fluids are poured into a reservoir to be processed in a continuous manner. Figure 4.10 shows an animation of a three-dimensional discrete/continuous simulation of pipe casting developed with a Brooks Automation AutoMod simulation package.

Mathematically, we can represent a model state using a random variable X, where $X(t)$ is a state variable that changes over time t. For example, $D(t)$ is a continuous state variable that represents the delay of the ith customer in a bank. Another example of a discrete-state variable is the number of customers $\{Q(t)\}$ in a queue at time t in an M/M/1 queuing system.

Furthermore, models are either *deterministic* or *stochastic* depending whether they model randomness and uncertainty in a process or not. A stochastic process is a probabilistic model of a system that evolves randomly in time and space. Formally, a stochastic process is a collection of random variables $\{X(t)\}$ defined on a common sample (probability) space. Examples of stochastic models operating with random variables include interarrival times of customers arriving at a bank and service or processing times of customers

Figure 4.10 Discrete/continuous AutoMod simulation model.

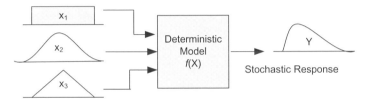

Figure 4.11 Deterministic model with stochastic response.

requests or transactions, variable cycle times, and machines' time to failure and time to repair parameters.

Deterministic models, on the other hand, involve no random or probabilistic variables in its processes. Examples include modeling fixed-cycle-time operations, such the case of automated systems and arrivals with preset appointments to a doctor or a lawyer's office. The majority of real-world operations are probabilistic. Hence most simulation studies involve random generation and sampling from certain probability distributions to model key system variables. Variability in model inputs lead to variability in model outputs. As shown in Figure 4.11, therefore, a deterministic model will generate stochastic response (Y) when model inputs (X_1, X_2, and X_3) are stochastic.

Finally, and based on the nature of model evolvement with time, models can be *static* or *dynamic*. In static models the system state does not change over time. For example, when a certain resource is always available in a manufacturing process, the state of such a resource is fixed with time. Every time this resource is required or needed, there will be no expected change in its status. Monte Carlo simulation models are time-independent (static) models that deal with a system of fixed states. In such spreadsheet-like models, certain

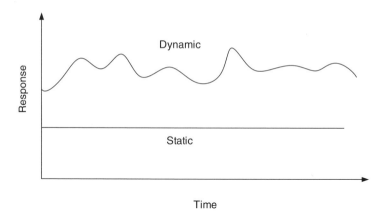

Figure 4.12   Static and dynamic model responses.

variable values are changed by random generation, and a certain measure or more are evaluated as to such changes without considering the timing and the dynamics of such changes. Most operational models are, however, dynamic. System state variables often change with time, and the interactions that result from such dynamic changes do affect the system behavior. Hence, in DES simulation, the time dimension is live. As shown in Figure 4.12, a model can involve both static and dynamic responses. This again makes DES models the most common among simulation types since they model variables that change their state at discrete points in time.

Dynamic simulation models are further divided into terminating and non-terminating models based on run time. Terminating models are stopped by a certain *natural* event, such as a doctor who sees a limited number of patients per day, a workshop that finishes all quantity in a certain order, or a bank that operates for 8 hours a day. In such systems, the length of a simulation run depends on the number of items processed, or on reaching a specified time period, with the model terminating upon completion. Models that run for a specified period often include initial conditions such as cleaning the system, filling bank ATMs, resetting computers in a government office, or emptying buffers in an assembly line.

Nonterminating models, on the other hand, can run continuously since the impact of initialization is negligible over the long run and no natural event has been specified to stop them. Most plants run on a continuous mode, where production lines start every shift without emptying the system. Hence, the run time for such models is often determined statistically to obtain a steady-state response, as we discuss later.

Figure 4.13 presents the taxonomy of different simulation types with highlighted attributes of DES. As noted in the figure, DES models are digital (discrete), stochastic, and dynamic computer models of terminating or nonterminating (steady-state) response. Such three characteristics often

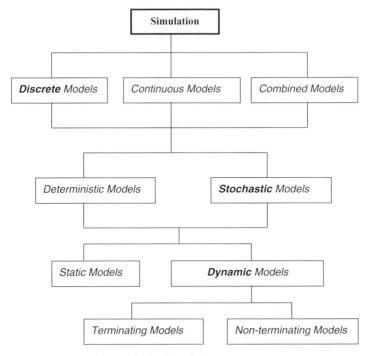

Figure 4.13   Simulation taxonomy.

resemble the actual behavior of many real-world systems and transactional processes. For example, in manufacturing systems, the flow of raw material, assembly components, and products can be modeled as discrete, dynamic, and stochastic processes. Similarly, many service facilities often deal with discrete entities that run dynamically in a stochastic manner.

4.4   THE ROLE OF SIMULATION

After understanding the various meanings and aspects of the term *simulation modeling*, it is necessary to clarify the role that simulation plays in modern industrial and business firms. In this section we clarify the role of simulation by first justifying the use of simulation both *technically* and *economically* and then presenting the spectrum of simulation applications to various industries in the manufacturing and service sectors. It is also worth mentioning that using simulation in industrial and business application is the most common but not the only field in which simulation is utilized; it is also used for educational and learning purposes, training, virtual reality applications, movies and animation production, and criminal justice, among others.

4.4.1 Simulation Justified

The question of why and when to simulate is typical of those that cross the minds of practitioners, engineers, and managers. We simply simulate because of simulation capabilities that are unique and powerful in system representation, performance estimation, and improvement. Simulation is often the analysts' refuge when other solution tools, such as mathematical models, fail or become extremely difficult to approximate the solution to a certain problem. Most real-world processes in production and business systems are complex, stochastic, and highly nonlinear and dynamic, which makes it almost impossible to present them using physical or mathematical models. Attempts to use analytical models in approaching real systems usually require many approximations and simplifying assumptions. This often yields solutions that are unsuitable for problems of real-world applications.

Therefore, analysts often use simulation whenever they meet complex problems that cannot be solved by other means, such as mathematical and calculus-based methods. The performance of a real system is a complicated function of the design parameters, and an analytical closed-form expression of the objective function or the constraints may not exist. A simulation model can therefore be used to replace the mathematical formulation of the underlying system. With the aid of the model and rather than considering every possible variation of circumstances for the complex problem, a sample of possible execution paths is taken and studied.

In short, simulation is often utilized when the behavior of a system is complex, stochastic (rather than deterministic), and dynamic (rather than static). Analytical methods, such as queuing systems, inventory models, and Markovian models, which are commonly used to analyze production and business systems, often fail to provide statistics on system performance when real-world conditions intensify to overwhelm and exceed the system-approximating assumptions. Examples include entities whose arrival at a plant or bank is not a Poisson process, and the flow of entities is based on complex decision rules under stochastic variability within availability of system resources.

Decision support is another common justification of simulation studies. Obviously, engineers and managers want to make the best decisions possible, especially when encountering critical stages of design, expansion, or improvement projects where the real system has not yet been built. By carefully analyzing the hypothetical system with simulation, designers can avoid problems with the real system when it is built. Simulation studies at this stage may reveal insurmountable problems that could result in project cancellation, and save cost, effort, and time. Such savings are obtained since it is always cheaper and safer to learn from mistakes made with a simulated system (a computer model) than to make them for real. Simulation can reduce cost, reduce risk, and improve analysts' understanding of the system under study.

The economic justification of simulation often plays a role in selecting simulation as a solution tool in design and improvement studies. Although simulation studies might be costly and time consuming in some cases, the benefits and savings obtained from such studies often recover the simulation cost and avoid much further costs. Simulation costs are typically the initial simulation software and computer cost, yearly maintenance and upgrade cost, training cost, engineering time cost, and other costs: for traveling, preparing presentations with multimedia tools, and so on. Such costs are often recovered with the first two or three successful simulation projects. Further, the cost and time of simulation studies are often reduced by analyst experience and become minuscule compared to the long-term savings from increasing productivity and efficiency.

4.4.2 Simulation Applications

A better answer to the question "why simulate?" can be reached by exploring the wide spectrum of simulation applications to all aspects of science and technology. This spectrum starts by using simulation in basic sciences to estimate the area under a curve, evaluating multiple integrals, and studying particle diffusion, and continuer by utilizing simulation in practical situations and designing queuing systems, communication networks, economic forecasting, biomedical systems, and war strategies and tactics.

Today, simulation is being used for a wide range of applications in both manufacturing and business operations. As a powerful tool, simulation models of manufacturing systems are used:

- To determine the throughput capability of a manufacturing cell or assembly line
- To determine the number of operators in a labor-intensive assembly process
- To determine the number of automated guided vehicles in a complex material-handling system
- To determine the number of carriers in an electrified monorail system
- To determine the number of storage and retrieval machines in a complex automated storage and retrieval system
- To determine the best ordering policies for an inventory control system
- To validate the production plan in material requirement planning
- To determine the optimal buffer sizes for work-in-progress products
- To plan the capacity of subassemblies feeding a production mainline

For business operations, simulation models are also being used for a wide range of applications:

- To determine the number of bank tellers, which results in reducing customer waiting time by a certain percentage

TABLE 4.1   Examples of Simulation Applications

Manufacturing Industries	Service Industries
Automotive	Banking industry
Aerospace	Health systems
Plastics industry	Hotel operations
Paper mills	Communication services
House appliances	Computer networks
Furniture manufacturing	Transportation systems
Chemical industry	Logistics and supply chain
Clothing and textile	Restaurants and fast food
Packaging	Postal services
Storage and retrieval	Airport operations

- To design distribution and transportation networks to improve the performance of logistic and vending systems
- To analyze a company's financial system
- To design the operating policies in a fast-food restaurant to reduce customer time-in-system and increase customer satisfaction
- To evaluate hardware and software requirements for a computer network
- To design the operating policies in an emergency room to reduce patient waiting time and schedule the working pattern of the medical staff
- To assess the impact of government regulations on different public services at both the municipal and national levels
- To test the feasibility of different product development processes and to evaluate their impact on company's budget and competitive strategy
- To design communication systems and data transfer protocols

Table 4.1 lists examples of simulation applications in the manufacturing and service sectors.

To reach the goals of the simulation study, certain elements of each simulated system often become the focus of a simulation model. Modeling and tracking such elements provide attributes and statistics necessary to design, improve, and optimize the underlying system performance. Table 4.2 presents examples of simulated systems with model main elements in each system.

4.4.3   Simulation Precautions

Like any other engineering tool, simulation has limitations. Such limitations should be dealt with as a motivation and should not discourage analysts and decision makers. Knowing limitations of the tool in hand should emphasize using it wisely and motivate the user to develop creative methods and establish the correct assumptions that benefit from the powerful simulation capa-

TABLE 4.2 Examples of Simulated Systems

Simulated System	Examples of Model Elements (E-R-I-C-O)
Manufacturing system	Parts, machines, operators, conveyors, storage
Emergency Room	Patients, beds, doctors, nurses, waiting room
Bank	Customers, bank tellers, ATMS, loan officers
Retail store	Shoppers, checkout cash registers, customer service
Computer network	Server, client PCS, administrator, data protocol
Freeway system	Cars, traffic lights, road segments, interchanges
Fast-food restaurant	Servers, customers, cars, drive-through windows
Border crossing point	Cars, customs agents, booths, immigration officers
Class registration office	Students, courses, registration stations, helpers
Supply chain/logistics	Suppliers and vendors, transportation system, clients

bilities and preclude simulation limitations from being a damping factor. However, certain precautions should be considered in using simulation to avoid the potential pitfalls of simulation. Examples of issues that we should pay attention to when considering simulation include the following:

1. The simulation analyst or decision maker should be able to answer the question of when not to simulate. A lot of simulation studies are considered to be design overkill when conducted for solving problems of relative simplicity. Such problems can be solved using engineering analysis, common sense, or mathematical models. Hence, the only benefits from approaching simple systems with simulation is being able to practic modeling and to provide an animation of the targeted process.

2. The cost and time of simulation should be considered and planned well. Many simulation studies are underestimated in terms of time and cost. Some decision makers think of simulation study as model-building time and cost. Although model building is a critical phase of a simulation study, it often consumes less time and cost than does experimental design or data collection.

3. The skill and knowledge of the simulation analyst. Being an engineer is almost essential for simulation practitioners because of the type of analytical, statistical, and system analyses skills required for conducting simulation studies.

4. Expectations from the simulation study should be realistic and not overestimated. A lot of professionals think of simulation as a "crystal ball" through which they can predict and optimize system behavior. It should be clear to the analyst that simulation models themselves are not system optimizers. While it should be asserted that simulation is just a tool and an experimental platform, it should also be emphasized that combining simulation with appropriate statistical analyses, experimental design, and efficient search engine can lead to invaluable system information that benefits planning, design, and optimization.

5. The results obtained from simulation models are as good as the model data inputs, assumptions, and logical design. The commonly used phrase *garbage-in-garbage-out* (GIGO) is very applicable to simulation studies. Hence, special attention should be paid to data inputs selection, filtering, and assumptions.

6. The analyst should pay attention to the level of detail incorporated in the model. Depending on the objectives of the simulation study and the information available, the analyst should decide on the amount of detail incorporated into the simulation model. Some study objectives can be reached with macro-level modeling, whereas others require micro-level modeling. There is no need for the analyst to exhaust his or her modeling skills trying to incorporate details that are irrelevant to simulation objectives. Instead, the model should be focused on providing the means of system analysis that yields results directly relevant to study objectives.

7. Model validation and verification is not a trivial task. As discussed later, model validation focuses on making sure that a model behaves as required by the model-designed logic and that its response reflects the data used into the model. Model verification, on the other hand, focuses on making sure that the model behavior resembles the intended behavior of the actual simulated system. Both practices determine the degree of model reliability and require the analyst to be familiar with the skills of model testing and the structure and functionality of the actual system.

8. The results of simulation can easily be misinterpreted. Hence, the analyst should concentrate efforts on collecting reliable results from the model through proper settings of run controls (warm-up period, run length, and number of replications) and on using the proper statistical analyses to draw meaningful and accurate conclusions from the model. Typical mistakes in interpreting simulation results include relying on a short run time (not a steady-state response), including in the results biases caused by initial model conditions, using the results of one simulation replication, and relying on the response mean while ignoring the variability encompassed into response values.

9. Simulation inputs and outputs should be communicated clearly and correctly to all parties of a simulation study. System specialists such as process engineers and system managers need to be aware of the data used and the model logic in order to verify the model and increase its realistic representation. Similarly, the results of the simulation model should be communicated to get feedback from parties on the relevancy and accuracy of results.

10. The analyst should avoid using incorrect measures of performance when building and analyzing model results. Model performance measures should be programmed correctly into the model and should be represented by statistics collected from the model. Such measures should also represent the type of information essential to the analyst and decision maker to draw conclusions and inferences about model behavior.

11. The analyst should avoid the misuse of model animation. In fact, animation is an important simulation capability that provides engineers and decision makers with a great tool for system visualization and response observance. Hence, it is true that "a picture is worth a thousand words." Such a tool is also useful for model debugging, validation and verification, and presentation to executives and customers. However, a lot of people misuse model animation and rely on their observation to draw conclusions as to model long-term behavior. Given that simulation models are stochastic and dynamic in nature, it should be clear to the analyst that a model's status at a certain time does not necessarily reflect its long-term behavior. Instead, model statistics are a better representation of model response.

12. The analyst needs to get the support of upper management and decision makers to make a simulation study fruitful and successful.

13. Finally, the analyst should select the appropriate simulation software tools that fit the analyst's knowledge and expertise and that are capable of modeling the underlying system and providing the simulation results required. The criteria for selecting the proper simulation software tools are available in the literature and are not the focus of this book. It should be known, however, that simulation packages vary in their capabilities and inclusiveness of different modeling systems and techniques, such as conveyor systems, power and free systems, automated guided vehicle systems, kinematics, automated storage and retrieval systems, human modeling capabilities, statistical tools, optimization methods, animation, and so on.

4.5 SIMULATION SOFTWARE

Simulation software tools are in common use among simulation and six-sigma practitioners in various types of applications and for different purposes. The major benefit that simulation packages provide to six-sigma practitioners is the ability to model dynamic systems under close-to real-world conditions. Processes of a transactional nature in both manufacturing and services often behave in a complex, dynamic, and stochastic manner. Such capability therefore allows analysts to better express new concepts and designs for transactional processes, to measure process performance with time-based metrics, to conduct statistical analyses, and to improve or optimize new and current systems.

Indeed, many simulation software tools were initially built by practitioners to model particular real-world systems, such as material handling systems, health systems, and assembly lines. They were built from entities and processes that mimic the objects and activities in the real system. To further meet the needs of six-sigma practitioners and engineers in general, simulation vendors developed and integrated many modules into their software products to simplify model building, facilitate model customization, allow for the creation of

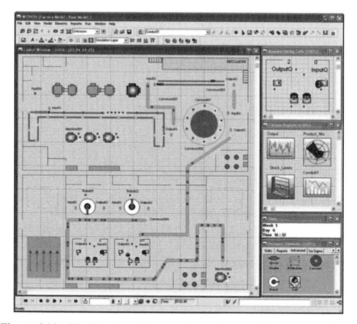

Figure 4.14 Flexible simulation software for modeling and analyses.

impressive animation, and enable analysts to conduct statistical analyses and run optimization searches. Some of these modules include six-sigma definitions and calculators and provide links to spreadsheet and statistical software to run experimental designs and six-sigma analyses directly. Figure 4.14 shows an example of a flexible simulation environment developed with the WITNESS simulation package. WITNESS includes a module for six-sigma calculation, a module for optimization, and a direct link to the MINITAB statistical package.

The spectrum of simulation software currently available ranges from procedural languages and simulation languages to special simulators. The vast amount of simulation software available can be overwhelming for new users. Common software packages include AutoMod, WITNESS, Simul8, and Arena. In Chapter 11 we provide a comprehensive list of simulation software tools and vendors and provide further details on simulation software features, applications, and comparisons.

4.6 SUMMARY

Basic concepts of simulation modeling include the system, model, and simulation concepts. Systems include inputs, entities, relationships, controls, and outputs. System elements should be tuned toward attaining an overall system goal. A system model is a representation or an approximation of a real-world system. Models can be physical, graphical, mathematical, and computer models. The goal of modeling is to provide a tool for system analysis, design, and improvement that is cheaper, faster, and easier to understand. Simulation is the art and science of mimicking the operation of a real-world system on the computer. It is aimed at capturing the complex, dynamic, and stochastic characteristics of a real-world process, where other types of models fall short. Based on the type of state variables, computer simulation models can be discrete, continuous, or combined. They can also be deterministic or stochastic, based on randomness modeling. Finally, they can be static or dynamic, based on the changes of system state. The focus of this book is on discrete event simulation models, which represent systems' complex and stochastic behavior dynamically in terms of discrete state variables. DES is justified by its growing capability, software tools, and wide spectrum of real-world applications. DES simulation is the backbone of simulation-based lean six-sigma and design for lean six-sigma.

5

DISCRETE EVENT SIMULATION

5.1 INTRODUCTION

As discussed in Chapter 4, computer simulation, is a technique that imitates the operation of a real-world system as it evolves over *time*. (See Section 4.2.1 for the definition of *system* used in this book.) It is considered an experimental and applied methodology that seeks to describe the behavior of systems, construct hypotheses that account for observed behavior, and predict future behavior. A simulation is therefore the execution of a model, represented by a computer program that gives information about the system being investigated.

Process simulation, often used to model production and business processes in both the manufacturing and service sectors, is referred to as *discrete event simulation* (DES). Discrete event systems are dynamic systems that evolve in time by the occurrence of events at possibly irregular time intervals. Since this resembles the nature of the majority of real-world production and business systems, DES models are used widely in real-world applications. Examples include traffic systems, manufacturing systems, computer-communications systems, call centers, bank operations, hospitals, restaurants, production lines, and flow networks. Most of these systems can be modeled in terms of discrete events whose occurrence causes the system to change from one state to another in a stochastic manner.

The *simulation modeling approach* for analyzing a production or business system is opposed to the analytical approach, where the method of analyzing the system is purely theoretical or an approximation of a theoretical model. As a more realistic alternative to the analytical approach, the simulation approach gives more flexibility to the six-sigma team and convenience to the decision maker by utilizing certain effective mechanics. Events activation, random number generation, and time advancement are the main DES mechanics. Events, which are activated at certain points in time, affect the overall state of the system (system variables, resources, entity flow, etc.). The points in time at which an event is activated are usually randomized using a random number generation technique that is built in the model software tool so that no input from outside the system is required. Simulation events exist autonomously in a discrete manner. Time (a simulation clock) is therefore advanced to a next scheduled event since nothing will happen between the execution of any two consecutive events. In addition to powerful DES mechanics, fast computations on today's high-speed processors, along with growing graphics capability, have contributed greatly to the effectiveness and animation or visualization capability in DES models. This chapter provides a deeper understanding of the DES process, components, and mechanisms. Examples of manual simulations are used to clarify DES functionality.

5.2   SYSTEM MODELING WITH DES

By utilizing computer capabilities in logical programming, random generation, fast computations, and animation, DES modeling is capable of capturing the characteristics of a real-world process and estimating system performance measures at different settings of its design parameters. This is particularly important in six-sigma DAMIC and DFSS applications, where system performance is optimized and improved. To measure such performance, DES imitates the stochastic and complex operation of a real-world system as it evolves over time and seeks to describe and predict the system's actual behavior.

The DES approach has numerous powerful capabilities that can provide the six-sigma team with information elements that quantify the system characteristics, such as design or decision parameters and performance measures (CTQs/CTSs). Such information makes a DES model an effective decision support system. For example, DES can be used to estimate the number of resources needed in a system (e.g., how many operators, machines, trucks). The model can also be used to arrange and balance resources to avoid bottlenecks, cross-traffic, backtracking, and excessive waiting times or inventories.

Toward this end, system modeling with DES includes mimicking the structure, layout, data, logic, and statistics of the real-world system and representing them in a DES model. An abstraction of system modeling with DES is illustrated in Figure 5.1. Abstracting the real-world system in a DES model

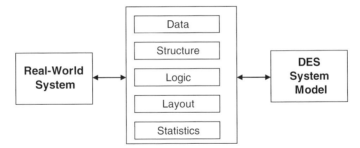

Figure 5.1   Elements of system modeling with DES.

can be approached by precise understanding and specification of the details of the five system modeling elements shown in the figure.

5.2.1   System Structure

A system DES model is expected to include the structure of the actual system being simulated. This structure is basically the set of system elements in terms of physical components, pieces of equipment, resources, materials, flow lines, and infrastructure. Depending on the nature of the system being modeled and the objective of the simulation study, the six-sigma team often decides on what elements to include in the simulation model. Elements of a manufacturing system are different from the elements of a business system. Whereas manufacturing systems are modeled using machines, labor, workpieces, conveyors, and so on, business systems are modeled using human staff, customers, information flow, service operations, and transactions.

Modeling such elements thoroughly is what makes a model realistic and representative. However, the level of details and specifications of model structural elements depends primarily on the objective and purpose for building the model. For example, the details of machine components and kinematics may not be helpful in estimating its utilization and effectiveness. Hence, basic graphical representations of structural elements are often used in DES models for animation purposes. On the other hand, physical characteristics such as dimensions, space, distances, and pathways that often affect flow routes, cycle times, and capacity should be part of the model structure. For a plant or a manufacturing system, Table 5.1 shows examples of structural elements and their impact on model performance factors.

5.2.2   System Layout

A system layout is simply the configuration plan for a system's structural elements. The layout specifies where to locate pieces of equipments, aisles, repair units, material-handling systems, storage units, loading and unloading docks,

TABLE 5.1   Examples of Structural Elements in DES

Structural Element Modeled	Model Performance Factor Affected
Conveyor length	Conveyor capacity
Unit load dimensions	Number of units stacked
Buffer size	Buffer capacity
Length of aisles and walkways	Walking distance and time
Size of automated guided vehicle AGV	Number of AGV carriers
Length of monorail	Carrier traveling time
Dog spacing of power and free system	Power and free throughput
Dimensions of storage units	Storage and retrieval time

and so on. Similar to system structure, placing and sizing model elements according to the layout specified results in a more representative DES model. Sticking to layout specifications helps capture the flow path of material or entities within the system. Hence, flow diagrams are often developed using system layouts. When designing new systems or expanding existing ones, the layout often plays an important role in assessing design alternatives. Examples of questions answered in a system layout include: What is the system structure? What does the system look like? How close is the packaging department to the shipping dock? How large is the storage area?

Facility planning is the topic under which the layout of a plant or a facility is designed. Department areas and activity-relationship charts are often used to provide a design for a facility layout. Locations of departments, distances between them, and interdepartmental flow need to be captured in the DES model to provide accurate system representation.

5.2.3   System Data

Real-world systems often involve a tremendous amount of data while functioning. Data collection systems (manual or automatic) are often used to collect critical data for various purposes, such as monitoring of operations, process control, and generating management reports. DES models are data-driven; hence, pertinent system data should be collected and used in the model. Of course, not all real system data are necessary to build and run the DES model. Deciding on what type of data is necessary for DES is highly dependent on the model structure and the goal of simulation. Generally speaking, all system elements are defined using system data. Examples include parameter settings of machines, material-handling systems, and storage systems. Such parameters include interarrival times, cycle times, transfer times, conveying speed, time to failure (MTBF), time to repair (MTTR), product mix, defect and scrap rates, and others. Model performance and results are highly dependent on the quality and accuracy of such data, based on the commonly used term *garbage-in-garbage-out*.

TABLE 5.2 Data Collected for Various DES Elements

Element Modeled	Pertinent Simulation Data
Machine	(Load, cycle, unload) time, MTBF, MTTR
Conveyor	Speed, capacity, type, accumulation
Operator	Walk speed, work sequence, walk path
Buffer	Capacity, discipline, input/output rules
Automated guided vehicle	Speed, acceleration/deceleration, route
Power-and-free system	Speed, dog spacing, chain length
Part/load	Attributes of size, color, flow, mix

Thus, developing a DES model requires the six-sigma team to define precisely the data elements needed, the method of collecting such data, and how the data will be represented and used in the DES model. Chapter 4 has shed light on the key topic of input modeling and statistical distributions used in DES models. Table 5.2 presents the data that need to be collected for modeling various simulation elements.

5.2.4 System Logic

System logic comprises the rules and procedures that govern the behavior and interaction of various elements in a simulation model. It defines the relationships among model elements and how entities flow within a system. The programming capability of simulation languages is often utilized to implement the system logic designed into the DES model developed. Similarly, real-world systems often involve a set of simple or complex logical designs that control system performance and direct its behavior. Abstracting relevant logic into a DES model is a critical modeling task. It is probably the most difficult modeling challenge that faces simulation six-sigma teams, especially with limitations of some simulation software packages. It is worth remembering, however, that complex logic is a main driver behind the commonality of simulation modeling in production and business applications.

In a typical simulation model, it is often the case that several decision points exist within the model operations, such as splitting and merging points. At these decision points, certain scheduling rules, routing schemes, and operational sequences may need to be built into the DES model to reflect the actual performance of the underlying system. Table 5.3 provides examples of such logical designs.

5.2.5 System Statistics

System statistics are means of collecting run-time information and data from a system during run time and aggregating them at the end of simulation run time. During run time, such statistics are necessary to control the operation

TABLE 5.3   Examples of Model Logical Designs

Model Activity	Logical Design
Parts arriving at loading dock	Sorting and inspection scheme
Requesting components	Model mix rules
Producing an order	Machine scheduling rules
Material handling	Carrier routing rules
Statistical process control	Decision rules
Machining a part	Sequence of operation
Forklift floor operation	Drivers' dispatching rules
Storage and retrieval system (AS/RS)	AS/RS vehicle movement rules

TABLE 5.4   Examples of Model Statistics

Model Statistic	Value Measured
Jobs produced per hour	System throughput
Percent of machine busy time	Machine utilization
Number of units in system	Work-in-progress level
Time units spend in system	Manufacturing lead time
Number of defectives	Process quality
Number of machine failures	Maintenance plan
Number of units on a conveyor	Conveyor utilization
Number of units on a buffer	Buffer utilization

and flow of system activities and elements. At simulation end, these statistics are collected to summarize system performance at various system design and parameter settings. In a system DES model, therefore, statistics are collected and accumulated to provide a summary of results at the end of run time. Such statistics are used to model real-time monitoring gauges and clocks in a real-world system. Because of model flexibility, however, some statistics that are used in the model may not actually be in the real-world system. This is because statistics do not affect model performance. Therefore, we can define statistics that are necessary to system operation and other statistics that may provide useful information during run time and summarize the results at the end of run time. Appendix A covers statistics collected from simulation and output analyses performed using such statistics. Table 5.4 provides examples of such statistics.

5.3   ELEMENTS OF DISCRETE EVENT SIMULATION

A DES model is built using a set of model components (building blocks) and is executed based on a set of DES mechanisms. The nature and functionality of these components and mechanisms may vary from one simulation package

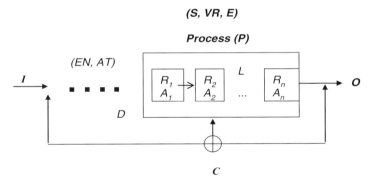

Figure 5.2   DES system elements.

to another. However, basic components of DES models are often defined using a common set of DES elements. These elements were defined in Chapter 4 in terms of system inputs, process (elements and relationships), outputs, and controls. Hence, DES elements include system entities (EN), often characterized by a set of attributes (AT). The system state (S) is described by state variables (VRs). The system state changes upon the occurrence of a set of events (E). During operation, different system activities (A) or tasks are performed by system resources (R). Finally, the model involves delays (D) that take place in the system and the logic (L) that governs the system operation. These elements are shown in Figure 5.2. In this section we discuss these elements and in the next section discuss key mechanisms used to execute DES models.

5.3.1   System Entities

Entities are items that enter a system as inputs, are processed through the system resources and activities, and depart the system as outputs. As DES dynamic objects, entities are model-traceable elements that are often of interest to system designers, managers, and six-sigma operatives such as green belts, black belts, and master black belts. Examples include parts or products in manufacturing, customers in banking, calls in a customer-service center, patients in health systems, letters and packages in postal services, documents in an office, insurance policies in an insurance company, data in an information system, and so on.

In DES, entities are characterized by attributes such as price, type, class, color, shape, ID number, origin, destination, priority, due date, and so on. Specific values of such attributes are tied to entities and can differ from one entity to another. Hence, attributes can be considered as *local* variables that are tied to individual entities. These attributes can be used at different locations and instances within the DES model to make various decisions for directing the flow of entities, assigning them to storage locations, activating resources, and so on.

Also, the type of entity is the basis for classifying DES systems between discrete and continuous. Discrete entities are modeled with discrete systems. Continuous entities such as flow of bulk materials, fluids, and gases are modeled with continuous systems. As mentioned earlier, our focus in this book is on discrete entities since they represent the majority of transaction-based systems in both manufacturing and service industries.

5.3.2 System State

The system state is a description of system status at any point in time during a simulation. It describes the condition of each system component in the DES model. As shown in Figure 5.2, these elements include system inputs (I), process (P), outputs (O), and controls (C). Hence, at any point in time, the system state defines the state of system inputs (i.e., types, amount, mix, specifications, arrival process, source, attributes). It also defines the system process in terms of type of activities (A), number of active resources (R), number of units in the flow, utilization of resources, time in state, delay time, time in system, and so on. Similarly, a system state defines the state of model outputs (i.e., types, amount, mix, specifications, departure process, destination, attributes, variability). Table 5.5 includes examples of system states that typically take place in different system elements.

DES records changes in a system state as it evolves over time at a discrete point in time to provide a representation of system behavior and to collect statistics and performance measures that are essential to system design and analysis. System state variables are used to quantify the description of a system state.

5.3.3 State Variables

DES models include a collection of variables that describe the system state at any specific point in time. Such variables contain the information needed to describe a model component status and measure its performance. Examples include the number of units in the system, the percentage of each resource status (i.e., % idle, % busy, % broken, % blocked), and the number of busy or idle operators.

TABLE 5.5 Examples of States of Different Model Elements

Model Element	System State
Machine	Busy, down, idle, blocked
Buffer	Empty, full, half-full
Conveyor	Empty, full, broken, blocked, starved
Labor	Idle, busy, walking
Vehicle	Moving, stopped, broken, parking

Control Variables   Response Variables

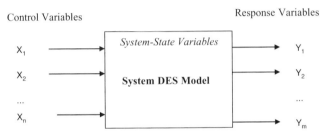

Figure 5.3   System state variables.

In addition to the overall system-level variables, model variables include input factors $(X_1, X_2, \ldots, X_n)$. They represent a set of independent variables whose changed values affect system behavior, and different settings of such factors often lead to different sets of model outcomes. Examples include inter-arrival times, number of operators and resources, service time, cycle time, capacity of buffers, and speeds of conveyors.

Model variables also include system response (output) variables $(Y_1, Y_2, \ldots, Y_n)$, which represent dependent measures of system performance. Examples include system throughput, average utilization of resources, and manufacturing lead time. Figure 5.3 is a schematic representation of state variables.

5.3.4   System Events

An event is an instantaneous occurrence that changes the system state. As discussed earlier, the system state is the description of system status at any time, which is defined by a set of state variables. The event is the key element in DES models since the models are characterized by being event-driven. Updating the system state, collecting system statistics, and advancing the simulation clock take place at event occurrence. The set of events $(E_1, E_2, E_3, \ldots, E_n)$ that occur at certain corresponding times $(T_1, T_2, T_3, \ldots, T_n)$ are stored chronologically in an event list (discussed later in DES mechanisms). System states $(S_1, S_2, S_3, \ldots, S_n)$ are changed based on the events and their implications; hence, model behavior is referred to as event-driven. Figure 5.4 shows how a system state is updated at event occurrences.

Event occurrence in a DES model can be the arrival of an entity, the start of a coffee break, the end of a shift, the failure of a resource, a change in batch size, the start of a new production schedule, the departure of an entity, and so on. Since DES model elements are interrelated and DES environments are dynamic, the occurrence of such events often leads a series of changes to the system state. For example, the arrival of a customer at a bank increases the waiting line if the bank teller is busy, or changes the teller status from idle to busy if a teller is available. Similarly, when the customer departs the bank, the server is changed back to idle, another customer is requested from the waiting

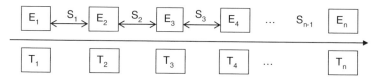

Figure 5.4   Event-driven system state.

TABLE 5.6   Activities in a Process Operations Chart

Symbol	Process Meaning	Example
◯	Processing operation	Drilling operation
➡	Transporting operation	Forklift transfers a unit load to storage
△	Storage operation	Finished goods are stored before shipping
D	Delay operation	Work-in-process units are delayed before final assembly
☐	Inspection operation	One of every 100 units is inspected at a quality control station

line, and the number of customers served is increased. Hence, the state variables affected are updated, relevant statistics are accumulated, and the simulation clock is advanced to the next event in the event list.

5.3.5   System Activities

An activity is a task performed in a model that has a specified time duration. Activities are determined simply by establishing a sequence for the operations needed to process an entity. Examples of such operations include receiving, directing, transferring, cleaning, machining, packaging, and shipping. Such activities either process the entity (value-added) directly, such as cutting and serving, or indirectly, such as material handling, inspection, and storage activities (non-value-added).

A process map often includes a sequence of all activities or operations required to process an entity, along with specifications and the classification of different types of operations. A process chart often classifies operations into process (a circle), transport (an arrow), storage (a triangle), delay (D), and inspection (a square), as shown in Table 5.6. Similarly, activities in a DES model can be classified to provide a better understanding of the process flow and a deeper knowledge of process operations.

System activities can be also classified as value-added or non-value-added:

1. *Value-added* (VA) *activities*. Value is added through activities that transform the characteristics of an entity from one form to another. Value is defined by the customer. By making changes to entities, such activities increase their values. Hence, the price of a ton of reinforcement steel is much higher than the price of a ton of steel billet. Steel billets are heated, formed, quenched, cooled, and cut before becoming reinforcement steel. Such operations gradually increase the value of steel entities.

2. *Non-value-added* (NVA) *activities*. Many operations performed on entities may not add value to them but they are still needed to complete the process. Examples include transporting and transferring operations, storing and delaying materials, and quality control inspections. The reader is encouraged to revisit Chapter 2 for the lean six-sigma perspectives about NVAs.

The time duration for system activities is specified in three ways, depending on the nature of the activity:

1. *Fixed time duration*. Time assigned to an activity has a fixed value and possesses no variability. Typical examples include fixed cycle times in automatic operations such as CNC machining, planned stoppage times, and timed indexing operations.

2. *Probabilistic time duration*. The time assigned in this case incorporates randomness and variability; hence, the activity time changes from one entity to another. Sampling from theoretical or empirical statistical distributions is often used to represent activity time duration.

3. *Formula-based time duration*. In this case the activity time is calculated using an expression of certain system variables. For example, loading time is determined based on the number of parts loaded, walking time is determined as a function of load weight, and machine cycle time is a function of cutting parameters.

5.3.6 System Resources

A system resource represents the tool or mean by which model activities are carried out. Examples include pieces of equipment, operators, personnel, physicians, repairpersons, machines, specialized tools, and other means that facilitate the processing of entities. Resource allocation and scheduling schemes are often implemented to provide best task assignments to resources. Table 5.7 includes examples of system resources in manufacturing and service applications.

Key factors that affect the performance of resources include capacity, speed, and reliability. Capacity affects resource utilization, which measures the percentage of resource use. Since resources consume capital, system designers and managers prefer to increase the utilization of resources through better sched-

TABLE 5.7   Examples of System Resources

Manufacturing System	Service System
Machines and machine centers	Physicians
Operators and general labor	Bank tellers
Inspectors and quality controllers	Hospital beds
Repairpersons and maintenance crews	Drivers
Assembly stations and tools	Drive-through windows

TABLE 5.8   Factors Affecting System Resources

Resource Factor	Performance Measure	Unit Metric	Example
Capacity	Utilization	Percent busy	M_2 is 85% utilized
Speed	Throughput	Units produced per hour (UPH)	M_2 produces 60 UPH
Reliability	Uptime	Mean time between failures	M_2 has a MTBF value of 200 hours

uling and resource allocation. Resource speed determines the productivity of the resource, often measured as throughput or yield. Similarly, resource throughput is a larger-the-better measure. Eliminating waste in time and inefficiencies in operation increases throughput. Finally, resource reliability determines the percentage of resource uptime (availability). It is always required to increase the uptime percentage of resources through better maintenance and workload balancing. Table 5.8 summarizes the three resource factors:

1. *Resource capacity*: a factor that is often measured by how many entities are able to access or enter a resource. Most resources, such as machines, bank tellers, and physicians, often treat one entity at a time. Some, however, perform batch processing of identical or different entities, such as filling machines, assembly stations, testing centers, and group-based activities. Capacity limitations often affect the utilization of resources, where system designers often strive to strike a balance between measures. Therefore,

$$\% \text{ utilization} = \frac{\text{actual capacity}}{\text{designed capacity}}$$

2. *Resource speed*: a factor that determines the throughput or yield of a resource. Fast resources often run with a short processing time and result in processing more entities in the time unit. Examples include the speed of a transfer line, the cycle time of a machining center, the service time at a bank teller, and the diagnostic time at a clinic. Units produced per hour (UPH) or

per shift (UPS) are common throughput measures in manufacturing systems. Cycle time is often the term that is used to indicate the speed of machines and resources in general. Machine throughput is therefore, the reciprocal of cycle time. That is,

$$\text{resource throughput} = \frac{1}{\text{cycle time}}$$

For example, a machine with a cycle time of 60 seconds produces 1/60 unit per second or 1 unit per minute. Takt time is often used to indicate the speed of automatic production lines.

3. *Resource reliability:* a resource factor that determines the resource uptime or availability. It is often measured in terms of mean time between failures (MTBF), where a resource failure is expected to occur at MTBF time units. Repair time, after a failure occurs, is measured by mean time to repair (MTTR). The resource uptime percentage is determined by dividing MTBF by available time (both MTBF and MTTR):

$$\text{resource uptime } (\%) = \frac{\text{MTBF}}{\text{MTBF} + \text{MTTR}}$$

5.3.7 System Delay

A system delay is an activity that takes place within the system but does not have a specified time duration. The duration is determined during run time based on dynamic and logical interactions among system elements. Examples include customer waiting time, delays in unit sequencing, and delays caused by logical design. Measuring delays by accumulating the time of delay occurrences is the principle advantage of DES models. Statistics important to decision makers can be collected using this capability: for example, the average waiting time for bank customers; manufacturing lead time; and time span from the start to the end of part manufacturing, including delays.

5.3.8 System Logic

System logic controls the performance of activities and dictates the time, location, and method of their execution. DES models run by executing logical designs such as rules for resources allocation, parts sequencing, flow routing, task prioritization, and work scheduling. Route sheets, work schedules, production plans, and work instructions are examples of methods used to implement the system logical design. Table 5.9 summarizes examples of system logic.

Simulation packages often provide built-in routing and dispatching rules for simulation entities. They also provide an English-like syntax to program

TABLE 5.9   Examples of System Logic

Logical Design	Implementation Method	Example
Material flow	Route sheet	Machining sequence
Workforce scheduling	Work schedule	Nursing shifts in hospitals
Product-mix scheduling	Production plan	Production lot size
Task prioritizing	Shop floor controls	Expedited processing
Standard operations	Work instructions	Product assembly

the model logic. Simulation packages often make use of industrial and business terminology as syntax: for example, process, part, labor, machine, conveyor, carrier, and so on. However, simulation software tools often vary in the capability of their logical programming platform. While some tools provide open editors for writing complicated logic in a C-like or VB-like programming structure, others provide limited programming capability for making the model building easier and faster.

Like any other programming language, writing the logic in simulation requires developing a flowchart of the logical design and translating the logical design into code using the software syntax. Since simulation logic is typically developed to mimic real-world operational logic, a thorough and practical understanding of the underlying production or business process is an essential ingredient for developing representative simulation logic.

5.4   DES MECHANISMS

As discussed earlier, DES models are dynamic, event-driven, discrete in time, stochastic (randomized and probabilistic), and computer-animated. Such characteristics are established in the DES model based on certain powerful mechanisms. Main DES mechanisms include the creation and updating of an events list, the time-advancement mechanism, the capability of sampling from probability distributions with random number generation, the capability of accumulating statistics over run time, and the power of two- or three-dimensional dynamic graphical representation with animation mechanism.

These mechanisms represent the backbone of DES functionality in simulation software tools since they are critical to creating operational models. Although simulation software tools vary in methods and algorithms to implement these mechanisms, the DES functionality often includes the creation of an event list, advancing the simulation clock (event-driven advancement), updating the event list, updating pertinent statistics, and checking for termination. Figure 5.5 is shows a flowchart of DES functionality.

Each DES mechanism involves many theoretical details, and in this section we discuss the key characteristics and methods used in these mechanisms.

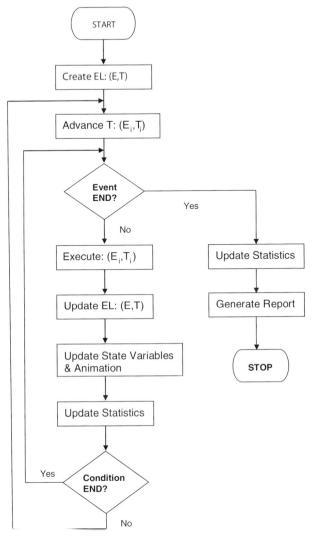

Figure 5.5 DES functionality.

5.4.1 Discrete Event Mechanism

Events in a DES are discrete since they take place at discrete points in time. DES functionality is, therefore, based on creating an event list and executing the events of the list chronologically. Such functionality is based on a discrete event and clock-advancement mechanism. The discrete event mechanism is the most distinctive feature of DES simulation models. As discussed earlier, an *event* is defined as an occurrence, a situation, or a condition that results in changing the *state* of the system model instantaneously. The state of a

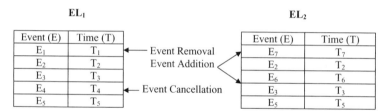

Figure 5.6   Event list operations.

system is represented by a collection of state variables that describe the system status.

Arrival and departure of entities such as parts and customers are events that change the state of a production system or a service station. Information about event types and their scheduled occurrence times are stored in an event list (EL). An event list comprises a set of events (E) and event–time (T) combination 1. That can be expressed as follows:

$$EL = \{(E_1, T_1), (E_2, T_2), (E_3, T_3), \ldots, (E_n, T_n)\}$$

At each event occurrence, two actions take place in the model: The EL is updated in terms of content and time, and the statistics collected are updated. Upon the occurrence of an event, the event executed will be removed from the list and the next most imminent event will top the list. Other events may enter or leave the list accordingly. Updating an event list includes two main operations (Figure 5.6):

1. *Event removal.* Upon the occurrence of an event, the list may be updated by removing one or more events from the list, resulting in changing the chronological order of event execution. Some events may be deleted from the list before their execution. An event may be deleted for many reasons: for example, when the occurrence of one event precludes the occurrence of another. Event removal can take place in two locations in an eventlist:

   a. *From the top of the EL.* When an event is removed from the top of an EL, it is done or executed. For example, event E_1 in Figure 5.6 was processed; hence the event is removed from the top of EL_1 when updating the list to become EL_2.

   b. *From any other location within the EL.* When an event is removed from any other location in the EL other then the top of the list, it is canceled. For example, event E_4 in Figure 5.6 was canceled; hence the event is removed from EL_1 and did not show up in EL_2.

2. *Event addition.* Upon the occurrence of an event, the list may be updated by adding one or more events to the list, resulting in changing the chronological order of event execution. Event addition can take place in two locations in an EL:

a. *To the top of the EL.* When an event is added to the top of the EL, it is considered as the most imminent event to take place. For example, event E_7 in Figure 5.6 was added to the top of EL_2 and will be processed first. E_7 was not a member of EL_1 when updating the list to become EL_2.

b. *To any other location within the EL.* When an event is added to any location in an EL other then the top of the list, it is just being added as a future event. For example, event E_6 in Figure 5.5 was added to EL_2 and will be processed right after E_2.

Thus, with the discrete event mechanism, we view a system as progressing through time from one event to another rather than changing continuously. As mentioned earlier, many real-world systems (e.g., plants, banks, hospitals) operate in a discrete manner (transaction- or event-based). Since these systems often involve waiting lines, a simple discrete event mechanism is often modeled using queuing models. DES models can be viewed as queuing networks that may or may not be amenable to queuing theory. Discrete events occur in the simulation model similar to the way they occur in the queuing system (arrival and departure of events). For instance, a customer joins a queue at a discrete instant of time, and at a later discrete instant the customer leaves the bank. Time is accumulated upon the occurrence of events at discrete points in time. The customer is either in or out of the queue. This view results in a computationally efficient way of representing time. The ability to represent time in an event-driven manner is the greatest strength of DES since it captures the dynamic behavior of real-world production and business systems.

5.4.2 Time-Advancement Mechanism

The time associated with the discrete events $(T_1, T_2, T_3, \ldots, T_n)$ in a simulation model is maintained using a *simulation clock* time variable. This variable is updated through the *next-event time-advancement mechanism*, which advances the simulation clock to the time of the most imminent event in the event list. For example, Figure 5.7 shows an event list of 10 events $(E_1, E_2, E_3, \ldots, E_{10})$ ordered chronologically as $T_1, T_2, T_3, \ldots, T_{10}$. In DES, the time periods between events (e.g., E_1 and E_2) are skipped when executing the model, resulting in a *compressed* simulation time. Hence, the total time required to process the 10 events in the DES computer model is much shorter than the actual clock time.

The time-advancement mechanism simply states that if no event is occurring in a time span, there is no need to observe the model and update statistics, so the time is simply skipped to the next event scheduled in the EL. This event-driven time-advancement process continues until some specified stopping condition is satisfied. For example, if a bank teller starts serving a customer in a bank, there is no need to track every second during the service time. Instead, the clock is advanced to the next event, which may be the end of the service time if no other event is expected to take place during the service time.

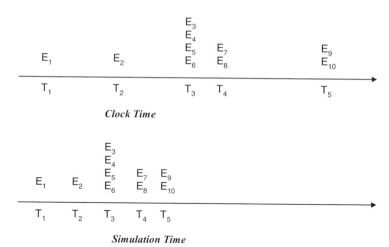

Figure 5.7   Time advancement and compression in DES.

5.4.3   Random Sampling Mechanism

The stochastic nature of simulation models is established by means of sampling from probability distributions using random generation techniques. *Stochastic*, which means random, is incorporated into the model using mathematical methods to generate streams of pseudorandom numbers. A random number generation (RNG) mechanism allows for capturing the random variations in real-world systems in DES models. The ability to allow for randomness is one of the great strengths of DES.

From a practical point of view, random numbers are basic ingredients in the simulation models of most real-world systems. Machine cycle times, customer service times, equipment failures, and quality control tests are examples of randomness elements in DES models. Random model data inputs often lead random model outputs. Thus, simulation outcomes, representing real-world systems, are often stochastic in nature and lead to inconsistent performance levels.

The probabilistic nature of the time period that separates successive events when executing a DES model represents the merit of randomness in simulation models. Events (e.g., customer arrival, machine failure, product departure) can occur at any point in time, resulting in a stochastic simulation environment. To generate successive random time samples ($T = T_1, T_2, \ldots$) between events, *sampling* from probability distributions is used. Hence, most simulation languages, as well as generic programming languages, include an RNG engine to generate event times and other random variables modeled. For example, interarrival times of certain entities to a system are generated randomly by sampling from a continuous exponential distribution while the number of entities arriving per unit time is randomly generated from a discrete Poisson distribution. Modeling randomness in discrete event simulation requires RNG and a method for selecting data points randomly from statistical distributions. A linear congruential generator (LCG) is used for RNG, and the inverse method is used for sampling from statistical distributions.

RNG with LCG The purpose of an RNG machine in DES is to produce a flow of numbers from a continuous uniform distribution between 0 and 1. Such numbers represent random and independent observations of a certain random factor that is uniformly distributed because of the equal probability of the uniform random variable. These random numbers can also be used to sample from probability distributions other than a uniform distribution.

In addition to the LCG method, common RNG techniques include combined linear congruential generators. LCG is, however, the most common arithmetic operation for generating uniform (0,1) random numbers. The randomness of sampling is almost guaranteed using a *congruential method*. Each random number generated using this method will be a decimal number between 0 and 1. Given parameters Z_0, a, c, and m, a pseudorandom number R_i can be generated from the following formula:

$$R_i = \frac{Z_i}{m} \quad \text{where} \quad Z_i = (aZ_{i-1} + c) \bmod(m) \quad i = 1, 2, 3, \ldots$$

where Z_i values are integers between 0 and $m - 1$, but values of the random numbers generated that are used in simulation are decimal numbers between 0 and 1. The operation $\bmod(m)$ means to divide $(aZ_{i-1} + c)$ by m and return the remainder of the division as the next Z_i. The initial value Z_0 is referred to as the *seed* of the random generator. The values of a, c, and m constants are selected properly to produce a good flow of random numbers. For example, if $m = 20$, $a = 10$, $c = 5$, and $Z_0 = 50$, the recursion generating of Z_i values is $Z_i = (10Z_{i-1} + 5) \bmod(20)$, which results in the following values:

$Z_0 = 50$

$Z_1 = [10(50) + 5] \bmod(20) = 505 \bmod(20) = 5$ and $R_1 = 5/20 = 0.25$

$Z_2 = [10(50) + 5] \bmod(20) = 55 \bmod(20) = 15$ and $R_2 = 15/20 = 0.75$

.
.
.

An RNG mechanism generates numbers that are not truly random because they can be generated in advance and because of the recursive nature of LCG, where a similar sequence of random numbers can be repeated again and again. For this reason, RNG methods are often called *pseudo-RNG*. Tables of ready-to-use random numbers are available, especially for Monte Carlo simulations, where random sampling is used to estimate certain experimental output. Also, most simulation software packages have the capability of automatic random number generation. Appendix B presents a table of random numbers.

Sampling with the Inverse Method   Using an RNG with uniform(0,1) values, successive random samples can be generated from probability distributions using three methods:

1. *Inverse method:* used for sampling from exponential, uniform, Weibull, and triangular probability distributions as well as empirical distributions.
2. *Convolution method:* used for sampling from Erlang, Poisson, normal, and binomial distributions.
3. *Acceptance–rejection technique:* used for sampling from beta and gamma distributions.

The inverse method involves probability distributions with a closed-form cumulative density function (CDF), such as the exponential and uniform distributions. The method consists of two steps: random number generation of R and computing the random variable (x) from the CDF that corresponds to R. Since the R values are all between 0 and 1, the CDF values, or $f(x) = P(y \leq x)$ are also between 0 and 1, where $0 \leq f(x) \leq 1$. The value of x is computed by determining the inverse of CDF at R or $x = F^{-1}(R)$. The probability density function (PDF) of the exponential distribution is

$$f(x) = \lambda e^{-\lambda x} \qquad \text{where} \quad x > 0$$

To determine a random sample (x) from $f(x)$, we first determine the CDF of the exponential distribution by integrating $f(x)$ from 0 to x, which results in

$$f(x) = 1 - e^{-\lambda x} \qquad \text{where} \quad x > 0$$

The random sample is then determined as $x = F^{-1}(R)$, where R is generated with an RNG. This results in the following:

$$x = -\frac{1}{\lambda} \ln(1 - R)$$

For example, five customers ($\lambda = 5$) enter a grocery store each hour, and for $R = 0.75$, the time period until the next customer arrival is $t = -(1/5) \ln(1 - 0.75) = 0.277$ hour or 16.62 minutes.

The convolution method expresses the random sample as a statistical sum of other easy-to-sample random variables. Examples include the probability distributions of a sum of two or more independent random variables, such as Poisson and Erlang, whose sample can be obtained from the exponential distribution sample. For example, the *m*-Erlang random variable is defined as the statistical sum of m independent and identically distributed exponential random variables. Hence, the m-Erlang sample is computed as

$$x = -\frac{1}{\lambda} \ln(R_1, R_2, \ldots, R_m)$$

Sampling from a normal distribution can also be achieved with convolution methods, since as the central limit theorem states, the sum of n independent and identically distributed random variables becomes asymptotically normal as n becomes sufficiently large.

The acceptance–rejection method is developed for distributions with complex PDFs, where the complex PDF $f(x)$ is replaced by a more analytically manageable proxy PDF $h(x)$. Random samples are then taken from $h(x)$.

5.4.4 Statistical Accumulation Mechanism

Statistics comprise a set of performance measures and monitoring variables that are defined within the DES model to quantify its performance and to collect observations about its behavior. Those statistics are defined in the model for various reasons:

1. *Monitoring simulation progress.* Values of the set of model statistics can be defined to be part of model animation. Such values are updated discretely through model run time. The six-sigma team can observe the progress of simulation by monitoring the changes that occur to the values of the statistics defined. For example, the six-sigma team can observe continuously the changes that occur to the number of units or customers in the system.

2. *Conducting scheduled simulation reviews.* In many simulation studies it is often necessary to review the system state at certain points in simulation run time. Those could be the end of one week of operating, the completion of a certain order, or at a model-mix change. With model statistics defined, the six-sigma team can halt simulation at any point in time reached by the simulation clock, and review simulation progress by checking the values of such statistics. For example, the six-sigma team can stop the model at the end of the day shift at a fast-food restaurant and review the number of customers who waited more than 10 minutes at the restaurant drive-through window.

3. *Summarizing model performance.* At the end of the simulation run time, the model generates a report that includes averages (point estimates) of the statistics defined, along with variability measures such as variance, range, standard deviation, and confidence intervals. Those statistics can be used as a summary of the overall system performance. For example, the average throughput for a production line is found to be 55.6 units per hour.

Collecting statistics from a DES model during run time and at the end of simulation is achieved using a statistical accumulation mechanism. Through

TABLE 5.10 Examples of Bank and Plant Statistics

Statistics in a Bank Example	Statistics in a Plant Example
Average customer waiting time	Average throughput per hour
Average time spent in a system	Number of defectives per shift
Percent of time a bank teller is idle	Average manufacturing lead time
Number of customers served per day	Percentage of machine utilization
Maximum length of the waiting line	Average number of units in buffer

this mechanism, the model keeps track of the statistics defined, accumulates their values, and provides averages of their performance. This mechanism is performed at the occurrence of model events. As shown in Figure 5.5, model statistics are updated after the execution of each event as well as at the end of simulation.

For example, when defining customer waiting time as a statistic in a DES model of a bank, a waiting-time statistical accumulator updates the overall customer waiting time upon the arrival or departure of customers. As a simulation run ends, the total accumulated waiting time is divided by the number of customers (processed in the bank during run time) to obtain the average waiting time for bank customers. In manufacturing, a plant statistic can be the average number of units produced per hour. Other statistics in a bank and a plant example are shown in Table 5.10.

5.4.5 Animation Mechanism

Animation in system simulation is a useful tool to both the simulation modeler and the decision maker. Most graphically based simulation software packages have default animation elements for system resources, buffers, operations, labor, and so on. This is quite useful for model debugging, validation, and verification. This type of animation comes with little or no additional effort and gives the modeler additional insight into how the model works and how to test different scenarios. Further, the more realistic the animation, the more useful the model becomes to the decision maker in testing scenarios and implementing solutions.

Simulation models can run and produce results without animation. The model logic is executed, model statistics are accumulated, and a summary report is produced at the end of the model run time. Simulation-based six-sigma teams used to rely on the results of the model program to validate the model and correct errors in logic programming and model parameters. Only the simulation modeler will be able to verify model correctness and usefulness. Decision makers review the model results and suggest what-if scenarios and experiments to be performed. Nowadays, however, with animation being a mechanism of most simulation packages, decision makers can watch the model run (in a two- or three-dimensional graphical representation), track

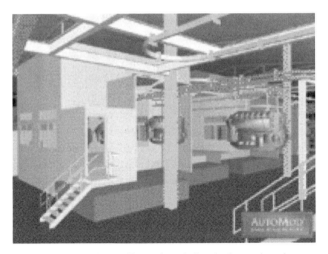

Figure 5.8 Three-dimensional simulation animation.

model entities, observe the impact of changes, and test model behavior at any selected point in the process flow. Such capability has helped modelers sell new concepts, compare alternatives, and optimize performance. Finally, animated model can be used for training purposes and as a demonstration tool in various types of industries.

Along with animation capability, DES models are often combined with good model management tools. User-friendly graphical user interfaces for graphics editing with libraries of graphical tools have made model animation an easy and enjoyable process. Some of those tools have been developed combined with a simulation database to store models, data, results, and animations. Modules of both graphical and parametric representation for basic model components such as parts, machines, resources, labor, counters, queues, conveyors, and many others are available in most simulation packages. Figure 5.8 shows an example of three-dimensional simulation animation built with AutoMod software.

5.5 MANUAL SIMULATION EXAMPLE

Discrete event simulation models are driven by the mechanisms of discrete event (transaction-based), time advancement, random sampling, statistical accumulation, and animation, as discussed in Section 5.4. Such mechanics can be better understood by analyzing a queuing system. Queuing systems are closely related to simulation models. Queuing models are simple and can be encountered analytically to compare the results to simulation model results. Other analytical models that can be analyzed with simulation include inventory models, financial models, and reliability models.

Since we have closed-form formulas to analyze simple queuing models, the need to approach such models with simulation is often questionable. Simulation is often used in such cases to clarify the application of DES mechanisms. Also, when assumptions that are required for developing analytical queuing systems do not apply, simulation is used to analyze the systems. Events arrival and departure, determining each event's time of occurrence, and updating the event list are examples of those mechanics. Based on the mechanics of modeling discrete events, the modeling state variables and statistics collected are determined at different instants when executing the model.

To clarify the mechanics that take place within a DES model, an example of a discrete event simulation model that represents a simple single-server queuing model (M/M/1) is presented. In this model, cars are assumed to arrive at a single-bay oil-change service station based on an exponentially distributed interarrival time (t) with a mean of 20 minutes. Two types of oil changes take place in the station bay: a regular oil change that takes about 15 minutes on average, and full service, which takes about 20 minutes on average. The station history shows that only 20% of customers ask for a full-service oil change. A simulation model is built to mimic operation of the service station. Model assumptions include the following:

- There are no cars in the oil-change station initially (the queue is empty and the oil-change bay is empty).
- The first car arrives at the beginning of simulation (clock time $T = 0$).
- Cars interarrival times are distributed exponentially (this assumption is essential in analyzing the model as a queuing system), whereas the oil-change time is distributed discretely, with a 15-minute service time having a probability of 80% and a 20-minute service time having a probability of 20%.
- The move time from the queue to the oil-change bay is negligible.
- Cars are pulled from the queue based on first-in-first-out (FIFO) discipline.
- No failures are expected to occur at the oil-change bay.

The logic of the simulation model is performed based on the following mechanics: In a discrete event mechanism, entities (cars) arrive and depart the service station at certain points in time. Each arrival or departure is an event, and each event is stored chronologically in an event list according to the following formulas:

next arrival time = current simulation clock time + generated
interarrival time
next departure time = current simulation clock time + generated service time

For each car arrival, certain logic is executed based on the discrete event and time-advancement mechanisms. If a bay is idle (empty), the car arriving

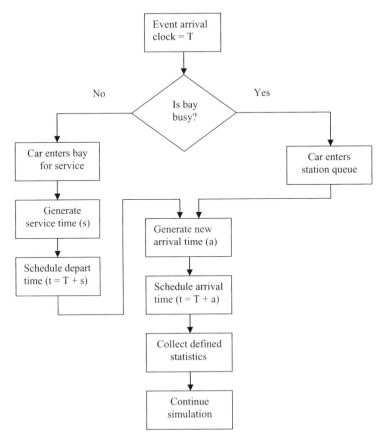

Figure 5.9   Execution of a car arrival–departure process.

at time T enters the bay for service. An RNG is used to sample randomly a service time s and schedule the car departure to be at time $t = T + s$. In case the bay is busy, the car enters the station queue and waits for the bay to be empty based on a FIFO schedule. Once the car finishes service, an interarrival time (a) is sampled randomly using the RNG, and a new arrival is scheduled at time $t = T + a$. Statistics are accumulated and collected and simulation continues similarly for another arrival. The execution of event arrival is shown in a flow chart as depicted in Figure 5.9.

Cars interarrival time is generated randomly from a continuous exponential distribution (mean = 20 minutes), while the service time is generated randomly from a discrete distribution (15 minutes with probability 80% and 20 minutes with probability 20%). Sampling from the exponential distribution with a CDF of $f(t) = 1 - e^{-\lambda t}$, $t > 0$, results in $t = -(1/\lambda) \ln(1 - R)$, where $R = f(t)$. Values of R used to obtain successive random time samples (t) are selected from a uniform (0,1) distribution using the LCG method. Similarly, R is generated for sampling from the discrete distribution to determine the service time (s), where $s = 15$ minutes if $0 \leq R \leq 0.8$ and $s = 20$ minutes if $0.8 < R \leq 1.0$.

Therefore, when starting the simulation ($T = 0$, the time at the simulation clock), we assume that the first car (customer 1) arrives at $T = 0$. The arrival of customer 1 is the event that changes the state of the service station from idle to busy. The service time assigned for the first customer (customer 1) using $R = 0.6574$ is $s = 15.00$ minutes. Thus, customer 1 departure time is scheduled as ($0 + 15.00 = 15.00$ minutes).

While customer 1's car is receiving a regular oil change, another car (customer 2) may be driven into the service station. The arrival time of customer 2 is determined using $R = 0.8523$ as $t = -(20) \ln(1 - 0.8523) = 38.24$ minutes. Hence, the arrival time for customer 2 is scheduled as ($0 + 38.24 = 38.24$ minutes).

Based on this mechanic, the first event list (EL) is formed with two scheduled events on the simulation clock:

- (E_1, t_1): departure of customer 1; the departure time is $T = 15.00$ minutes.
- (E_2, t_2): arrival of customer 2; the arrival time is $T = 38.24$ minutes.

Apparently, customer 1 will depart the service station before customer 2 arrives. E_1 will be removed from top of the EL and the list will be reduced to E_2 alone. The simulation clock will be advanced to the next event (E_2) time ($T = 15.00$ minutes) in order to execute E_2. The departure of customer 1 is the event that changes the state of the service station from busy to idle.

The service time assigned to customer 2 using $R = 0.8867$ is $s = 20.00$ minutes. Hence, the departure time of customer 2 is scheduled as ($38.24 + 20.00 = 58.24$ minutes). After the arrival of customer 2, another car (customer 3) is scheduled to arrive at $t = -(20) \ln(1 - 0.2563) = 5.92$ minutes using $R = 0.2563$. The arrival of customer 3 is scheduled as $38.24 + 5.92 = 44.16$ minutes.

Based on this mechanic, the event list is updated with the two scheduled events (E_3: departure of customer 2 and E_4 = arrival of customer 3). The two events are scheduled on the simulation clock as follows:

- E_3: $T = 58.24$ minutes
- E_4: $T = 44.16$ minutes

Apparently, customer 3 will arrive at the service station before the departure of customer 2, resulting in a waiting time of ($58.24 - 44.16 = 14.08$ minutes).

Simulation run time continues using the mechanism described until a certain terminating condition is met at a certain number of time units (e.g., three production shifts, 1 year of service, 2000 units produced, and so on). A simulation table that summarizes the dynamics of the first 20 cars arriving at the service station is shown in Table 5.11, where the following symbols are used:

TABLE 5.11   Simulation Table for the First 20 Cars

A	B	C	D	E	F	G	H	I
1	0.00	0.00	15.00	0	0.00	15.00	15.00	0.00
2	38.24	38.24	20.00	38.24	0.00	58.24	20.00	23.24
3	5.92	44.16	20.00	58.24	14.08	78.24	34.08	0.00
4	31.40	75.56	15.00	78.24	2.68	93.24	17.68	0.00
5	13.25	88.81	20.00	93.24	4.43	113.24	24.43	0.00
6	28.12	116.93	15.00	116.93	0.00	131.93	15.00	3.69
7	14.35	131.28	15.00	131.93	0.65	146.93	15.65	0.00
8	15.22	146.50	15.00	146.93	0.43	161.93	15.43	0.00
9	21.87	168.37	20.00	168.37	0.00	188.37	20.00	6.44
10	13.98	182.35	15.00	188.37	6.02	203.37	21.02	0.00
11	36.54	218.89	15.00	218.89	0.00	233.89	15.00	15.52
12	9.95	228.84	15.00	233.89	5.05	248.89	20.05	0.00
13	10.54	239.38	15.00	248.89	9.51	263.89	24.51	0.00
14	23.64	263.02	20.00	263.89	0.87	283.89	20.87	0.00
15	11.70	274.72	15.00	283.89	9.17	298.89	24.17	0.00
16	15.90	290.62	15.00	298.89	8.27	313.89	23.27	0.00
17	28.70	319.32	15.00	319.32	0.00	334.32	15.00	5.43
18	25.65	344.97	15.00	344.97	0.00	359.97	15.00	10.65
19	22.45	367.42	20.00	367.42	0.00	387.42	20.00	7.45
20	12.50	379.92	15.00	387.42	7.50	402.42	22.50	0.00
Sum	379.92		330.00		68.66		398.66	72.42

- A: customer number (1,2, . . . , 20)
- B: randomly generated interarrival time in minutes for each customer
- C: arrival time of each customer
- D: randomly generated service time in minutes for each customer
- E: time at which service (oil change) begins
- F: time that each customer waits in the queue waiting for service to begin
- G: time at which service (oil change) ends
- H: total time a customer spends in the system (waiting plus service)
- I: idle time for the station bay

Several statistics can be calculated manually from data in the simulation table. These statistics are collected by accumulating state variables and by averaging the accumulated values over the run time. In computer simulation, model state variables are tracked dynamically to accumulate times and occurrences of events over run time and to provide averages of statistics of interest at the end of the simulation run time. The following calculations show examples of key statistics that are of interest to the analyst in the service station example:

- Average customer waiting time = (total waiting time in queue)/(number of customers) = 68.66/20 = 3.43 minutes.
- Average car service time = (total service time)/(number of customers) = 330.00/20 = 16.50 minutes.
- Average interarrival time = (sum of interarrival times)/(number of customers) − 1 = 379.92/(20 − 1) = 19.99 minutes.
- Average waiting time for waiting customers = (total waiting time)/(number of customers who wait) = 68.66/12 = 5.72 minutes.
- Average time-in-system = (total time in system)/(number of customers) = 398.66/20 = 19.93 minutes. This can also be calculated by summing the average waiting time and average service time (i.e., 3.43 + 16.5 = 19.93).
- The probability that a customer has to wait = (number of customers who wait)/(number of customers) = 12/20 = 60%.
- The utilization of the station bay = [1.00 − (total idle time/total run time)] = 100% [1.00 − (72.42/402.42)] = 82.00%.

5.6   COMPUTER DES EXAMPLE

In this section we present a DES computer simulation model for a small health care clinic. Simulation is utilized to build a clinic model that can be used to assess operational alternatives for maximizing the number of patients served while minimizing the patient time-in-system (clinic). Through this simple example we describe how DES mechanics are utilized to represent clinic elements and patient flow and to generate simulation data from relevant sampling distributions. We briefly discuss model assumptions, model construction, and model validation and verification. Finally, a model analysis sample will be presented to demonstrate how DES provides answers to key questions on the behavior of the clinic system.

The clinic treats both urgent and acute care patients. Most patients require acute care. Most urgent care patients are assessed by clinic doctors and sent to the hospital. The clinic admits two types of patients: new patients (i.e., walk-ins) and return patients (i.e., patients with appointments). A patient arrives at the clinic's waiting room and waits at the registration counter. Patients then wait for the medical assistant to admit them to the clinic. The medical assistant calls patients, takes their blood pressure, checks their temperature, and ask them relevant questions before they enter the examination room. The two physicians in the examination rooms diagnose patients, release them, or direct them to the lab or the pharmacy. If the patient is sent to the lab he or she returns with the lab results to the examination room to see the same physician that he or she saw earlier. Patients who have been released leave the clinic. Figure 5.10 shows patient flow in the clinic. Potential patient waiting points are included on the map.

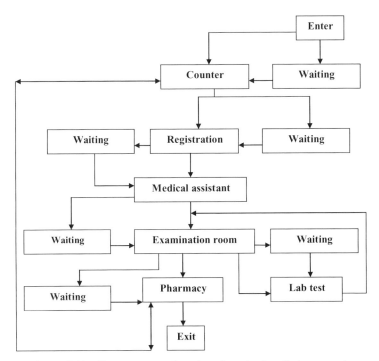

Figure 5.10   Process map of patient flow in the clinic example.

Unlike manual simulation, the clinic logic and flow shown in Figure 5.11 is not easy to model manually (analytically). A network of queues can be used to approximate the solution with many assumptions. Hence, computer DES with AutoMod software is used to model the clinic, estimate its performance metrics, and select the best operating strategy.

The student version of the AutoMod software package is used to simulate the clinic example. Even with the limited number of simulation entities in the student version, AutoMod provides a flexible model building and logic programming platform that makes it easy to develop three-dimensional structures, incorporate complex logic, and insert process data. It also estimates many simulation statistics and provides an excellent module to analyze simulation outputs with AutoStat. The following assumptions are used in building the clinic simulation model:

- The patients' interarrival time is distributed exponentially with a mean of 15 minutes. Arriving patients register and wait to be called. The clinic waiting room has a maximum capacity of 50 seats.
- The clinic operation pattern is 8 hours per day from 8:00 A.M. to 5:00 P.M. with a 1-hour lunch, which is set initially 12:00 to 1:00 P.M. Due to the

TABLE 5.12   Distribution of Clinic Staff

Resource	Staff
Counter	1 clerk
Registration	2 clerks
Medical care	2 assistants
Examination	2 physicians
Test lab	1 technician
Pharmacy	1 pharmacist

nature of the clinic work, however, staff members take their lunch break based on their available idle time.

- Twenty-five percent of patients are required to take a lab blood test, which takes 25 minutes on average. Patients who take their lab test must return to the physician who requested the test. Patients returning from the lab test have a different processing times by physicians.
- Sixty percent of patients are required to reschedule appointments with a clinic physician for further treatment.
- Fifty percent of patients are sent to the pharmacy to get prescription drugs.
- Fifteen percent of patients are treated by a clinic physician and released from the clinic in their first visit.
- The medical assistant performs three functions: taking blood pressure, checking temperature, and asking patients questions about their health.
- Clinic staff are distributed as shown in Table 5.12.

To model the stochastic nature of the clinic operation, the simulation data that have been collected are fitted to standard distributions. Model inputs are generated randomly from AutoMod's built-in statistical distributions. In AutoMod, distributions are sampled using an inverse method with LCG RNG. Further details on modeling simulation inputs are presented in Chapter 6. Table 5.13 summarizes the sampling distributions used in the clinic model.

Since AutoMod is load-driven, clinic model building with AutoMod begins by defining a process for creating patients (loads). An initialization process can be used to create a dummy load and clone to various clinic processes. Patients are directed through clinic operations based on assigning load attributes. The two main load types defined are for walk-ins and scheduled (new or returning) patients. Queues and resources are then used to construct the clinic model. Each queue is defined in terms of capacity and location, and each resource is defined with a certain distribution cycle time. Syntax is written to direct patient flow and to control interactions among clinic processes. The model intended flow logic is used to verify the model, and the performance data expected or observed are used to validate the model. Figure 5.11 shows a simple animation of the AutoMod clinic simulation model.

TABLE 5.13   Sampling Distributions Used in the Clinic Example

Process	Patient Type	Distribution
Arrival	All patients	Exponential (15 min)
Counter	Just entering clinic To be released from clinic	Uniform (1.2, 0.6) min Exponential (4 min)
Registration	New and walk-ins Returning	Uniform (3.5, 1.5) Uniform (2, 0.5)
Medical assistant	Taking blood pressure Checking temperature Questionnaire	Normal (60, 5) sec Normal (60, 5) sec Triangular (2, 4, 10) min
Examination	All patients Returning from lab test	Triangular (10, 15, 40) Triangular (4, 6, 10) min
Lab test	Performing blood test	Normal (25, 3) min
Pharmacy	First time Second time	Triangular (1.6, 4, 8) min Triangular (1, 3, 5) min

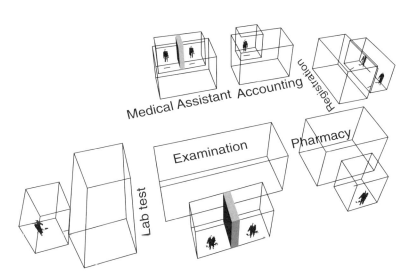

Figure 5.11   AutoMod simulation model of the clinic example.

Once validated and verified, the model can be used to collect clinic statistics automatically. Unlike manual simulation, however, computer simulation creates a simulation events list, tracks events, updates the event list, and develops an internal simulation table to accumulate simulation statistics. We obtain the results by running the model for 40 hours. The following statistics are generated from the model:

average time that patients spend in the clinic = 84.62 minutes

average service (total treatment) time =   56.1 minutes

average waiting time =   28.52 minutes

number of patients treated daily =   27 patients

We found that the patient examination process contributed the major delay in the clinic. The examination process takes on average 42.90 minutes, with a waiting time of about 20.97 minutes and a treatment time of about 21.93 minutes. Also, patients complain that physicians are pressured to complete treatment quickly. Still, however, due to routing patients, the two clinic examination physicians are utilized 93.6% and 89.5%. This is a relatively high utilization value.

The objective is to use the model to reduce waiting time and increase service time without further increasing the utilization of examination resources and without reducing the number of patients treated daily. It is therefore suggested as an alternative process configuration to minimize cost that one of the registration clerks be removed and the examination physicians be increased by one. The alternative clinic process structure includes changing clinic resources as shown in Table 5.14.

By running another simulation with the new clinic structure, the following statistics are generated from the model:

average time that patients spend in the clinic = 74.30 minutes

average service (total treatment) time =   60.12 minutes

average waiting time =   14.18 minutes

number of patients treated daily =   32 patients

The updated statistics show a reduction in the average time spent in the clinic, with an increase in the treatment time. The reduction is achieved

TABLE 5.14 Alternative Clinic Process Structure

Resource	Staff
Counter	1 clerk
Registration	1 clerk
Medical care	2 assistants
Examination	3 physicians
Test lab	1 technician
Pharmacy	1 pharmacist

primarily in waiting time. Examination physicians spend a longer time with patients. The total number of patients treated daily has increased to an average of 32. Average utilization of the three physician resources is also reduced to an average of 75% (74.8%, 75.5%, and 73.4%). The updated process configuration is also validated with subject matter experts and recommended for implementation. Improvement was also justified with a cost–benefit analysis.

5.7 SUMMARY

The DES techniques presented in this chapter are most capable and power-ful tools for business process simulation of transactional nature within DFSS and six-sigma projects. A DES provides modeling of entity flows with capa-bilities that allow the simulation modeler to see how flow objects are routed through the system. DES witnesses growing capability, software tools, and a wide spectrum of real-world applications.

Several considerations should prepare the team for planning a transaction-based process simulation study and for addressing the suitability of their choice of adopted simulation package. First, a clear classification of the project business processes is needed as a product development and production process for problem solving, including distribution processes or service-based processes. Process cycle times and resource requirements are usually handled by simulation, producing more accurate results. Many DES simulation pack-ages are capable of representing highly variable steps with tightly coupled resources. Modeling shifts, downtime, overtime, and learning curves with multiple replications for resources are some of the important considerations in building a valid simulation model.

In the context of six-sigma, discrete event simulation is quite suitable for production process problem solving, where outputs are produced in a batch or continuous-flow mode in relatively high volumes. Tasks such as assembly, disassembly, setup, inspection, and rework are typical steps in production processes with queuing rules and downtime modeling. Such processes are usually modeled to obtain steady-state behavior past the warm-up period.

In modeling distribution processes in a production environment, it is impor-tant to define attributes for flow entities in order to keep track of unique char-acteristics such as value-adds, cost, and distance traveled. Due to the transient nature of distribution processes, the simulation model warm-up period is usually longer than other production processes.

In a service-based industry, processes present a major area for employment of simulation studies. These processes are typically characterized by a non-value-added time that exceeds the value-added time (processing). The simu-lation of service processes represents a challenge because the entities and processing resources are both typically human. Modeling human behavior is complex and unpredictable. For example, customers calling a service center may hold or hang up. Modeling adaptation is required to model such situations. Also, processing times are highly variable and customer arrivals are random and add to the model complexity. Model accuracy demands representative probability distributions. Otherwise, a steady state may not be reached.

6

THE SIMULATION PROCESS

6.1 INTRODUCTION

The set of techniques, steps, and logic followed when conducting a simulation study is referred to in the context of this book as a *simulation process*. The details of such a process often depend on the nature of the six-sigma project, the project objectives, the simulation software used, and even on the way the team handles simulation modeling. Although the tactics of such a process often varies from one application to another and from one simulation project to another, the overall structure of the simulation process is common. It is necessary to follow a systematic method when performing the simulation process. This chapter is focused on analyzing the various aspects of the simulation process, the process followed by a six-sigma team carrying out a complete simulation study.

Simulation projects are conducted primarily for one of three purposes: system design, problem solving, and continuous improvement. The three purposes can, however, be achieved using a generic systematic approach. This approach starts with a clear definition of the design challenge, the problem, or the improvement opportunity. Alternatives of system design, problem solving, and improvement opportunity are then developed. These alternatives are evaluated using simulation-based performance criteria chosen by the team. The alternative with the best estimated performance is finally selected and implemented.

Simulation is used as an effective tool for diagnosing the system and defining problems, challenges, and opportunities. It is also used for developing and evaluating alternatives as well as for assessing the performance of each alternative. Hence, a complete simulation study often includes problem definition; setting simulation objectives; developing a conceptual model; specifying model assumptions; collecting pertinent model data; building, verifying, and validating the simulation model; analyzing model outputs; and documenting the project findings.

6.2   CATEGORIES OF SIMULATION STUDIES

Based on the principal goal of the project, simulation studies can be categorized into three classes: system design, problem solving, and continuous improvement. These categories are shown in Figure 6.1. The tactics and techniques used in a simulation study often vary depending on these three categories. Finally, it is worth mentioning that simulation is the environment of simulation-based six-sigma, lean six-sigma, and design for six-sigma methods. Following is a discussion of the three categories of simulation studies.

6.2.1   Simulation-Based System Design

Many simulation studies are used to assist the design process. This is often aligned with a typical objective of a DFSS project. However, the focus here is on a event-driven or transaction-based process and service design rather than the product design (see, e.g., Yang and El-Haik, 2003). Process compatibility studies can benefit greatly from simulation-based design. Applications of simulation projects include a wide spectrum of projects, such as the development of new facilities, major expansions of a manufacturing system, a new clinic, a new bank, a new vehicle program, and a new transportation network.

Designing a new system or process with or without simulation often involves testing new concepts, building new structures or layouts, and developing new logic. Typical objectives of design simulation studies include developing a new system design, assessing and validating a proposed design, and comparing two or more design alternatives. We can conclude that some

Figure 6.1   Categories of simulation studies.

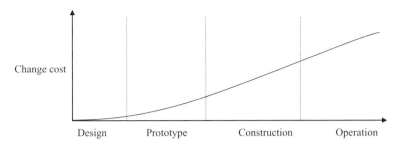

Figure 6.2 Cost of making changes at various project stages.

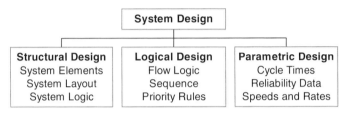

Figure 6.3 System design with simulation.

synergy and alignment can be leveraged by drawing on the similarities between system design and the simulation process. This is explored further in later chapters of this book.

The flexibility of simulation modeling provides an excellent platform for developing and testing design alternatives as well as for analyzing a variety of process schemes. The relatively low cost and high efficiency of simulation often promotes the use of simulation studies, solely on such merits. Saving achieved from simulation-based design studies often come from testing design feasibility, enhancing the system structure, and recommending process changes up front (see Section 8.4). Design changes have low cost and resource implications when carried out at the concept stage (see Chapter 8 for a theoretical context and Chapter 10 for an application case study). Figure 6.2 shows the project cost relation to the stage of the project. As shown in the figure, the cost of engineering changes increases exponentially with time in any project. Expensive changes often take place at the construction and operation stages of the project, where the project is subjected to costly delays, operation interruptions, and difficult physical changes.

As shown in Figure 6.3, the design of such systems includes system structural design, logical design, and system parametric design.

1. System structural design includes developing a system structure in terms of elements, layout, and representation. In a plant example, this may include selecting machines, pieces of equipment, and material-handling systems, and developing an appropriate layout of the system structure. The layout would

include departmental spacing, workspace design, aisles, material flow diagram, and so on. An activity relationship diagram is often used to guide and structure the system layout. Finally, representation is made using graphical tools and methods to present the model structure defined.

2. System logical design includes developing the logic that governs the behavior of the underlying system. The system functionality and the way through which system elements work on system entities are defined through the system logical design. Examples of logical design include flow logic, sequence of operation, priority schemes, making decisions, scheduling rules, and so on.

3. Parametric design includes setting parameters (control factors) to values or levels that lead to *best* performance. These parameters are defined as settings for model structural elements. Examples include machine cycle times and reliability data, material handling and transfer speed, buffer capacity, and so on. Different levels of such parameters are tested using the simulation model through experimental design and optimization search to arrive at the parameter settings that lead to best system performance. For example, different parameters in a manufacturing process (e.g., number of operators, machine cycle times, conveyor speeds, buffer capacities) are set so that the throughput of a plant is maximized.

System structural, logical, and parametric design affects system performance (model response) directly in terms of system throughput, lead time, inventory levels, and so on. In designing systems, such responses[1] are used to guide the design process. For example, one layout may result in longer material flow and consequently, longer manufacturing lead time. Delays created because of improper logical design also affect lead time and throughput. Finally, a longer cycle time or a high percentage of downtime at a certain workstation may creat a system bottleneck that causes delays and results in throughput. The design for six-sigma (DFSS) method is applied to enhance the system design so that best performance can be achieved.

6.2.2 Problem-Solving Simulation

Production and business systems often face challenges that affect their operation and performance. The impact of such challenges varies from reduced efficiency or frequent delays and failures to major shutdowns and catastrophes. Hence, a big portion of the work of production managers, system engineers, and operations managers is often focused on monitoring the system performance, tackling operational problems, and attempting to prevent future problems. In addition, many of these problems may be concluded from customer complains, market feedback, and actual sales numbers.

Diagnosing such problems and providing the proper solution is not always a trivial task. Diagnostic analysis, testing what-if scenarios, and validating the

---

[1] We also used critical-to-quality characteristics (CTQs) as an alternative six-sigma terminology.

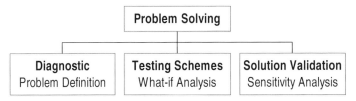

Figure 6.4   Problem solving with simulation.

feasibility of proposed fixes are typically required. Various engineering analyses that are based on analytical modeling or engineering sense are often utilized to solve problems. Examples include inventory models, queuing models, reliability models, failure mode and effect analysis, root-cause analysis, fault-tree analysis, and so on. However, as discussed in Chapter 4, the assumptions and analytical limitations of such tools limit their application for solving problems of considerable complexity. DES capabilities provide a flexible problem-solving platform. Hence, as shown in Figure 6.4, simulation modeling is utilized to perform the main problem-solving techniques of diagnosing systems, testing operational schemes, and validating prescribed solutions. These techniques fit the objective of six-sigma DMAIC projects.

As shown in Figure 6.4, solving problems with simulation includes problem diagnostic and definition, testing solution schemes, and validating problem solution.

1. Problem diagnostic with simulation is not always an easy task. It depends heavily on the credibility of the model built. The model has to reflect the current system state (including the impact of the problem). In some cases, the problem may be found while the model is being built. This is because, as discussed in Chapter 4, system structure, layout, data, and logic are explored thoroughly when building the model. Once the model is built, observing the model performance during run time may also lead to zoom-in on the problem or problem area. The quality of animation and graphical representation often play a major role in problem diagnostic by watching the model. In other cases it is necessary to perform some tests, inserting some logic, or studying the simulation report in order to diagnose the source of the problem.

2. After diagnosing the problem, simulation can be used to test a variety of problem solution schemes. Such schemes are often tested using what-if analysis, where the impact of each scheme on model performance is tested and analyzed. The most feasible scheme that leads to a comprehensive fix of the problem (without adding another problem) is selected as a solution to the problem defined. Model flexibility and efficiency allow for testing large number of solution schemes and running various what-if scenarios.

3. Finally, the viability and robustness of the solution scheme proposed for the underlying problem is tested. Sensitivity analysis can be used to test

solution robustness to various sources of variations. Not all solution schemes are able to function properly with other system elements under system constraints and work conditions. Small variations in such conditions may lead to drastic changes in system performance. Hence, simulation is used to test solution sensitivity and robustness before considering the solution scheme to be final.

6.2.3 Continuous Improvement Simulation

It is often asserted that the success of production and business systems in sustaining a certain level of performance depends on effort in establishing and implementing plans for continuous improvement. Companies do not always wait until a problem arises to take correction and improvement actions. Managers and engineers often believe that there is always a window for improvement in the way that companies produce products or provide services. Through this window, system managers and planners often foresee opportunities for making the system better and more prepared to face future challenges. As discussed in Chapter 2, this matches the philosophy of lean manufacturing and lean six-sigma (LSS) studies. The Japanese term *Kaizen* is used widely to refer to continuous improvement methods and techniques that companies follow to boost efficiency and sustain high performance. Hence, continuous improvement effort is often focused on defining improvement opportunities, proposing improvement plans, and testing the validity of the improvement actions proposed. A lot of those actions arise from developing future operational plans for expansion and business growth.

Opportunities of improvement can be defined by thinking of better ways to do things, brainstorming on cost and waste reduction measures, and benchmarking competition. Simulation can be utilized to search for such opportunities and for testing different improvement plans. Some of those plans may not be feasible; hence, the simulation can test plan feasibility and applicability. Finally, the model can be used to validate the improvement plan selected and to check its robustness and viability. Figure 6.5 shows the utilization of simulation in aiding the continuous improvement effort.

As shown in the figure, continuous improvement with simulation includes defining opportunities, testing improvement plans, and validating the plan selected.

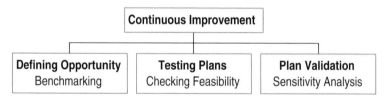

Figure 6.5   Continuous improvement with simulation.

1. Defining an improvement opportunity can be approached in different ways. Throwing ideas for improvement into routinely scheduled meetings, holding brainstorming sessions, and benchmarking competition are examples of techniques adopted to generate ideas and opportunities for improvement. Having a representative simulation model that can be watched and analyzed by a continuous improvement team can be beneficial in this regard. Changes suggested and ideas raised by team members can be tested and evaluated quickly at virtually no cost.

2. After a certain improvement opportunity is defined, it often necessary to provide a plan and mechanism for achieving the defined improvement. Different methods and techniques can be used to improve performance, cut cost, and reduce waste. Lean manufacturing techniques are commonly used by engineers to achieve improvement plans. In lean manufacturing, the focus is on cost and waste reduction, reducing manufacturing lead time and company response time, and increasing equipment and labor effectiveness. Simulation can aid in applying and testing the impact of the improvement plans proposed.

3. Finally, the viability and robustness of the improvement action or plan proposed is tested. Sensitivity analysis can be used to test the plan's robustness to various sources of variations. Using the simulation model, improvement actions are implemented and performance is evaluated to quantify the improvement. The sensitivity of the plan to a variety of operating conditions is then tested to make sure that there no negative impact will occur on other system functions and that the improvement level can be sustained.

6.3 SYSTEMATIC SIMULATION APPROACH

The approach followed for applying each of the three categories of simulation studies discussed in Section 6.2 has a specific nature and requirements. For the three categories, however, we can follow a generic and systematic approach for applying a simulation study effectively. This approach consists of common stages for performing the simulation study, as shown in Figure 6.6.

The approach shown in the figure is a typical engineering methodology for the three categories of simulation studies (i.e., system design, problem solving, and system improvement) that puts the simulation process into the context of engineering solution methods. Engineers and simulation analysts often adopt and use such an approach implicitly in real-world simulation studies without structuring it into stages and steps. The approach is an engineering methodology that consists of five iterative stages, as shown in Figure 6.6: Identify the simulation problem, develop solution alternatives, evaluate solution alternatives, select the best solution alternative, and implement the solution selected.

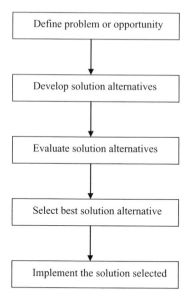

Figure 6.6   Systematic simulation approach.

6.3.1   Identification of the Problem or Opportunity

Any simulation study should start by defining the problem to be solved and analyzed through simulation. System design, performance, and improvement challenge are three types of simulation problems. In continuous improvement studies, the problem is viewed as an improvement opportunity if there is no specific problem to be solved.

The simulation problem is defined in terms of study scope, study objectives, and model assumptions.

1. Defining the scope of a simulation study starts by describing the system structure, logic, and functionality. A process map or flowchart can be used to define the process structure and flow. The study scope also includes a clear description of the system problem or opportunity. Describing the challenges, limitations, and issues with the current state and explaining why such issues take place provide the simulation analyst with a satisfactory definition of the simulation problem. If possible, the problem can be structured mathematically or by using a schematic diagram to make it easy for the analyst to understand various aspects of the problem.

2. Based on the problem scope, the simulation study objectives are set and expressed in the most concise way possible. Together with the overall objective of the study, which is often aimed at solving the problem that has been defined, the study subobjectives provide the criteria and mechanism for solving the problem. For example, the overall objective might be to increase customer satisfaction, and the specific objectives might be to reduce both

customer waiting time and processing time. It is always better to define quantifiable metrics that measure the objectives that have been defined.

3. Model assumptions are defined to narrow down the scope of the problem and to sharpen the study objectives. Model assumptions also help in directing model analysis and in selecting proper methods and techniques to solve the problem defined. Most model assumptions can be related to the surrounding system conditions, the constraints on system functionality, and the availability of monetary and technical resources. For example, the plant operating pattern might be a restriction on its daily productivity. Model assumptions are also focused on the source and type of data used in the model, such as reliability data, processing times, and decision-making rules. Statistical methods and distributions are always utilized in this regard.

6.3.2   Development of Solution and Improvement Alternatives

Once a problem or an opportunity has been identified, solution alternatives can be explored and developed. Such alternatives are explored by analyzing the problem variables, objectives, and constraints and attempting to define the parameter setup, the structural changes, or the logical design necessary to meet the problem objectives without violating its physical and logical constraints. Solution alternatives are then developed by structuring the set of methods and actions within each alternative explored. It is important to pay attention to the way in which we present the set of solution alternatives that has been developed. The analyst should be able to pinpoint the differences among solution alternatives. Tables, graphs, and summary sheets can be used to present and compare the solution alternatives developed.

1. Defining the set of solution alternatives is approached by exploring the solution domain and listing the potential solution methods. In addition to a thorough understanding and solid experience in the underlying system, a combination of creativity and engineering skills is required to generate a set of solution alternatives. Brainstorming sessions and idea-generating techniques can be used in this regard in a manner similar to that used for generated improvement ideas. With simulation, however, solution ideas can be generated in a much easier way. By providing a close-to-reality representation of the system, the model can be used to observe the system's dynamic behavior, promote idea generation, and predict the performance for each idea generated.

2. Exploring the solution domain of the underlying problem often leads to defining a set of potential solution ideas. Those ideas are transformed into a set of solution alternatives by structuring each idea in terms of a method and a set of actions and changes. The feasibility of each idea is checked and validated in the process of developing the set of solution alternatives. Each feasible solution idea is put in terms of a concise and specific plan that is aimed at

making the change required to eliminate the problem or to improve system performance. Structuring an idea into a solution alternative requires knowledge in engineering systems, representation methods, and best practices.

6.3.3 Evaluation of Solution Alternatives

In this stage, the set of solution alternatives selected are evaluated based on the objectives defined, and solution alternatives are ranked accordingly. Evaluating solution alternatives is essential to compare their performance and make it easier for decision makers to select the best solution alternative. Evaluating solution alternatives includes defining a set of performance criteria and evaluating each solution alternative for the set of criteria defined. Simulation plays a primary role in performance evaluation under complex, dynamic, and stochastic behavior of real-world systems.

1. Defining a set of performance criteria includes providing the proper quantitative metrics that measure an aspect of the system performance. This includes monetary criteria such as cost, profit, and rate of return, as well as technical and operational criteria such as throughput, effectiveness, and delivery speed. This may also include some sociotechnical criteria that may be difficult to measure, such as appeal, reputation, and customer satisfaction. It is important to select, as much as possible, a set of measurable criteria that can be assessed using the simulation model and can be used as a basis to compare solution alternatives. In addition to the simulation capability of flexible programming, model counters, tallies, and statistics provide the analyst with a variety of techniques to measure performance criteria. Some of those criteria, such as throughput, lead time, and inventory level, can be taken at a system level. Others, such as utilization, effectiveness, and reliability, can be taken at the process or operation level. Finally, some of the measures, such as quality, cost, and number of units processed, can be related to the system entity.

2. Once the set of performance measures are defined, a simulation model can be used to evaluate the set of solution alternatives. Each simulation-based evaluation provides a set of performance measures. Solution alternatives are then compared and the alternative with the best performance is selected. The simulation model is set to quantify comparison criteria (a set of performance measures) without the need for a closed-form definition of each model response. For example, in a manufacturing system model, system throughput is a function of buffer sizes, machine cycle times and reliability, conveyor speeds, and so on. It may not be possible analytically to provide a closed-form definition of the throughput function. However, randomness often combines the response signals of simulation models. Variability encompassed in the simulation model leads to variability in the model outcome. Hence, the stochastic nature of simulation outcomes should be included in an alternative comparison. Multiple simulation replications are often used to reflect

variability in model outcomes, and statistical measures such as output means, variances, and confidence intervals are computed to represent such variability.

6.3.4   Selection of the Best Alternative

In this stage, the best solution or improvement alternative is selected based on the simulation evaluation in the preceding step. The best solution in terms of values of performance measures is selected by comparing its overall performance to the rest of the solution alternatives. Comparing the performance of solution alternatives based on multiple objectives, however, is not an easy task.

Selecting the best solution alternative is a decision-making process that is based on an overarching assessment of solution alternatives. If the comparison is made based on a single objective, the alternative with best performance is selected. When multiple performance measures are used, multicriteria decision making (MCDM) is required to assess solution alternatives based on multiple decision criteria. If one alternative dominates other alternatives for all performance criteria, the nondominated solution is selected. Otherwise, MCDM techniques such as goal programming and analytical hierarchy process are used to support the multicriteria decision. Such methods are based on both subjective and objective judgments of a decision maker or group of experts in weighting decision criteria and ranking solution alternatives. An overall utility function [often referred to as a multiattribute utility function (MAUF)] is developed by combining criteria weights and performance evaluation. MAUF is used to provide an overarching utility score to rank solution alternatives.

Statistical comparative analysis and hypothesis testing are also used to compare solution alternatives. It is worth mentioning, however, that other factors that may not be evaluated using the simulation model, such as economical factors, safety, and environmental concerns, are often included in the selection process. The solution selected is then recommended for implementation in the real-world system as a design for a new system, as a solution to an existing problem, or as an improvement plan for an existing system.

6.3.5   Implementation of the Alternative Selected

Finally, the solution alternative selected is considered for implementation. Depending on the nature of the problem, implementation preparations are often taken prior to the actual constructional changes on the floor. Like any other project, implementing the solution recommended by the simulation study is performed into phases. Project management techniques are often used to structure the time frame for the execution plan and to allocate the resources required. Although some of the parameters and logic recommended by the simulation may be altered during the installation stage, the model can still be used as a tool for guiding the implementation process at its various stages.

Milestone meetings for progress reviews can benefit from simulation results and simulation animation of the system structure.

6.4   STEPS IN A SIMULATION STUDY

In this section we present a procedure for conducting simulation studies in terms of a step-by-step approach for defining the simulation problem, building the simulation model, and conducting simulation experiments. This procedure is a detailed translation of the systematic simulation approach presented in Figure 6.6. Figure 6.7 is a flowchart of the step-by-step simulation procedure.

The simulation systematic approach shown in Figure 6.7 represents the engineering framework of the simulation study. The steps may vary from one analyst to another because of factors such as the nature of the problem and the simulation software used. However, the building blocks of the simulation procedure are typically common among simulation studies.

The simulation procedure, often represented by a flowchart, consists of the elements and the logical sequence of the simulation study. It also includes decision points through which the concept and model are checked, validated, and verified. Iterative steps may be necessary to adjust and modify the model concept and logic. Finally, the procedure shows steps that can be executed in parallel with other steps.

6.4.1   Problem Formulation

The simulation study should start with a concise definition and statement of the underlying problem. The problem statement includes a description of the situation or the system of the study and the problem that needs to be solved. Formulating the problem in terms of an overall goal and a set of constraints provides a better representation of the problem statement. A thorough understanding of the elements and structure of the system under study often helps in developing the problem statement.

Formulating a design problem includes stating the overall design objective and the constraints on the design process. For example, the goal might be to design a material-handling system that is capable of transferring a certain item from point A to point B. The constraints on the design process may include certain throughput requirements, budget limitations, unit load capacity, inclination and declination requirements, and so on. Thus, the design problem is formulated such that the goal defined is met without violating any of the constraints. Unlike linear programming, however, in simulation there is no need to express the problem goal and constraints mathematically. Both are modeled using different elements and resources of DES.

Similarly, formulating a problem in an existing system includes stating the overall problem-solving objective and the constraints on the solution pro-

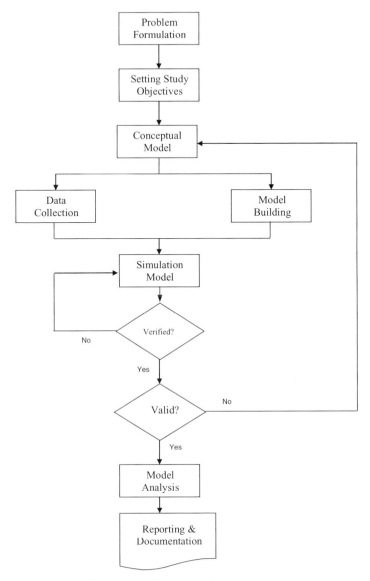

Figure 6.7 The simulation procedure.

posed. For example, the problem might be identified as a reduction in system throughput by a certain percentage. Hence, the simulation goal is set to boost the system throughput to reach a certain target. The constraints on the solution proposed may include the limited capacity of workstations, conveyor speeds, number of operators, product mix, and budget limitation, among others. Thus, the problem is formulated such that the goal defined is met without violating any of the constraints.

Finally, the formulation of an improvement problem may include stating the overall improvement objective in terms of multiple and often competing goals while meeting process constraints. For example, the first goal might be to reduce the manufacturing lead time by a certain percentage in order to apply lean manufacturing principles, and the second goal, to include meeting certain throughput requirements. Both types of goals are often subject to a similar set of process constraints, such as budget limitations, variations in manufacturing operations, and flow requirements. Thus, the improvement problem is formulated such that the two goals are met without violating any of the constraints that have been defined.

6.4.2 Setting Study Objectives

Based on the problem formulation, a set of objectives can be set to the simulation study. Such objectives represent the criteria through which the overall goal of the study is achieved. Study objectives simply indicate questions that should be answered by the simulation study. Examples include determining current-state performance, testing design alternatives, studying the impact of speeding up the mainline conveyor, and optimizing the number of carriers in a material-handling system.

Specifying study objectives serves various purposes. First, we can decide if simulation is the best tool to use to solve the underlying problem. Do we have enough data to determine metrics for the objectives that have been defined? Can we use analytical methods to answer the questions that have been raised? Is the software tool capable of presenting and analyzing study requirements so that the study objectives are achieved? These types of questions can be better answered by stating the objectives of the simulation study unambiguously.

In terms of modeling, model elements and model logic are selected and designed to provide appropriate measures of performance that quantify the study objectives. For example, to meet study objectives, an accumulating conveyor is used as a material-handling system. Also, statistics that represent throughput, lead time, delays, and number of carriers are inserted into the model over time to accumulate the data required at the end of the run. Specifying study objectives provides a clear vision for analyzing model outputs so that the questions raised are answered by the simulation report.

6.4.3 Conceptual Modeling

Developing a conceptual model is the process through which the modeler abstracts the structure, functionality, and essential features of a real-world system into a structural and logical representation that is transferable into a simulation model. The model concept can be a simple or a complex graphical representation, such as a block diagram, a flowchart, or a process map that depicts key characteristics of the simulated system, such as inputs, elements, parameters, logic, flow, and outputs. Such a representation should eventually be programmable and transferable into a simulation model using available simulation software tools. Thus, a successful model concept is one that takes into consideration the method of transferring each abstracted characteristic, building each model element, and programming the conceptual logic using the software tool.

The art of conceptual modeling is a combination of system knowledge and model-building skills. The modeler starts by a establishing a thorough understanding of the system, whether it is a new or an existing system. The modeler studies system inputs, elements, structure, logic, flow, and outputs and abstracts the overall structure and the interrelationships of structure elements into a conceptual model.

The model concept is presented considering the components and capabilities of the simulation environment. For example, developing a concept that includes a power-and-free conveyor system should find out if there is a capability for modeling such a system in the simulation software tool being used. Key parameters of system elements are also specified as a part of the model concept. For example, the concept of using a conveyance system to transfer entities from point A to point B should include parameters of conveyor type, speed, reliability, and capacity. Such parameters guide data collection and element selection.

Finally, the model concept is developed taking into consideration the problem as it was formulated and the objectives of the simulation study. Figure 6.8 shows the requirements for conceptual modeling.

6.4.4 Data Collection

Simulation models are data-driven computer programs that receive input data, execute the logic designed, and produce certain outputs. Hence, the data collection step is a key component of any simulation study. Simulation data can, however, be collected in parallel to building a model using the simulation software. This is recommended since data collection may be time consuming in some cases, and building the model structure and designing model logic can be independent of the model data. Default parameters and generic data can be used initially until the system data are collected.

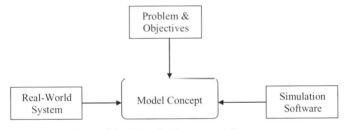

Figure 6.8 Developing a model concept.

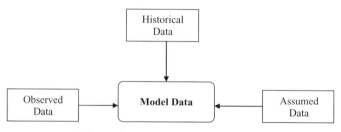

Figure 6.9   Collecting model data.

The quality of data used in the models drives the overall model quality and validity and affects the accuracy of the model results and statistics collected. Hence, the term *garbage-in-garbage-out* (GIGO) is often common in modeling. For a model to be representative, it has to be driven by representative data in terms of data integrity, accuracy, and comprehensiveness.

Data elements required for constructing a simulation model are often determined based on the model concept, model structure, and the nature and type of model outcomes and statistics to be collected. For example, typical data collected for modeling a bank operation include customer interarrival times and the service time distribution.

Depending on the nature of the simulation study, model data are collected by reviewing historical data, by observing and monitoring system operations, and by using benchmark data assumptions. The three types of data collected in model development are shown in Figure 6.9.

Historical data are often used when modeling existing systems that have been in operation for a certain time. Examples of historical data include actual production data, maintenance data, inventory records, customer feedbacks, and weekly and monthly reports on operations performance. Statistical methods of input modeling (discussed briefly in Appendix A), such as descriptive statistics, removing outliers, and fitting data distributions play an important role in analyzing historical data.

In case historical data are not available for a sufficient period of time, the actual performance of system can be observed to collect pertinent data. Time studies using a stopwatch are common data collection methods in this regard. A time study uses a standard form to collect data and starts by monitoring system behavior. System monitoring includes watching the operation of a certain system element, understanding its functionality, and deciding on the parameters to be collected and the times of collecting data. The data collected should be statistically representative and should be distributed to cover the entire spectrum of system operation. Selecting the appropriate sample size and times of observations are taken is essential to obtain representative data.

Finally, when no historical data are available and it is not permissible to collect data using time studies, simulation data may be benchmarked or assumed. Educated guesses, benchmark data, and theoretical statistical models can be used to develop the data required for simulation. This often occurs when modeling new systems and testing proposed design alternatives. Experience in both modeling and system analysis often equips the simulation analyst with the knowledge and skill that is essential to provide educated guesses of a variety of model data. Data from a similar business or manufacturing process can be used in the model, at least as a starting point. Finally, theoretical statistical distributions with estimated parameters can be used for model data. All types of assumed data are subject to modification and alteration as the model building progresses and more insight is gained into model behavior.

Input Modeling   The part of simulation concerned with selecting proper data representation for deterministic model parameters and stochastic model variables is referred to as *input modeling*. Deterministic model data are typically easier to predict and set since their values will be fixed in the model over the run time. Examples include constant demand, scheduled deliveries, fixed cycle times, fixed buffer capacity, and constant number of labor or machine resources. Values for these parameters are fixed initially if known for certain or can be viewed as decision or design variables (control factors) that are set to their optimum levels through experimental design or optimization methods.

Stochastic model inputs, on the other hand, can be viewed as noise (random) factors that change over time randomly or according to certain probability distributions. Over the simulation run period, values for these variables are sampled from theoretical or empirical probability distributions. Hence, it is first required to fit a certain probability distribution to the data collected, insert into the model the distribution selected, and using the sampling mechanism discussed in Chapter 5, generate samples of the variable over the run time. If values change randomly, uniform random numbers can be used to generate values for the variable. As discussed earlier, in addition to the variability of adaption over the run time, different random number streams often result in different sets of data sampled, which leads to stochastic variability in model outcomes. Hence, multiple simulation replications are often used to estimate each model performance measure in terms of a mean and a variance.

Many statistical methods are typically used for modeling simulation inputs. However, the basic concept in these methods is simple. If we compute the mean of a sample of 20 numbers, the value obtained will generally be much different from the population mean (i.e., the sampling distribution is not a good approximation of the data distribution). But if we sampled many sets of 20 numbers over and over again, computed the mean for each set, and constructed a relative frequency distribution of all sets, we would eventually find that the distribution of means is a very good approximation to the sampling distribution. The sampling distribution of the mean is a theoretical distribution that is approached as the number of samples in the relative frequency distribution increases. As the number of samples approaches infinity, the relative frequency distribution of actual data approaches the sampling distribution in the simulation model.

The sampling distribution of the mean for a certain sample size is just an example; there is a different sampling distribution for other sample sizes. There are also a number of statistical tests designed to assess whether a given distribution describes a data set faithfully. Although input modeling can be approached differently, a generic procedure for input modeling (fitting data collected to probability distributions) can be developed. This procedure is also built in as a module in many simulation packages. The procedure includes the following steps:

- *Step 1:* Plot the data.
  - Use a histogram and summary statistics to determine the general characteristics of the underlying distribution.
- *Step 2:* Select a family of distributions.
  - Use the results of *step 1* to select a set of "reasonable" distributions.
  - Fit each distribution to the data observed and estimate the distribution parameters.
- *Step 3:* Select the best distribution.
  - Determine which of the fitted distributions best represents the data observed using one or more appropriate statistics.
- *Step 4:* Check the distribution quality.
  - Determine the distribution goodness of fit:
    - Chi-square test
    - Kolmogorov–Smirnov test
    - Anderson–Darling test

Many simulation books include a detailed description of common probability distributions in simulation studies. (Many of these are discussed in Appendix A), which includes a summary of parameters, applications, and graphical representations of probability distributions that are commonly used in conjunction with simulation. Simulation books also provide suggestions for selecting probability distributions for different data situations. Table 6.1 presents a summary of some commonly used distributions in simulation studies of various data situations in manufacturing applications.

6.4.5 Model Building

Data collection and model building often consume the majority of the time required for completion of a simulation project. To reduce such time, the modeler should start building the simulation model while data are being

TABLE 6.1   Summary of Commonly Used Distributions in Simulation Studies

Simulation Variable	Probability Distribution	Key Distribution Parameters
Machine cycle time	Uniform	(min., max.)
	Triangular	(min., max., mode)
Speed for conveyor, automated guided vehicle (AGV), fork truck, etc.	Uniform	(min., max.)
Travel time for AGV, fork truck, etc.	Triangular	(min., max., mode)
	Lognormal	(mean, standard deviation)
Downtime (time to repair)	Triangular	(min., max., mode)
Uptime (time between failures)	Erlang	(mean, $K = 3$)
	Exponential	$\lambda = 1/MTBF$
Scrap rate	Binomial	% scrapped
Interarrival time	Erlang	(mean, $K = 2$)
	Exponential	$\lambda = 1/MTBA$
	Lognormal	(mean, standard deviation)
Part type assignment	Binomial	% product mix
	Empirical discrete	% A, % B, etc.

collected. The conceptual model can be used to construct the computer model using assumed data until the data collected become available. The overlap between model building and data collection does not affect the logical sequence of the simulation procedure. Constructing model components, entity flow, and logic depends mostly on the model concept and is in most cases independent of model data. Once the model is ready, model input data and parameter settings can be inserted into the model later. Also, since a large portion of a simulation study is often spent in collecting model data, building the model simultaneously reduces significantly the overall duration of the simulation study and provides more time for model analysis and experimentation.

There is no standard procedure for building a simulation model. The procedure is often based on the modeler approach and on the simulation software tool used. However, a generic procedure for building a simulation model effectively is expected to include key basic steps. This procedure is shown in Figure 6.10.

The model-building procedure starts by importing a computer-aided design (CAD) file of the system layout. The CAD file is a two- or three-dimensional layout file developed in any commercial drafting and design software tool. Although the layout has no impact on model performance, it is a useful graphical representation of the system that provides realistic system animation, especially a three-dimensional layout. Because of the large size of

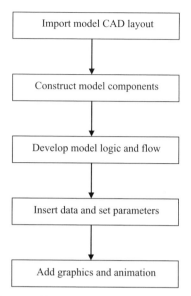

Figure 6.10   Model-building procedure.

three-dimensional files, most simulation models are built on a two-dimensional layout, which provides locations and spaces for various system elements, such as machines, pieces of equipment, conveyors, storage, and loading–unloading decks. It also provides a physical reference to the entity flow by showing physical system elements such as aisles, walkways, walls, empty spaces, and offices. Thus, when running the model, the entity flow within different model elements will match the flow of parts or customers in the real-world system.

Different model elements and components are built using the set of modules or building blocks defined in the simulation software tool used. Examples include resources or machines, queues or buffers, conveyors, and loads or parts. Most simulation tools have built-in, ready-to-use, and drag-and-drop modules of such components with some graphical representation. Constructing a model component includes selecting the simulation element that best represents the component and locating the element on its actual locations in the system's CAD layout. At this stage, model elements are not related to each other.

Element interrelationships, decision points, and entity flow are defined by developing the model logic. Depending on the simulation tool used, the model logic can be developed by defining in and out routing rules at each simulation element, the model logic can be written and debugged in a certain editing environment, or the two may be combined. Although most new simulation software tools strive to reduce the programming effort, writing code is still necessary to implement the model logic.

Once the model logic is developed, we start running the model. However, the model performance and results may not reflect the behavior of the system of interest without inserting representative data into the model and setting the parameters of various system components. Such data should be collected while the first three steps in the model-building procedure are being executed. As the model data become available, we start inserting the data in the model following the instructions suggested by the simulation tool used. Generally, this is an easy and quick step in the model-building procedure.

The last step in building a simulation model is to add the animation and graphical representations of model complements and the surrounding environment. Typical examples include developing animation that reflects a plant environment, a banking system, a restaurant, or others. Although animation does not really add to the quality of the simulation results, it helps greatly in model validation and verification. It is also a great selling and presentation tool, especially if developed with three-dimensional graphics. Some simulation environments allow for only two- or $2\frac{1}{2}$-dimensional graphical representation. In addition to providing a graphics editing environment, simulation software tools include libraries of ready-to-use graphics that suit different systems simulated for both manufacturing and service applications.

6.4.6   Model Verification

Model verification is the quality control check that is applied to the simulation model built. Like any other computer program, the simulation model should perform based on the intended logical design used in building the model. Although, model logic can be defined using different methods and can be implemented using different programming techniques, execution of the logic when running the model should reflect the initial design of the programmer or modeler. Different methods are used for debugging logical (programming) errors as well as errors in inputting data and setting model parameters. Corrected potential code and data discrepancies should always be verified by careful observation of changes in model behavior.

To verify a model, we simply check whether the model is doing what it is supposed to do. For example, does the model read the input data properly? Does the model send the right part to the right place? Does the model implement the production schedule prescribed? Do customers in the model follow the queuing discipline proposed? Does the model provide the right output? And so on. Other verification techniques include applying rules of common sense, watching the model animation periodically during run time, examining model outputs, and asking another modeler to review the model and check its behavior. The observations made by other analysts are valuable since the model builder will be more focused on the programming details and less focused on the implication of different programming elements. When the model logic is complex, more than one simulation analyst may have to work on building the model.

6.4.7   Model Validation

Model validation is the process of checking the accuracy of the model representation to the real-world system that has been simulated. It is simply about answering the following question: Does the model behave similarly to the simulated system? Since the model will be used to replace the actual system in experimental design and performance analysis, can we rely in its representation of the actual system?

Knowing that the model is only an approximation of the real-world system, key characteristics of actual system behavior should be captured in the model, especially those related to comparing alternatives, drawing inferences, and making decisions. Hence, necessary changes and calibrations that are made to the model to better represent the actual system should be returned to the model concept. The model concept represents the modeler's abstraction of the real-world system structure and logic. Thus, if the model were not fully valid, the model concept needs to be enhanced and then translated into the simulation model.

Several techniques are usually followed by modelers to check the validity of the model before using it for such purposes. Examples include checking the data used in the model and comparing them to the actual system data, validating the model logic in terms of flow, sequence, routing, decisions, scheduling, and so on, vis-à-vis the real-world system, and matching the results of the model statistics to those of actual system performance measures.

Cross-validation using actual system results and running certain what-if scenarios can also be used to check model validity. For example, last year's throughput data can used be to validate the throughput number produced by the model for the same duration and under similar conditions. We can also double the cycle time of a certain operation and see if the system throughput produced is affected accordingly or if the manufacturing lead time data reflect this increase in cycle time.

6.4.8   Model Analysis[2]

Having a verified and validated simulation model provides analysts with a great opportunity since it provides a flexible platform on which to run experiments and to apply various types of engineering analyses effectively. With the latest advances in computer speed and capacity, even large-scale simulation models of intensive graphics can be run for several replications in a relatively short time. Hence, it takes only a few minutes to run multiple simulation replications for long periods of time in most simulation environments.

As shown in Figure 6.11, model analysis often includes statistical analysis, experimental design, and optimization search. The objective of such methods is to analyze the performance of the simulation model, compare the performance of proposed design alternatives, and provide the model structure and

---

[2] We provide further insight on analyzing simulation outputs in Chapter 7 and Appendix A.

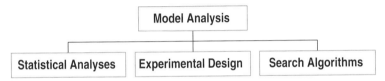

Figure 6.11 Continuous improvement with simulation.

parameter setting that will lead to the best level of performance. Statistical analyses include representing performance with descriptive statistics such as mean, variance, and confidence intervals. It also includes performing various statistical tests for hypothesis testing and comparative analysis. Experimental design with simulation includes conducting a partial or full factorial design of experiments to provide the best settings to model control variables. Finally, search algorithms include the application of optimization search methods such as exhaustive search, genetic algorithm, simulation annealing, and tabu search to optimize an objective function.

The speed and flexibility of the simulation model also facilitate conducting experimental design and running optimization algorithms. Also, the availability of different methods and analyses in commercial simulation software tools facilitates the use of output analyses. Most full versions of simulation packages are equipped with modules for statistical analyses and add-ins of experimental design and optimization methods.

6.4.9   Study Documentation

The final step in a simulation study is to document the study and report its results. Proper documentation is crucial to the success of a simulation study. The simulation process often includes communicating with many sides, writing complex logic, encountering enormous amounts of data, conducting extensive experimentation, and going through several progress reviews and milestones. Thus, without proper documentation, the analyst loses track of and control over the study and cannot deliver the required information or meet the study expectations. This often results in an inaccurate simulation model with poor results, inability to justify model behavior and explain model results, and loss of others' confidence in study findings and recommendations.

Documenting the simulation study is the development of a study file that includes the details of each simulation step. Comprehensive documentation of a simulation study comprises three main elements: detailed documentation of the simulation model, the development of an engineering simulation report, and the presentation of simulation results to customers and partners of the simulation project. Figure 6.12 presents the three elements of documenting a simulation study.

The first element in documenting a simulation study includes documentation of the simulation model. This includes documenting both the concept

Figure 6.12   Documenting a simulation study.

model and the simulation program. The documentation of the model concept includes documenting the system process map, flowcharts, block diagrams, and sketches that explain the model concept, model elements, modeling approach, and model logical design. Simulation model documentation includes a description of model structure, program details and flowcharts, code and explanation of routines using within-code comments and notes, and an explanation of the scheduling and sequencing rules, routing rules, and the decision-making process followed at decision points within the model. Model reporting also includes a statement of model assumptions and a description of model inputs and outputs. By reviewing such documents, we should be able to understand how the model reads inputs, processes the logic, uses available simulation elements, and provides the required outputs. Such documentation facilitates making model changes, explaining the data used in the model, debugging code and logic, understanding model behavior, and interpreting model results.

The major deliverable of the simulation study is the development of a simulation report. This includes a formal engineering report and supplementary materials of the simulation process. Such a report constitutes the communication medium between the simulation analyst and the other parties involved in the simulation study. Some of those parties may not even see the model or talk to the simulation analyst; hence, the report should be comprehensive, unambiguous, and understandable by people of various skills and backgrounds.

The elements of a formal simulation report include a description of the underlying system, formulation of the simulation problem, a statement of study objectives and assumptions, sources of model data, a description of the model structure, and a summary of simulation results, findings, conclusions, and recommendations. Supplementary materials include the data inputs used, system and model graphs, sketches and drawings, details of experimental design and output analysis, and a printout of simulation results. It is also common among simulation engineers to provide a one-page summary of key information provided by the simulation study. This type of executive review provides concise and focused information for those decision makers, managers, and firm executives who may not be interested in the details of model inputs, structure, and results analysis. A simulation report includes the following elements:

1. The System Being Simulated
    a. Background
    b. System description
    c. System design
2. The Simulation Problem
    a. Problem formulation
    b. Problem assumptions
    c. Study objectives
3. The Simulation Model
    a. Model structure
    b. Model inputs
    c. Model assumptions
4. Simulation Results
    a. Results summary
    b. Results analysis
5. Study Conclusion
    a. Study finding
    b. Study recommendations
6. Study Supplements
    a. Drawings and graphs
    b. Input data
    c. Output data
    d. Experimental design
    e. Others

The simulation study is often concluded by presenting the simulation results to customers, managers, and other parties to the simulation project. Such a presentation often includes a summary of the simulation results (summary of study steps, results and findings, conclusions, and recommendations). The presentation should include running animations, movies, or snapshots of the simulation model in a variety of situations. Such animations help explain the study results and aid the analyst in selling the design proposed, comparing design alternatives, and securing management support through proposed solutions and plans of action. The communication and presentation skills of the simulation analyst play a major role in developing a successful presentation and gaining managerial support to implement the recommendations of the simulation project.

6.5 EXAMPLE: APPLYING THE SIMULATION PROCESS TO A HOSPITAL EMERGENCY ROOM

In this section we apply the simulation process to an emergency room (ER) simulation. The main objective here is to use simulation to study the ER system, improve the ER operation by reducing total time in the system, and predict the effect of changes on long-term performance. ER simulation study requirements include the following:

- Identifying measurement points necessary for system evaluation
- Selecting and applying the appropriate simulation approaches and techniques to represent the system processes
- Identifying ER operational weak points and bottlenecks
- Suggesting improvements to the current situation
- Constructing a representation of the ER system configuration proposed
- Verifying the system proposed

ER Process Description As depicted in Figure 6.13, the ER example consists of three stations: the triage station, where patients record their names and fill out the triage form; the reception station, where the station attendant enters their data into a computer and creates medical files for new patients; and the treatment room, for physician diagnosis. If a patient needs further treatment, he or she will be sent to a specialist doctor (i.e., internist, cardiologist, gynecologist, ophthalmologist, etc.) For serious medical emergencies (i.e., accidents, heart attacks, parturitions, etc.), patients are treated directly by doctors without passing through the triage and data-entry stations.

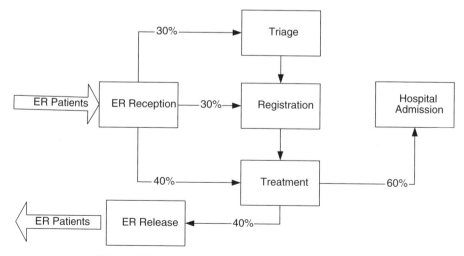

Figure 6.13 Conceptual model of the ER example.

An ER system can be viewed as a queuing system with a single channel (i.e., admission, treatment, release). The patient flow within the ER is sequential and can be combined into a one-stop service in order to apply the formulas of the single-server queuing model. In practice, however, different ER resources provide service at different locations.

The ER is not a typical first-in-first-served system; the ER operates using specific rules to route patients. The priority for a patient to enter the ER depends on his or her medical case and the risk involved in making the patient wait. Although more complex queuing models (e.g., a network of queues) can be used to approximate the system, simulation will be a better alternative, given the study requirements mentioned. The system can still be described by the rate of patient arrivals and the service mechanism, which includes the following three stages:

1. Passing by triage
2. Entering a patient's related data in computer records
3. Physical treatment

In the ER DES model, two main events cause an instantaneous change in the state of the ER system:

1. The entry of a patient into the system (patient arrival)
2. The release of a patient upon completion of treatment (patient departure)

Based on information obtained from subject matter experts, observation, and historical data, the following logic governs the flow of patients in the ER:

- The order of treatment priority is high risk, medium risk, and normal cases, respectively.
- ER patients are classified as follows:
  - *Risky case:* a probability of 40%
  - *Medium-risk case:* a probability of 30%
  - *Normal case:* a probability of 30%
- When a risky case arrives at the ER, it does not pass through the triage or data-entry stages; the person is treated directly by a physician.
- When a medium-risk case arrives, the person passes through triage but does not pass through the data-entry area.
- When a normal case arrives, the person must pass through all three stages: triage, data entry, and treatment.
- One physician can treat up to two patients in the emergency room.
- Patients leaving the emergency room are routed as follows:
  - *Those who will leave the hospital:* a percentage of 40%

- *Those who will continue the treatment in other specialist units in the hospital:* a percentage of 60%

Data Collection and Input Modeling   After getting permission from the hospital administration, the project team started their work by observing the situation in the ER and collecting pertinent data, including counting the queue of patients, measuring the service time at various process stages and the interarrival time of patients, tracking patient flow, and so on. Also, historical data from the ER were reviewed and summarized.

The team was faced with a lack of historical data (records of patient arrivals and service durations) essential to estimating the service and interarrival times. Thus, it was decided to collect data using a stopwatch. Pertinent data for 59 samples (patients) were collected. Table 6.2 summarizes the data collected in terms of patient arrival times and the time for end of service at the three ER stations. Using the data collected, values are estimated for each patient's interarrival and service times at the three ER stations.

To fit data collected to their corresponding statistical distributions, the input modeling module in an ARENA simulation package was used. Table 6.3 summarizes the results for interarrival times.

Model Building, Verification, and Validation   Using AutoMod simulation software, the base model is built as shown in the AutoMod animation shown in Figure 6.14. The model logic was first verified and the model behavior was validated using the data collected and observations of ER subject matter experts. AutoMod model elements used include:

- *Entity:* patients
- *Attributes:* health (classification of ER patients)
- *Activity:* treatment
- *Exogenous event:* arrival of patients
- *Endogenous event:* completion of treatment (departure of patients)
- *State variable:* number of patients waiting, number of busy doctors, etc.

Model Results   The model is set to run 10 days continuously (240 hours, assuming that the ER operates three shifts a day) after a warm-up period of 8 hours. As shown from the queue statistics in Figure 6.15, a total of 1001 patients were treated at the ER, of which 373 patients were classified as serious. A total of 628 patients went through the three ER stations, starting with triage. The average ER patient spent about 652 seconds waiting, about 122 seconds at triage, about 790 seconds at the reception registration, and about 2772 seconds in treatment. The average total time-in-system is about 3644 seconds.

TABLE 6.2 ER Data Collected

Patient Number	Time of Arrival at Triage (1)	Time of Arrival at Reception (2)	Time of Starting the Treatment	Time of Patient Departure (3)	Time Between Arrivals (min)	Service Time 1 (min)	Service Time 2 (min)	Service Time 3 (min)
1	8:30	8:31	8:35	9:55	0	1	4	30
2	9:00	9:01	9:05	9:10	30	1	4	5
3	9:20	9:25	9:30	9:45	20	5	5	15
4	9:50	9:51	9:55	10:05	30	1	4	10
5	10:15	10:16	10:18	10:30	25	1	2	12
6	10:20	10:21	10:30	11:10	5	1	9	40
7	10:22	10:24	10:30	11:30	2	2	6	60
8	10:35	10:36	10:38	11:00	13	1	2	22
9	10:40	10:41	10:42	12:10	5	1	1	88
10	10:45	10:46	10:47	11:45	5	1	1	58
11	10:45	10:46	10:48	12:05	0	1	2	77
12	11:00	11:01	11:05	2:12	15	1	4	127
13	12:00	12:01	12:05	12:10	60	1	4	5
14	12:10	12:11	12:12	12:15	10	1	1	4
15	12:30	12:31	12:35	12:38	20	1	4	3
16	12:30	12:31	12:36	12:40	0	1	5	4
17	12:30	12:31	12:38	12:40	0	1	7	2
18	12:35	12:36	12:36	12:40	5	1	1	4
19	12:40	12:41	12:45	12:50	5	1	5	5
20	12:50	12:51	12:55	13:00	10	1	5	5
21	12:50	12:51	12:58	13:10	0	1	8	12
22	12:50	12:51	12:56	13:15	0	1	6	19
23	12:50	12:51	12:55	13:00	0	1	5	5
24	13:20	13:21	13:22	13:30	30	1	2	8
25	13:30	13:31	13:35	14:00	10	1	5	25
26	13:30	13:31	13:35	13:40	0	1	5	5
27	13:35	13:36	14:00	14:30	5	1	25	30
28	14:00	14:01	14:10	14:12	25	1	10	2
29	14:00	14:01	14:05	14:20	0	1	5	15

(*Continued*)

TABLE 6.2 *Continued*

Patient Number	Time of Arrival at Triage (1)	Time of Arrival at Reception (2)	Time of Starting the Treatment	Time of Patient Departure (3)	Time Between Arrivals (min)	Service Time 1 (min)	Service Time 2 (min)	Service Time 3 (min)
30	14:45	14:46	14:48	15:30	45	1	3	42
31	15:30	15:31	15:35	16:00	35	1	5	25
32	8:29	8:33	8:43	8:55	0	4	10	10
33	8:57	8:59	9:00	9:05	18	2	1	5
34	9:05	9:06	9:36	9:40	8	1	30	5
35	9:07	9:08	9:16	9:21	2	1	8	4
36	9:40	9:41	10:00	10:20	33	1	19	20
37	9:50	9:51	10:20	10:30	10	1	29	10
38	9:55	9:56	10:30	11:00	5	1	34	30
39	10:00	10:01	10:40	11:00	5	1	39	20
40	10:00	10:01	10:50	11:40	0	1	49	50
41	10:10	10:11	10:12	11:00	10	1	1	48
42	10:35	10:36	10:40	11:20	25	1	4	40
43	10:35	10:36	10:41	11:00	0	1	5	19
44	10:40	10:41	10:45	11:30	5	1	4	45
45	10:55	10:56	10:58	12:10	15	1	2	72
46	11:00	11:05	11:10	12:30	5	5	5	80
47	11:15	11:16	11:20	12:00	15	1	4	40
48	11:30	11:31	11:35	12:00	15	1	4	25
49	11:40	11:41	11:45	13:00	10	1	4	75
50	11:40	11:41	11:50	13:00	0	1	9	70
51	11:45	11:46	11:48	12:10	5	1	2	22
52	11:45	11:46	11:55	12:20	0	1	9	25
53	12:00	12:01	12:10	12:40	15	1	9	30
54	12:10	12:11	12:20	13:15	10	1	9	55
55	13:30	13:31	13:34	14:00	80	1	3	26
56	13:40	13:41	14:00	14:15	10	1	19	15
57	14:00	14:01	14:20	14:30	20	1	19	10
58	14:05	14:06	14:10	15:00	5	1	4	50
59	16:20	4:21	16:30	17:10	125	1	9	40

TABLE 6.3   Fitting Data Collected to Statistical Distributions[a]

Patients Interarrival Time

Distribution summary
  Distribution: exponential
  Expression: −0.001 + EXPO(14.6)
  Square error: 0.002905
Data summary
  Number of data points = 59
  Min. data value = 0
  Max. data value = 125
  Sample mean = 14.6
  Sample std. dev. = 21.2

Goodness of fit:
  Chi-square test
    Number of intervals = 2
    Degrees of freedom = 0
    Test statistic = 0.154
    Corresponding p-value < 0.005

Service Time at the Triage Station

Distribution summary
  Distribution: lognormal
  Expression: 0.5 + LOGN(0.654, 0.347)
  Square error: 0.013047
Data summary
  Number of data points = 59
  Min. data value = 1
  Max. data value = 5
  Sample mean = 1.22
  Sample std. dev. = 0.832

Goodness of fit:
  Chi-square test
    Number of intervals = 2
    Degrees of freedom = −1
    Test statistic = 5.42
    Corresponding p-value < 0.005

Service Time at the Registration Station

Distribution summary
  Distribution: lognormal
  Expression: 0.5 + LOGN(8.19, 12.7)
  Square error: 0.044657
Data summary
  Number of data points = 59
  Min. data value = 1
  Max. data value = 49
  Sample mean = 8.46
  Sample std. dev. = 9.96

Goodness of fit:
  Chi-square test
    Number of intervals = 7
    Degrees of freedom = 4
    Test statistic = 16.9
    Corresponding p-value < 0.005

Service Time at the Treatment Station

Distribution summary
  Distribution: Weibull
  Expression: 2 + WEIB(15.8, 0.64)
  Square error: 0.001592
Data summary
  Number of data points = 34
  Min. data value = 2
  Max. data value = 127
  Sample mean = 23.1
  Sample std. dev. = 28.7

Goodness of fit:
  Chi-square test
    Number of intervals = 2
    Degrees of freedom = −1
    Test statistic = 0.122
    Corresponding p-value < 0.005

[a] See Appendix A for additional statistical substance.

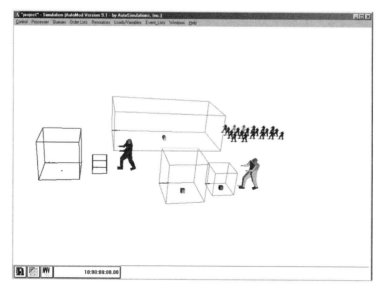

Figure 6.14   AutoMod animation of the ER model.

Name	Total	Cur	Average	Capacity	Max	Min	Util	Av_Time	Av_Wait
Space	1004	1	1.00	Infinite	2	0	—	860.56	—
Q_tria_wait	628	0	0.00	50	2	0	0.000	5.65	0.00
Q_tria	628	0	0.09	1	1	0	0.088	121.38	0.00
Q_rec_wait	628	1	0.47	40	7	0	0.012	641.41	0.00
Q_rec	627	1	0.57	1	1	0	0.574	790.29	0.00
Q_treat_wait	1001	3	3.21	40	10	0	0.080	2771.53	0.00

Time: 10:00:00:00.00

Figure 6.15   Summary of AutoMod queue statistics.

Figure 6.16 shows the AutoMod summary of resource statistics. The model is set to use one resource (teller) for reception, one for computer data entry, and a high number of resources for physicians since the number of physicians used at any point is based on the number of patients in the ER room. Utilization is a key resource statistic that can be used in directing ER improvement actions.

Based on model observations and model statistics, the following measures were taken to improve ER operations. Improvement efforts are focused on reducing patient waiting time before service, registration time, and waiting for

Resource Statistics

Update

Time: 10:00:00:00.00

Name	Total	Cur	Average	Capacity	Max	Min	Util	Av_Time	Av_Wait	State
R_nurse	628	0	0.09	1	1	0	0.088	121.38	0.00	----
R_comp	627	1	0.57	1	1	0	0.574	790.29	0.00	----
R_doctor(1)	427	0	0.88	1	1	0	0.880	1781.47	1124.32	----
R_doctor(2)	368	1	0.77	1	1	0	0.766	1797.31	1090.81	----
R_doctor(3)	169	1	0.35	1	1	0	0.354	1810.63	574.75	----
R_doctor(4)	31	0	0.06	1	1	0	0.062	1723.74	148.50	----
R_doctor(5)	3	0	0.01	1	1	0	0.006	1862.21	0.00	----
R_doctor(6)	2	0	0.00	1	1	0	0.004	1688.28	0.00	----
R_doctor(7)	0	0	0.00	1	0	0	0.000	0.00	0.00	----
R_doctor(8)	0	0	0.00	1	0	0	0.000	0.00	0.00	----
R_doctor(9)	0	0	0.00	1	0	0	0.000	0.00	0.00	----
R_doctor(10)	0	0	0.00	1	0	0	0.000	0.00	0.00	----
R_doctor(11)	0	0	0.00	1	0	0	0.000	0.00	0.00	----
R_doctor(12)	0	0	0.00	1	0	0	0.000	0.00	0.00	----
R_doctor(13)	0	0	0.00	1	0	0	0.000	0.00	0.00	----
R_doctor(14)	0	0	0.00	1	0	0	0.000	0.00	0.00	----
R_doctor(15)	0	0	0.00	1	0	0	0.000	0.00	0.00	----
R_doctor(16)	0	0	0.00	1	0	0	0.000	0.00	0.00	----
R_doctor(17)	0	0	0.00	1	0	0	0.000	0.00	0.00	----
R_doctor(18)	0	0	0.00	1	0	0	0.000	0.00	0.00	----
R_doctor(19)	0	0	0.00	1	0	0	0.000	0.00	0.00	----
R_doctor(20)	0	0	0.00	1	0	0	0.000	0.00	0.00	----

Figure 6.16 Summary of AutoMod resource statistics.

treatment time. This will significantly reduce the overall time spent in the system. Other measures are aimed at increasing the effectiveness of ER operations by optimizing the number of physicians and increasing the utilization of reception and triage resources. Other micro measures can also take place to improve work procedures, especially for preparing patient files and entering patient information.

Following is a summary of ER improvement actions:

- The long time spent in the reception stage was reduced first since a long waiting time may affect patient health negatively (i.e., may convert it from normal or medium to serious). The reception and registration time is improved by developing a more effective work procedure for preparing patient files and entering patient information. A time and motion study was recommended for this purpose. An integrated hospital database often provides a substantial time saving in this regard.
- Time spent while patients wait for physicians at the treatment station was reduced by changing the rule of "one doctor to two patients" and allowing the immediate assignment of another available doctor to the second patient.
- The final number of physician resources is set to three based on the utilization of physician resources (see Figure 6.16).
- The triage and registration resources were combined, due to low utilization of the triage resource.

Changes in the total time-in-system were observed through a confirmation simulation run with five replications, as shown in Figure 6.17. The new average

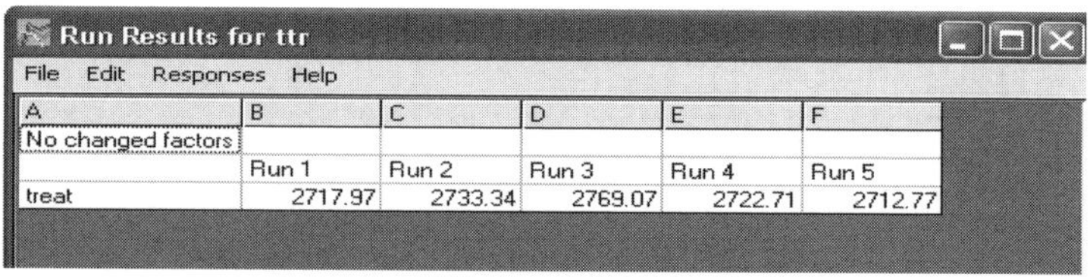

Run Results for ttr					
File Edit Responses Help					
A	B	C	D	E	F
No changed factors					
	Run 1	Run 2	Run 3	Run 4	Run 5
treat	2717.97	2733.34	2769.07	2722.71	2712.77

Figure 6.17 Five replications of the ER time-in-system.

time-in-system is 2732 seconds. Compared to an average time-in-system of 3644 seconds in the initial ER model, improvement actions resulted in about a 25% reduction in average patient time-in-system.

6.6 SUMMARY

In this chapter we presented a detailed description of the simulation process, including the details of simulation procedures for project scoping, data collecting, model building, model analyses, and model documentation. This process is later integrated into the six-sigma project charter in the 3S-LSS and 3S-DFSS approaches.

7

SIMULATION ANALYSIS

7.1 INTRODUCTION

Simulation models are often viewed as logic and data-driven statistical models that represent the behavior of a real process in terms of statistical measures such as performance values, average, variance, range, and percent time in a given state. Model outputs are analyzed and interpreted statistically to draw inferences on various aspects of model behavior. Simulation projects achieve their objectives by conducting statistical experiments on a computer model representing a certain real-world process. The skills necessary for building, running, and analyzing simulation models are highly dependent on understanding the statistical nature of simulation modeling as well as on mastering the essential statistical concepts and methods. The compatibility of statistical methods used in building and analyzing simulation models are critical to enter the correct data into the model and to draw the most accurate conclusions about model behavior. As discussed earlier, simulation models are driven by historical or observed data, and both model performance and model outputs are functions of data accuracy and representation. Similarly, providing the correct interpretation of simulation results and arriving at the right decisions and conclusions is a function of the accuracy of output performance and the viability of the output analysis techniques that are used.

In this chapter we focus on the main issues involved in analyzing simulation outputs: distinguishing between terminating and steady-state simulation,

Simulation-Based Lean Six-Sigma and Design for Six-Sigma, by Basem El-Haik and Raid Al-Aomar
Copyright © 2006 John Wiley & Sons, Inc.

understanding the stochastic nature of simulation outcomes, determining the simulation run controls (warm-up period, run length, and number of replications), and selecting the appropriate model output analysis method.

Output analysis methods include both descriptive and analytical statistical analysis to draw the correct conclusions from model outputs. Examples include performance estimation (point and interval estimators), comparing alternatives, and hypothesis testing. The reader is encouraged to consult Appendix A regarding these concepts. Output analysis may include optimizing model performance by determining the parameter setting that leads the best performance level. Experimental design (partial and full factorial design of experiment) and specialized optimization search methods such as genetic algorithms and simulated annealing are often utilized for this objective.

7.2   TERMINATING VERSUS STEADY-STATE SIMULATION

Based on the model stopping condition, simulation models can be terminating or nonterminating (steady-state). A terminating condition is determined based on how representative the profile of the underlying simulation response is compared to the real world. In *terminating simulation* the model is run for a certain period of time or until a certain event takes place or a certain condition is met, regardless of the profile of model performance output. In *nonterminating simulation* the simulation model is run long enough to pass the transient behavior of simulation response and reach a steady-state performance profile. Each type has its own real-world applications and output analysis methods.

7.2.1   Terminating Simulation

Terminating simulation refers to running a simulation model for a certain time period or until a certain event occurs or a certain condition is met. The objective here is to describe the behavior of the process over a particular period of time. The simulation records the performance at all points in time within the defined period, including the impact of initial states which encompass transient periods within the simulation time. The transient behavior, however, is part of the normal operation in a terminating system, and hence it cannot be discarded or ignored in simulation output analysis.

Typically, terminating simulations represent real-world processes that start empty, receive entities gradually for a certain run time, reach a peak performance, and empty out again before shutting down. The system may repeat cycles of such performance during run time. Examples include a bank that operates 8 hours a day or a job shop that produces several items of 100 units each according to a certain production schedule. Table 7.1 lists several applications of termination simulation. Figure 7.1 shows a bank simulation termination example.

TABLE 7.1 Examples of Simulation Termination

Application	Terminating Time/State/Condition
Bank operation	Completing an 8-hour business day
Job shop operation	Scheduling the production of 100 units of three product types
Roadway traffic	Simulating intersection traffic during rush hour
Inventory control	Pulling parts from storage until a certain reorder point is reached
Restaurant operation	Modeling drive-through window during lunchtime

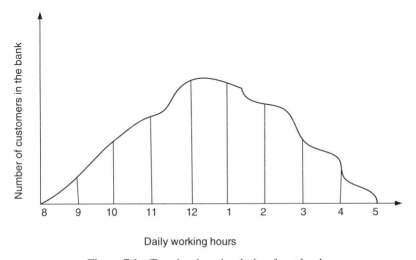

Figure 7.1 Termination simulation for a bank.

As shown in the figure, the number of customers in the bank increases working hours gradually and reaches its maximum by midday. The high load modeled does not hold for a long time. The number of customers gradually decreases toward the end of the business day. In this example, no specific workday interval can be considered a transient period or a steady state period; that is, terminating simulation includes both transient and steady-state behavior. Therefore, average performance measures are of little meaning. Generally, the terminating simulation is not intended to estimate the long-term (steady-state) performance of the underlying system. Instead, the focus in the analysis of terminating simulation models is on performance measures such as:

- Determining maximum and minimum performance values
- Observing and determining changes in behavior patterns over the run time
- Determining the utilization of resources
- Determining the number of servers, operators, resources, and so on

Terminating simulation models are generally highly sensitive to initial model conditions and to the terminating state selected. Usually, the main issues involved in experimenting with terminating simulation include the following:

- Selecting the initial model state
- Selecting the terminating time, event, or condition
- Determining the number of simulation replications

Selecting the Initial Model State   The initial state of the model represents the model status at the beginning of simulation when the simulation clock time equals zero. Although most terminating simulations start with the model empty, some cases require initializing the model. For example, in a production process model, we may start with a full warehouse and use the model to estimate the time it takes for the assembly line to use all items in storage and start to feel a component shortage. This is often decided based on the state of the process modeled. For example, in a plant simulation, we need to check if the plant starts with machines ready to work or if the machines require initial setup or configuration. We also need to check if the production line starts with full, half-full, or empty buffers. In some cases we may be interested in beginning a plant simulation at a certain point in time where the initial model state is defined to represent model behavior at the starting point selected for the simulation.

Because of the limited run time, the model initial state typically affects the behavior recorded and the statistics collected in terminating simulations. For example, if you start a bank simulation at 8:00 A.M. when the bank is full of customers, the data observed and collected on bank teller utilization will be higher than when the bank starts empty and fills with customers gradually. Thus, we must be careful in defining the proper initial conditions and state and should be aware of the implications and consequences of such definitions. We should also be able to interpret the results obtained from the model while taking into consideration the initial state defined.

Selecting the Terminating Time   Selecting the terminating event (time, state, or condition) is another key issue that affects the behavior (and results) of terminating simulation. Observing the operating patterns of simulated systems often makes it easy to define the terminating event. For example, a doctor's clinic working hours may include being open from at 8:00 A.M. to 12:00 noon, closing for lunch from noon to 1:00 P.M., and being open again from 1:00 to 5:00 P.M. A plant may operate based on a 12-hour shift starting from 6:00 A.M. with two 15-minute coffee breaks at 9:45 A.M. and 2:45 P.M. and lunchtime from noon to 1:00 P.M. In such systems, the terminating time is defined to reflect the operating pattern. Running the clinic or plant model for several days at the same initial conditions is expected to produce replications of each day's performance.

The terminating time can also be determined contingent on reaching a certain state. This state needs to be selected carefully so that certain key questions are answered from the model behavior observed and the results obtained. For example, a terminating cause may include reaching a state of serving the 100th customer in a fast-food restaurant or when all machines are broken in a shop. Studying the model behavior up to the state selected is expected to represent some meaningful milestone or turning point in the model behavior. We may be interested in using the first 100 customers served as a sample for a marketing study, or we may be interested in seeing how the shop would recover were all the machines to be broken. This also applies to selecting terminating time based on meeting a certain condition, such as when all parts in the warehouse are used and the inventory reorder point is reached.

Determining the Number of Simulation Replications Experimenting with terminating simulation often includes running several replications with different random number streams to provide a certain statistical confidence level in the results obtained. Since the objective here is not to estimate measures for the model's steady-state performance, multiple replications are used primarily to develop confidence intervals of model statistics and not to obtain average long-term performance estimates.

In general, more replications provide better estimation of model statistics. For example, running several days of bank operations at the same running conditions provides insight on the changes in service pattern from one day to another. Changing the seeds of random number generators in the model often results in changing the bank operating pattern from one simulation run to another. In Section 7.3 we discuss the details of determining sample size (number of simulation replications).

7.2.2 Steady-State Simulation

A steady-state or nonterminating simulation applies to systems whose operation continues indefinitely or for a long time period with no statistical change in behavior. Such a period includes a relatively short transient (warm-up) period and a relatively long steady-state period. The warm-up period is excluded from results analysis, and simulation statistics are collected from the steady-state period. Examples of nonterminating systems include a service station that runs around the clock or a plant that runs one or two shifts a day but where the work continues in the next shift exactly as it left off at the end of the preceding shift (without emptying parts from the system or changing its resources and parameter settings). As shown in Figure 7.2, the plant production rate continues to change over run time in a steady-state mode after passing the plant transient period. This represents the pattern in most plants and production systems.

A steady-state performance does not mean that the system performance is fixed after a certain period. In fact, model response will continue to exhibit

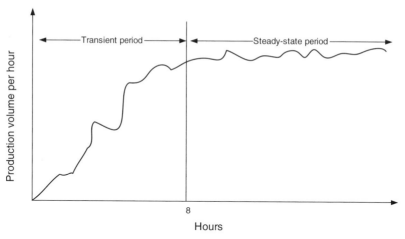

Figure 7.2   Nonterminating simulation for a plant.

variability throughout the total run time. The observed performance variability during the steady-state period will follow approximately the same distribution as the total run time. This perfectly reflects the stochastic nature of simulation models representing real-world processes. Changing model parameters at any point during the steady-state period may result in experiencing another transient period before reaching steady state and may well produce observations that demonstrate different distribution effects. Different streams of random numbers may also result in obtaining different distributions of the performance observed for the CTQs.

Experimenting with steady-state simulation requires determining the following:

- Determining the length of the warm-up (transient) period
- Determining the length of the data collection period (steady-state period)
- Determining the number of simulation replications required to collect representative sample observations

These decisions, often referred to as *model run controls*, are discussed in the next section. The objective of determining such controls is to obtain output observations that are identical (unbiased), independent, and normally distributed. Observations with such features facilitate the use of a wide range of statistical methods when analyzing model outputs.

7.3   DETERMINATION OF SIMULATION RUN CONTROLS

To perform statistical analysis of the simulation output, some conditions need to be established. In theory, the observations collected from any simulation experiment must be unbiased, independent, and normally distributed. This implies drawing independent observations (not correlated) from a stationary process (time invariant) whose outcomes are normally distributed (bell-shaped). Since none of the three conditions are satisfied in many real-world applications of simulation modeling, simulation output analysis, both descriptive and inferential, is performed while ensuring that the model remains feasible from a statistical point of view.

First, the run length of the model, the results collection period is increased to pass the transient period and reach a steady-state outcome. This is aimed at approaching a stationery outcome so that observations drawn are identical from a distribution perspective. Second, multiple simulation replications with different streams of random numbers are run to alleviate the problem of lack of independence among simulation observations. Third, the sample size (i.e., the number of simulation replications) is increased so that the central limit theorem[1] is applied on the averages of samples drawn, and the normal distribution assumption is met. This helps to restrict the manner in which simulation observations are collected so that reliable results are obtained from the model.

Model parameters that determine the length of the warm-up period and run time and the number of separate simulation replications are often referred to as model run controls. Most simulation packages require the modeler to provide settings for such parameters prior to running simulation experiments. The determination of such parameters is important for both terminating and nonterminating simulation. In terminating simulation, no warm-up period is required since the transient performance is part of the model behavior observed. Also, since no steady-state outcome is reached in terminating simulation, the assumption of identical (unbiased) observation does not hold. The method of determining run length also differs between the two types of simulation. The methods of determining the three model run controls – warm-up period, run length, and number of replications – are discussed below.

7.3.1 Determination of the Warm-up Period

To estimate long-term performance measures of a simulated system, nonterminating (steady-state) simulation is used. At the beginning of the

---

[1] Let $X_1, X_2, \ldots, X_N$ be a set of N independent random variates and let each X_i have an arbitrary probability distribution $f(x_1, x_2, \ldots, x_N)$ with a mean and a finite μ_i and a variance σ_i^2. Then the normal form variate

$$X = \frac{\sum_{i=1}^{N} x_i - \sum_{i=1}^{N} \mu_i}{\sqrt{\sum_{i=1}^{N} \sigma_i^2}}$$

has a limiting cumulative distribution function that approaches a normal distribution. Under additional conditions on the distribution of the addend, the probability density itself is also normal, with mean $\mu = 0$ and variance $\sigma^2 = 1$.

simulation run time, simulation outcomes exhibit erratic behavior caused by initial bias. An obvious remedy is to run the simulation for a period of time large enough to remove the effect of the initial bias and collect results from the steady-state period.

The initial run time within the transient period is referred to as the *warm-up period*. At the end of the warm-up period, different simulation outputs are collected for simulation analysis. Observations collected during the warm-up period are not reliable enough to draw any statistical inference. The results are simply thrown away automatically when producing statistics in simulation packages. The length of the warm-up period is determined to be sufficient to reach the steady-state condition. Graphically, the profile of the simulation response is observed to be somehow stable.

The length of warm-up period is a function of the initial model state, the amount of model activity, and activity times. Based on the nature of the process, it may take a few hours or several hundred hours to fill the system and reach steady state. We can see the impact of the warm-up period on the model behavior by observing the fluctuations in certain model response variables, such as production rate, in terms of hourly throughput. Starting the system elements "empty and idle" often results in low throughput numbers in the first few hours since there is not enough accumulation of entities downstream. Such low throughput does not represent the typical system behavior and should not be used to make decisions regarding the capacity of resources. Similarly, the lead time for the first few entities that leave the system may be relatively short, since system resources are idle and there is no queuing lines or congestion in the flow. Therefore, this lead time is not considered valid for a schedule of shipments and to set delivery dates with customers. Observations collected within the warm-up period are deleted since they reflect the impact of model initial conditions and they do not estimate its long-term performance.

A practical method for determining the warm-up period for a nonterminating system is based on experimentation of the distribution of the average system response as measured in a certain performance metric [e.g., units produced per hour (UPH) of a production system]. The warm-up period ends when trends and variations caused by initial bias stop and a steady pattern of random fluctuations dominates. A repeated distribution of a certain response average indicates a steady-state performance.

Welch's graphical procedure or Welch's moving-average procedure is another simple technique for determining the time step after which the transient mean converges to a steady-state mean. The method is applied when variation in model response is large and it is not possible to observe the start of the steady-state response. The erratic model response is first smoothed using a moving average before detecting the end of the transient period. The moving average is constructed by selecting a window of the latest data points and calculating the arithmetic mean. Each data point is the average response value of n replications. An indicator for the end of the transient period is when the moving-average plot appears to flatten out.

A practical way to determine the length of the warm-up period can be summarized as follows:

1. Run the model for a long time and observe the fluctuations in model response (Y). Make sure that model response reflects all sources of variations.
2. Determine the run time (t_m) in terms of m time steps that result in a representative steady-state response. For example, to claim a steady state, rare events in the system (such as machine breakdown) should occur several times (at least five times).
3. Run n simulation replications each of length t_m, where $n \geq 5$, using different streams of random numbers.
4. At each time step (t_j), determine the average of response values (Y_i), where $i = 1, 2, \ldots, n$, and obtain an average response measure ($\bar{Y}_j$).
5. Plot the values of the average response measure ($\bar{Y}_j$), where $j = 1, 2, \ldots,$ m.
6. Observe the time step (t_w) at which the system reaches statistical stability (i.e., obtaining a flat average response, obtaining a repeating pattern of average performance, or obtaining an approximately identical probability density distribution of Y_i).
7. When high variability of several sources causes the model's output response to be erratic, smooth the response by using Welch's moving average calculated for a selected window of the most recent average responses. The steady-state response is reached in this case when the moving average starts to flatten. As a precaution, extend the warm-up period (t_w) a little more to avert the risk of underestimating the warm-up.

Figure 7.3 shows the behavior of the model's average output response in both transient and steady-state periods. The warm-up period is marked by the end of the time step (t_w) at which the average response is $\bar{Y}_w$. As shown in the figure, the shape of distribution of the averages of the model response changes substantially from one time step to another within the transient period and becomes stable in the steady-state period. Also, the values of average response tend to flatten after the warm-up period is over (not necessarily a straight line as shown in Figure 7.3). All values of response averages in the warm-up period are excluded from the simulation results and the result collection period starts from t_w and ends at the end of run time at t_m.

To increase the simulation effectiveness, some modeling techniques can be used to shorten the warm-up period. In general, if the model starts out empty of simulation entities, it usually takes more time before it reaches steady state. Hence, initializing system conditions to conditions that are more typical than the default "empty and idle" state often results in shortening the warm-up

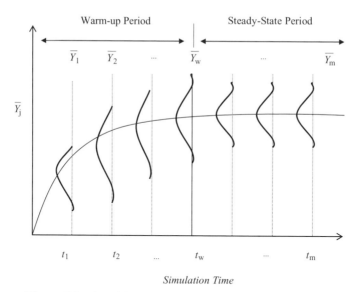

Figure 7.3   Graphical determination of a warm-up period.

period. Examples include starting the production system with full buffers and with a certain number of forklifts available for material handling.

For example, using the simulation model to estimate the throughput of a production system in terms of units produced per hour (UPH), a pilot simulation run of 25 hours resulted in the following throughput values:

21	28	33	44	50
51	53	56	60	57
59	60	58	60	61
60	61	59	60	61
58	61	59	61	60

As it can be seen from the results, the throughput starts relatively low (21 UPH) since the system is still filling up with units during the first hour. It was observed that for the first 39 minutes, no units have reached the end of the production line. The 39 minutes represents an estimate of the ideal production lead time with no queues and congestion. Units then accumulate in the system and gradually start to depart the system with fewer gaps. Comparing the throughput for the first 5 hours (21, 28, 33, 44, and 50) to the throughput for the next 5 hours (51, 53, 56, 60, and 57) shows that the model starts to produce a more stable response with significantly less variation. The throughput for the next 5 hours results in a similar pattern, and so on. Hence an overall run time of 25 hours is found to be satisfactory.

To detect the end of the warm-up period graphically, using the procedure discussed earlier, four more simulation replications are run, each of which is

TABLE 7.2 Throughput Distribution at Selected Run Times

Time, t_j	Sample Mean, $\overline{Y}_j$	Standard Deviation, s_j	95% CI_j
1	20.90	4.12	(18.45, 23.35)
5	49.50	3.45	(47.40, 51.60)
10	57.24	2.14	(55.29, 59.19)
15	60.10	1.64	(58.75, 61.45)
20	60.06	1.65	(58.67, 61.46)
25	60.07	1.64	(58.78, 61.44)

a 25-hour run time and with different seeds for the random number generator. The throughput mean, standard deviation, and 95% confidence interval (CI) were determined at selected points in time (1, 5, 10, 15, 20, and 25 hours) in order to check the changes in the distribution of throughput numbers. The results are shown in Table 7.2.

By checking the results in Table 7.2, we can see that the throughput distribution varies drastically within the first 5 hours and stabilizes slightly toward the end the tenth hour. Throughput mean, standard deviation, and 95% CI become essentially identical after the tenth hour. We can use a warm-up period of 10 hours for the throughput simulation and collect statistics after this transient period.

7.3.2 Determination of the Simulation Run Length

Data gathering from a steady-state simulation requires some kind of statistical assurance that the simulation model has reached the steady state. The warm-up period needs to be set so that the simulation model reaches a steady-state condition before observations and statistics are collected from the model. Subsequently, it is essential to determine the length of the result collection period. The simulation run length represents the time period in which model behavior is observed, statistics are accumulated and collected, and performance measures are recorded. Therefore, it is essential to set the model to run long enough past the warm-up period to collect representative data and arrive at correct conclusions about the model behavior.

In a terminating simulation, the determination of the run length is not an issue since there is a time point, a condition, or a natural event that defines the stopping point of the simulation run. Running for an 8-hour shift, serving 50 customers, or meeting a certain condition are examples of model stopping points. For steady-state simulation, the model could be run indefinitely and an end of run time is therefore required. However, simulation runs may be time consuming, especially when running large number of replications or when running large and complex simulation models. Regardless of the capability of the computer system, the effectiveness of simulation runs should be kept in mind and we should not run the model longer than the time thought statistically sufficient. That is, determination of the appropriate run length is a trade-off between the accuracy of the collected statistics and the efficiency (and cost) of simulation experiments.

The rule of thumb for determination of a simulation run length is simply to let the model run long enough past the warm-up period so that every type of event (especially rare ones) can take place several times (at least five). This is a function of the frequency and duration of these rare events. The longer the run time, the higher the value of sampled events and the higher the developed confidence in the results obtained from the model. The length of the interval between occurrences of the least frequently occurring event, multiplied by the number of occurrence intervals (samples) required, determines the lower bound on the length of the simulation run. For example, consider the failure of a certain machine to be a rare event in a production system simulation with a mean time between failures (MTBF) of 40 hours. The length of the simulation run time should be at least 200 hours to permit machine failure to take place at least five times.

Based on the method of sampling, two techniques are commonly used for determining the length of steady-state simulation: the method of batch means and the independent replication method. Another technique is the regenerative method, which is used primarily for its theoretical properties and rarely in actual simulation for obtaining the numerical results of a steady-state output.

Method of Batch Means (Subinterval Method)   This method involves one very long simulation run, which is suitably subdivided into an initial transient period and *n* equal batches. Each batch is treated as an independent run of the simulation experiment. The transient period is truncated and observations are collected from the run batches, where each batch (subinterval) represents a single observation. For example, after a relatively long warm-up period of 50 hours, the model can be run for 500 hours. The results form five subintervals each of 100-hour length, collected separately as five simulation replications.

The batch means method is used in practice to suit the nature of a certain system or when the warm-up period is long and it is desired to save the time of multiple replications by running one long simulation replication. Figure 7.4 shows the graphical representation of the method of batch means. Notice that all batch means are observed beyond the warm-up period. However, successive batches with common boundary conditions are often correlated, and statistical measures such as confidence intervals and standard variance cannot be established based on observations collected. To avert such an issue, a large batch interval size is used to alleviate the correlation among batches and lead effectively to independent batches (i.e., similar independent simulation runs). As an alternative, the model is set to change the seeds of the model random

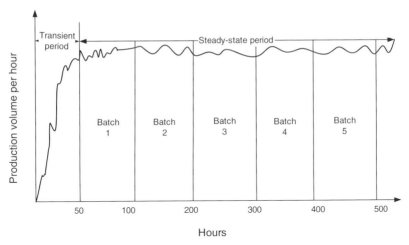

Figure 7.4 Method of batch means.

number generator at each batch (subinterval). Both actions are recommended for increased accuracy.

Independent Replication Method A more common approach is to collect the simulation results from multiple simulation replications. In this approach, the *independent replication method*, each observation is collected from an independent simulation run (i.e., no correlation exists among simulation runs) that has its own warm-up period. This method is the most popular steady-state technique for systems with a relatively short transient period. Figure 7.5 is a graphical representation of the independent replications method. After a relatively short warm-up period of 8 hours, the model is run for three independent replications each of 200 hours' run length.

In a manner similar to the method of batch means, the length of each run is determined simply by making sure that the least frequent event has taken place several times. For example, five occurrences of a machine failure with a 40-hour MTBF results in a run length of 200 hours. For each independent replication of the simulation run, the transient (warm-up) period is removed or truncated. For intervals observed after the transient period, data are collected and processed statistically for the point estimates of the performance measure and for its subsequent confidence interval.

Since randomness is incorporated in simulation models using random number generation (RNG), each simulation replication is run with a different random stream in order to develop independent runs of the simulation experiment. As discussed Chapter 4, the samples collected using RNG are not truly random (often referred to as *pseudorandom numbers*) since they can be predicted in advance, due to their recursive nature, which may result in a

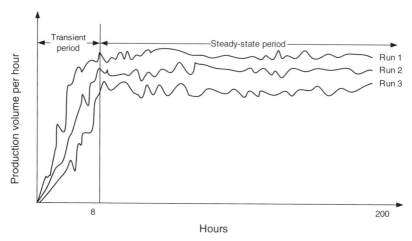

Figure 7.5   Method of independent replications.

repeating pattern. In some cases it is essential to watch for the pattern of random numbers used and to focus the effort on removing the excursiveness.

7.3.3   Determination of the Number of Simulation Runs

As discussed previously, to obtain representative performance estimation from stochastic simulation models, multiple simulation replications are run and the run time is increased. After determining the warm-up period and obtaining a steady-state simulation, it is then necessary to determine a statistically sufficient number of simulation runs (replications) at the planning stage of a simulation model. The determination of the number of simulation runs (n) is a critical decision since the confidence level of simulation outputs that are drawn from the simulation depends greatly on the size of the data set. Typically, the larger the number of runs, the higher the associated confidence.

More simulation runs, on the other hand, reduce the effectiveness of simulation experiments, especially for running extensive large-scale simulations. Hence, we often seek the smallest number of simulation runs that will provide the desirable confidence level with a desirable margin of error between the point estimate and (the unknown) true mean. To this end we first need to select a certain confidence level, assign the error margin, and estimate the amount of variability in the simulation response. A sample containing a certain number of pilot simulation runs with the selected warm-up period and run length is used to determine the variance (s^2) in the underlying simulation response. For large pilot simulation runs, the simplest way to determine the number of simulation runs (n) that satisfies the error margin selected at the confidence level selected is expressed as follows:

$$n = \frac{\left(Z_{\alpha/2}\right)^2 s^2}{\delta^2}$$

where $Z_{\alpha/2}$ is the standard normal score, δ is the desirable margin of error (i.e., the absolute error), which is the half-length of the confidence interval with a $100(1 - \alpha)\%$ confidence level, and s^2 is the variance obtained from the pilot run. For example, if we are using the simulation to observe the throughput of a production system in terms of average units produced per hour (UPH) and to assure at the 95% confidence level that the performance mean obtained from the simulation model is estimated within ±0.50 UPH of the true (unknown) throughput mean (μ), we use an initial sample size of 10 pilot runs to estimate the performance variance. The results of the initial simulation runs in terms of UPH are as follows:

| 60.50 | 58.40 | 62.35 | 60.00 | 63.20 | 59.70 | 58.45 | 61.34 | 57.40 | 60.35 |

Based on the results of the initial runs, a sample mean of 60.07 UPH and a standard deviation of 1.64 UPH is obtained. Using a significance level of $\alpha = 0.05$, $Z_{\alpha/2} = Z_{0.025} = 1.96$. Thus, the sample size (n simulation replications) is estimated as follows:

$$n = \frac{\left(1.96\right)^2 \left(1.64\right)^2}{\left(0.5\right)^2} = 41.34 \approx 42$$

That is, we need to run the model 42 times to measure system throughput with the given accuracy and confidence levels. Since we already have the results of 10 runs, we need to run 32 more simulation replications.

7.4   VARIABILITY IN SIMULATION OUTPUTS

When analyzing the outputs of simulation-based six-sigma models representing complex production and business processes, we are often interested not only in performance evaluation but also in sensitivity analysis, experimentation, and optimization. As discussed previously, stochastic simulation models generate inconsistent performance levels over time. Such variability presents some challenge to the effort to analyze simulation outputs. In this section we address the sources and implications of variability in simulation models.

We can classify the factors that contribute to variation in simulation outputs into controllable and uncontrollable (noise) factors. Model design factors such as buffer sizes, number of production resources, and speeds of conveyance systems are usually considered controllable factors. Different settings of process controllable factors may result in a different set of model outputs. Designers and engineers can control such factors and change their parameters to the benefit of the process performance. Once set to their levels, control factors do not change by time or by changes in running conditions. Experimental design and optimization methods are utilized to determine the best levels of control factors at which best model performance is achieved.

On the other hand, random factors are those uncontrollable factors whose individual values change over time, reflecting the actual behavior of the real-world process. These variables are presented in simulation models by sampling from probability distributions, as discussed in Chapter 6. Key distribution parameters are often estimated through statistical approximations based on fitting a collected set of empirical data to one of several commonly used distributions.

The behavior of DES models is usually driven by the stochastic nature of real-world processes such as the entity arrival process, in-system service process, and output departure process. Variability in those processes is caused by random factors such as arrival rate, servicing or processing rate, equipment time between failures and time to repair, percentages of scrap and rework, and so on.

Such variability in model processes results in stochastic variability in model critical-to-quality response (CTQs). By way of example, the throughput of a production system is subject to variability in material delivery, tool failure, and operator efficiency is represented by a DES model in a stochastic response. Based on the inherent fluctuations, the throughput could be a low value in one shift and higher in another. The amount of variability in model outputs depends on the amount of variability in model inputs along with the dynamics caused by the logical interactions among different model elements and processes. Thus, simulation (runs) yield estimates of such performance measures (throughput in this case) in terms of means and variances to reflect such variability. This variability also results in differences among the output averages of multiple simulation replications with different streams of random numbers.

Other noise elements may include uncontrollable internal sources of variation within the simulated system, such as changes in the workplace climate, team dynamics, union relations, and worker strikes. They also represent external factors such as the variation in the plant logistics, supply chain dynamics, and environmental changes. Although we can use DES to include certain assumptions that represent such noise factors, most practical simulation studies are focused on modeling controllable design factors (e.g., as decision variables) and random factors (e.g., noise factors). Random factors are integrated into the model by sampling from probability distributions, and model design factors are set through experimental design and optimization methods. By determining optimum settings to model controllable factors, statistical and simulation-based optimization methods aim at obtaining a CTQ, a system response, that meets design expectations with minimal sensitivity to variations in model random factors.

In addition to controllable and random factors, model artificial factors are another source of variation in simulation outputs. Artificial factors are

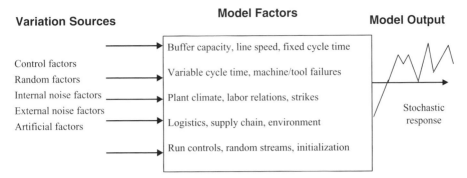

Figure 7.6   Sources of variation in a simulation model.

simulation-specific factors, such as the model initialization state, warm-up period, run length, termination condition, and random number streams. Changing the settings of such factors, from one simulation run to another, often results in changes in model outcomes. Hence, a key part of simulation analysis is devoted to running pilot simulations, testing the impacts of these factors in the model, and providing proper settings to different model run controls, as discussed earlier, prior to applying statistical and simulation-based optimization methods. Figure 7.6 presents a summary of different sources of variation in simulation models along with examples of their corresponding model factors.

7.4.1   Variance Reduction Techniques

It was shown in Section 7.4 that the stochastic nature of simulation often produces random outputs (random and probabilistic model inputs result in random model outputs). As the amount of input variability increases, the model response observed tends to be more random, with a high variance and a wide confidence interval around the response mean. As a result, less precise performance estimates are obtained from the simulation model. To remedy this problem, a set of variance reduction techniques (VRTs) are employed to reduce the variability involved in model outcomes. Reducing the response variance often results in reducing the width of the confidence interval around the response mean, which increases the precision of simulation estimation and leads to making better decisions based on simulation results.

   Several methods are used to reduce the variance of the simulation model response. The simplest method of variance reduction is to eliminate the sources of variations in model inputs, simply by using expected values of model parameters rather than using empirical or theoretical statistical distributions. Such a method results in a stable model response that makes it easier to plan various production and business resources. However, this may not be a good approach since it does not represent the actual behavior of the system being simulated.

Another method of reducing response variance without affecting the model representation of the system being simulated is to increase the run length of the steady-state simulation so that the sample size used in determining the response confidence interval is very large. Increasing the run length is particularly beneficial when using the batch means steady-state method. When using the independent replication method, the number of replications needs to be increased instead. For terminating simulation where the run length is limited by the nature of the system pattern of operation, the number of simulation replications can be increased.

Since a confidence interval of model response is typically used in practice to represent model variability, VRT methods often express variability reduction in terms of reduction made to the half-width of the confidence interval. Based on the formula used previously to determine the number of simulation replications (n), the following formula can be used to approximate the number of simulation replications (n') that are needed to bring the confidence interval half-width (hw) down to a certain specified error amount at a certain significance level (α):

$$n' = \left(\frac{Z_{\alpha/2} s}{hw} \right)^2$$

where s is the sample standard deviation.

An alternative way to control variability is to benefit from our control of random number generation (RNG). There is a class of methods that can work on RNG to reduce the model-produced variability. Using common random numbers (CRNs), for example, is one of the most popular VRTs. CRNs are based on using one exact source (stream) of RNG to all simulation replications instead of using a RNG for each independent simulation replication. The objective is to induce certain types of correlations (synchronization) in the random numbers produced, so that some variation is reduced. To maintain the independence of simulation replications, as a random sampling requirement, a CRN uses a different random number *seed* that are assigned at the beginning of each independent simulation replication. Care should be taken in picking the seed, so that segments of the random stream are not shared between replications. Simulation software tools provide a CRN option that automatically assigns seed values to each replication so that the possibility of sharing segments of random stream is minimized. Each stochastic element in the model, such as cycle times, arrival times, and failure times, can be assigned a different random number stream to keep random numbers synchronized across the system being simulated.

The employment of CRNs is particularly useful when comparing the performance of two or more simulated systems. This is because using CRN results in creating positive correlation among strategies being compared reduces the variance of sample differences collected from the model but does not reduce the variance of the simulation output. If no correlation, or a negative correlation, was created between systems, using CRNs will result in either no impact or in increasing the variance obtained when comparing simulated systems.

Several other VRTs can be used in addition to the CRN method to reduce variability, including antithetical variates, control variates, indirect estimation, and conditioning. All are beyond the scope of this chapter.

7.5 SIMULATION-BASED OPTIMIZATION

Simulation-based optimization methods are those methods that perform an optimization search algorithm using a discrete event simulation model as a representation of the real-world process. In such methods, responses or output variables of the objective function (either single or multiple) and constraints are evaluated by computer simulation. It is not necessary to provide an analytical expression of the objective function or the constraints as it is the case in mathematical optimization methods such as linear and nonlinear programming. The focus of simulation-based optimization is on finding the best set of values for decision variables of a simulation model from among all possible solution alternatives without explicitly evaluating each possibility.

One objective of simulation-optimization methods is to minimize the time and effort wasted in an exhaustive search, blind search, long and costly experimentation, or complicated mathematical approximations to the solution. This facilitates the use of a combination of an optimization search engine and a process simulator for numerous real-world applications. These applications range from finding the best buffer design for a certain production system to determining the best investment portfolio for financial planning.

A simulation-based optimization method consists of two main modules: a discrete event simulation system model and an optimization or search engine. Potential solution alternatives (factorial combinations) are generated from the search engine in terms of a set of model control factors. The DES model, in return, assesses the performance at each factorial combination provided by running multiple simulation replications at each potential solution point. The search engine receives the simulation evaluation and uses the assessments to guide its defined selection process. The selection process in optimization search engines is typically based on the relative performance of the solution alternatives being evaluated so that good solution alternatives are kept and weak solutions die off gradually. Iteratively, the simulation-optimization method is repeated until a certain convergence condition is met. The outcome would be the best combinatorial design that yields the best level of model performance. Figure 7.7 shows the architecture of simulation-based optimization method.

Different approaches have been used in the literature to optimize or draw inferences from the output of a simulation model. Depending on the size of the scope of the project, some methods are based on using the simulation model for a complete enumeration of solution combinations. Others utilize

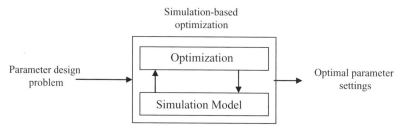

Figure 7.7   Architecture of simulation-based optimizatin.

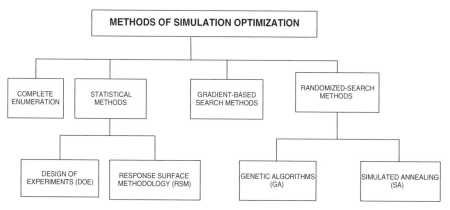

Figure 7.8   Methods of simulation-based optimization.

only samples of simulation evaluation in an attempt to estimate the gradient of the system response. Such methods are dependent on the problem dimensionality and may not be applied easily to real-world processes because of complexity, a large problem domain, and highly nonlinear and dynamic behavior. Figure 7.8 shows some of the major categories for simulation-optimization methods.

In this book we are limiting ourselves to simulation optimization by statistical means, which include mainly the application of design of experiments (DOE) and response surface methodology (RSM). Both methods are based on sampling from the process by conducting experiments on the process simulation model presented in Chapters 9 and 10.

In general, computer simulation experiments are conducted primarily for two purposes: (1) to determine the relationship between dependent and independent variables (factors), and (2) to find the configuration of independent variables that leads to the optimal solution.

By analyzing and comparing the results of simulation experimental design, DOE can identify which design variables are critical in affecting the process response. Further, DOE attempts to provide the optimum level of design variables that leads to the best process performance. The DOE-combined DES models can be an efficient process optimization methodology when approaching a simple system with a small and well-defined solution space. However, experimentation becomes costly and time consuming when more complex real-world systems are approached.

Since a simulation model can be thought of as a function of unknown form that turns input parameters into output performance measures, RSM attempts to develop the algebraic form (explicit formula) of this function to serve as a rough proxy for the full-blown simulation. Such a formula represents an algebraic model of the simulation model, so it is sometimes called a *metamodel*. The purpose of the metamodel is to represent an approximation to the response surface and how it would behave over various regions of the input factorial space. The least squares method is used to estimate the parameters in an approximate polynomials to the metamodel. The RSM approximation procedure requires a great deal of experimentation to develop the response surface for a complex system. Moreover, the hill-climbing procedure of RSM guarantees convergence to local optima only.

On the other hand, randomized-search methods are meta-heuristic approaches or generic optimizers that conduct direct and global searches in their optimization. Genetic algorithms (GAs), tabu search, and simulated annealing are examples of common randomized-search methods. GAs in particular have demonstrated an ability to operate with noisy observations, such as outputs generated by a simulation model. Therefore, the implementation of these generic optimizers for use with simulation models has been approached using different procedures. GAs are often noted for robustness in searching complex spaces and are, therefore, best suited for combinatorial problems.

Several researchers have described GA architectures for optimization through the use of simulation models. GA is selected as an optimization algorithm since it has proven to be robust in dealing with the level of complexity of these problems. Other applications include procedures that combine simulation with Tabu search to deal with problems of work-in-process inventory management. Another application involves the use of perturbation analysis, a gradient-based method, with inventory models where the demand has an associated renewal arrival process. An algorithm that combines simulated annealing and simulation is generally useful for applications with objectives such as finding an appropriate dispatching priority of operations, minimizing the total tardiness for a commercial flexible manufacturing system, or determining optimal parameter levels for operating an automated manufacturing system.

The challenges facing simulation optimization of complex systems can be classified into two categories: those from optimizing complex and highly nonlinear functions, and those related to the special nature of simulation modeling. Simulation optimization can be sought as a single-objective problem or as a multicriteria optimization problem. Four major approaches are often used to solve single-objective problems: gradient-based search, stochastic approximation, response surface, and heuristic search methods. For multicriteria simulation optimization, most of the work done in this area consists of slight modifications of the techniques used in operations research for generic multiobjective optimization. Among these is the approach of using one of the responses as the primary response to be optimized subject to other levels of achievement on the other objective functions. Another approach is a variation on the goal programming approach.

Most real-world problems require multicriteria simulation optimization. Traditional multicriteria mathematical optimization techniques such as multiattribute value function, multiattribute utility function, and goal programming were applied to multicriteria optimization problems. The simulation-optimization problem is different from the problems addressed by multiobjective mathematical programming algorithms in three important respects:

1. The relationships between output variables and decision variables are not of closed form.
2. The outputs may be random variables (stochastic as opposed to deterministic nature).
3. The response surface may contain many local optima.

These characteristics result in unique difficulties that may not be solvable by using conventional mathematical optimization methods.

PART III

SIMULATION-BASED SIX-SIGMA
AND DESIGN FOR SIX-SIGMA

8

SIMULATION-BASED SIX-SIGMA ROAD MAPS

8.1 INTRODUCTION

The objective of this chapter is to introduce the simulation-based design for six-sigma process, referred to as 3S-DFSS, and the simulation-based lean six-sigma process, referred to as 3S-LSS, to lay the foundations for the following chapters of the book. In this chapter we discuss the 3S-LSS road map and explain the various aspects of its application procedure. Chapter 9 presents a case study that illustrates the application of 3S-LSS to real-world problems, discusses the 3S-DFSS road map, and explains the various aspects of its application procedure. Chapter 10 presents a case study that illustrates 3S-DFSS application to real-world projects.

In general, DFSS combines design analysis (e.g., requirements cascading) with design synthesis (e.g., process engineering) within the framework of a service (product) development systems. Emphasis is placed on critical-to-quality (CTQ) characteristics or functional requirement[1] identification, optimization, and verification using the transfer function and scorecard vehicles. A *transfer function* in its simplest form is a mathematical relationship between

---

[1] A set of functional requirements is the minimum set of independent requirements for a product or process that completely characterizes the design objective (Suh, 1990). FRs can be defined from the CTQs by cascading using quality function deployment machinery (El-Haik and Roy, 2005).

---

a functional requirement (FR) and the critical design parameters. *Scorecards* help predict FR risks by monitoring and recording their mean shifts and variability performance. The 3S-DFSS approach builds on the proven capability of DFSS and extends these capabilities through the use of simulation modeling. The simulation model provides the DFSS process with a flexible process design optimization and verification vehicle.

On the other hand, LSS combines the use of six-sigma DMAIC process with lean manufacturing techniques. The primary focus here is on time-based performance measures such as productivity and lead time. DMAIC is a well-structured and data-driven six-sigma process that is aimed at tackling problems in existing systems. Its application often seeks problem solving of product- and process-related issues such as quality complains from customers, production problems, and delivery delays. Lean manufacturing involves both a philosophy and tool kits essential for improving effectiveness, reducing waste, and streamlining operations. Combined with DMAIC, lean techniques seek improvement to the structure and flow in a six-sigma performing process. Thus, the 3S-LSS proposed approach enables an effective and highly beneficial DMAIC–lean integration by utilizing simulation modeling as a flexible and inexpensive platform for DMAIC analyses and lean technique application.

Figure 8.1 presents a scoping flow of the DFSS and LSS approaches. Whereas DFSS projects usually target newly developed processes, LSS projects are often focused on problem solving and continuous improvement of existing processes. DMAIC is known as a problem-solving technique, and lean manufacturing is known as a mechanism for continuous improvement. In LSS, the DMAIC and lean techniques are combined and executed in conjunction with each other to reduce process variability and waste.

8.2   LEAN SIX-SIGMA PROCESS OVERVIEW

As presented by George (2002), whereas six-sigma focuses on defects and process variation, lean is linked to process speed, efficiency, and waste. As discussed in Chapter 2, lean techniques are aimed primarily at the elimination of waste in every process activity: to incorporate less effort, less inventory, less time, and less space. The main lean principles include zero waiting time, zero inventories, a pull rather than a push system, small batches, line balancing, and short process times. Basic lean-manufacturing techniques include the 5Ss (sort, straighten, shine, standardize, and sustain), standardized work, elimination of waste, and reducing unneeded inventory. This is attainable by eliminating the following seven wastes:

- Overproduction
- Inventory
- Transportation
- Processing

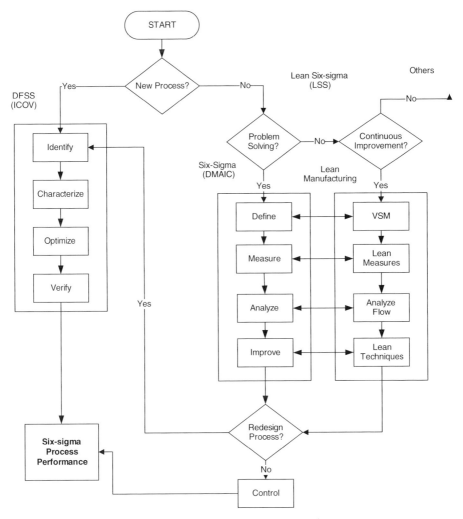

Figure 8.1 DFSS and LSS approaches.

- Scrap
- Motion
- Waiting or idle time

On the other hand, DMAIC is a problem-solving approach that ensures complete understanding of process steps, capabilities, and performance measurements by defining process CTQs/FRs, measuring CTQs/FRs, applying six-sigma tools and analysis to improve process performance, and implementing methods to control the improvement achieved. Thus, the cost reduction and continuous improvement focus of lean techniques complement the statistical

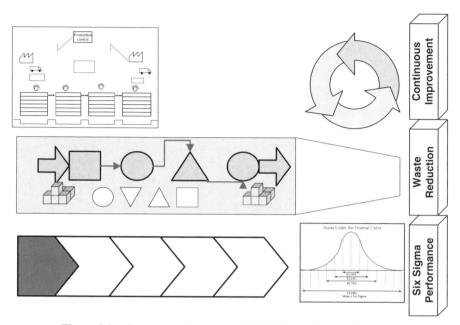

Figure 8.2   Concurrent impacts of DMAIC and lean techniques.

focus of DMAIC on solving quality-related problems by achieving a certain level of performance and implementing process controls to maintain performance and to minimize the chance of future problems. Such synergy leads to an enormous opportunity for process improvement.

Figure 8.2 provides a conceptual framework of the integration of lean and six-sigma into a comprehensive approach. Starting from a current-state value stream map, lean techniques are implemented systematically to reduce process waste (width) and to reduce process lead time (length). This results in enormous reduction in process cost and major improvement in process effectiveness. This can only be maintained through a continuous improvement mindset. DMAIC six-sigma application reduces process variability and increases its sigma rating. However, the use of both methods should not be parallel but should, instead, leverage their synergy. The outcomes of a DMAIC–lean integrated approach include reduced process cost, six-sigma performance, and continuous improvement plans, as depicted by Figure 8.2.

8.3   SIMULATION-BASED LEAN SIX-SIGMA ROAD MAP

The 3S-LSS road map is a graphical representation of integration of the standard DMAIC process and the relevant lean techniques on a simulation platform. Such a road map is aimed at providing six-sigma operatives (green belts, black belts, master black belts, and others) with a holistic understanding and an overall procedure for applying the 3S-LSS approach. The process proposed is made broad to a great extent, knowing that the specifics of each real-world application may require some change to the approach presented. The 3S-LSS road map involves the analysis and synthesis activities among DMAIC, lean, and simulation.

8.3.1   Simulation Role in DMAIC

The various DMAIC phases will benefit from the simulation model, as shown in Figure 8.3. The conceptual synthesis of simulation modeling over the DMAIC phases makes modeling a powerful platform for DMAIC application to production and business processes.

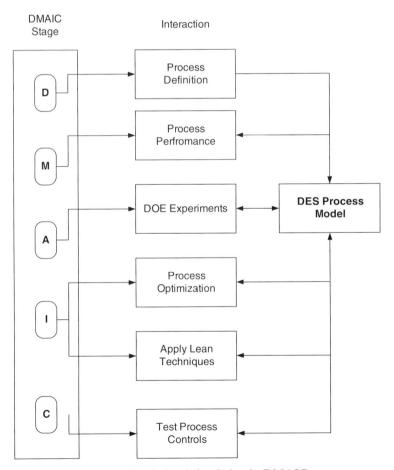

Figure 8.3   Role of simulation in DMAIC.

In the define phase, the process model can be used to better define the process in terms of structure, logic, and parametric data. Also, process-defined CTQs or FRs and variables are programmed into the model. In the measure phase, the model plays a central role in estimating process performance by measuring CTQs or FRs at various process locations. Given the complexity, stochastic, and dynamic nature of real systems, in many cases the model may be the only resort for system-level time-based performance assessment.

In the DMAIC analyze phase, the model is the backbone of experimental design due to its statistical nature. A plant or service process simulation model serves as an affordable alternative for conducting analysis effectively and in the most realistic fashion possible. Physical modeling of real-world processes is extremely costly, if not impossible. The flexibility of simulation models and the capability of random number generation and animation make the model a great environment for statistical analysis that benefits the decision-making process.

In the improve phase, the model is used for optimizing process parameters based on the statistical analysis conducted in the preceding phase. Further DOEs and other analysis forms can be run and a plethora of optimization methods can be applied using the simulation model. All other proposed problem-solving techniques and improvement actions can also be evaluated in this phase. Finally, in the control phase, control schemes and monitoring plans can be tested prior to deployment.

8.3.2 Integrating Lean Techniques into DMAIC

Typically, LSS refers to integrating lean techniques into the DMAIC process to form synergic and complementary problem-solving and improvement perspectives. In addition, the 3S-LSS approach proposed utilizes simulation as an integration environment. Since the DMAIC approach is split into five phases (define, measure, analyze, improve, and control), lean techniques are integrated into these five phases instead of being executed in parallel: hence the term *Lean six-sigma.*

In Chapter 2, several lean techniques were discussed. These techniques can be utilized at various phases of the DMAIC process to work in conjunction with DMAIC tools. Proper selection of lean tools at each DMAIC phase is critical to successful project application. Certain lean tools play a complementary role to six-sigma tools in a given DMAIC phase. Table 8.1 presents suggested examples of lean concepts and tools that can be utilized in various phases of the 3S-LSS approach.

As shown in Table 8.1, value stream mapping (VSM) (introduced in Chapter 2) is a key lean tool that aids the define phase of DMAIC. VSM provides a high-level graphical and time-based representation of process structure and flow. The process CTQs and variables defined can also be located in the current-state VSM. Problem-solving and improvement efforts will be assessed using the VSM and should be reflected in the future-state VSM. In

TABLE 8.1   Lean Tools Selected in Various DMAIC Phases

Define	Measure	Analyze	Improve	Control
Value stream mapping	Lead time	Work analyses	SMED	Visual controls
	Takt time	Flow analysis	JIT/Kanban	Standard work
	Inventory turns	Scheduling	Line balance	Kaizen

the measure phase, several lean measures, such as lead time, takt time, and throughput, can be used to guide improvement and benchmarking efforts. More business measures, such as inventory turns and on-time delivery rates, can also be used.

In the DMAIC analyze phase, lean techniques are focused on analyzing the process structure, configuration, and flow logic. This is achieved through the use of several lean techniques, such as work analyses, flow analysis, and scheduling. These analyses provide insight as to what needs to be changed in the process structure and operating procedures to reduce waste and streamline flow.

The role of lean techniques is further leveraged in the improve phase of DMAIC, where several lean techniques can be applied to improve the current-state VSM lean measures. The focus is on cutting or reducing all types of waste in time, material, and effort. Several tools, such as SMED, JIT/Kanban, and work balancing, can be used to achieve these objectives.

In the control DMAIC phase, several lean tools can be used to sustain improved performance and to implement control and monitoring plans successfully. Examples include Kaizen, work standards, and visual controls.

8.3.3   3S-LSS Road Map

Figure 8.4 provides a road map for applying the 3S-LSS process to real-world systems. For a comprehensive understanding, the road map should be contrasted with the system modeling presented in Chapter 5. We follow this road map in the case study discussed in Chapter 9. Using the Figure 8.4 road map, we will demonstrate the progression from phase 1: define through phase 5: control: the implementation of six-sigma improvement to a real-world system (Chapter 9) in several lean measures as reflected in their Z-score (sigma) gains.

As depicted in Figure 8.4, each DMAIC phase is split into three steps and integrated as a process flow that is coupled with a simulation-based assessment of performance. In the define phase, the 3S-LSS approach focuses on defining the scope of the improvement project and translating the project objectives into the most appropriate lean measures related to the project. This is an essential activity because it allows the six-sigma project team to verify applicability of the 3S-LSS approach to the underlying project scope. The key is to express the project CTQs or FRs as time-based lean measures that can

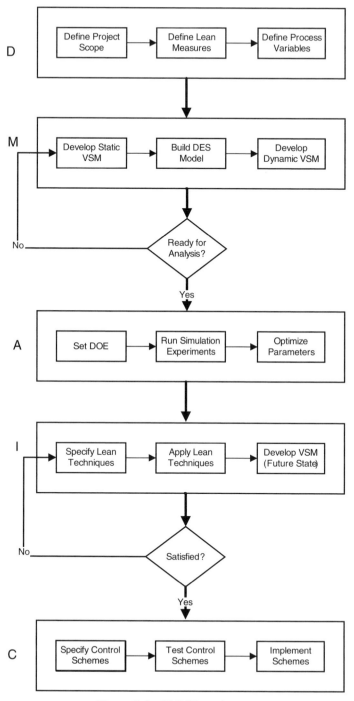

Figure 8.4 3S-LSS road map.

later be assessed using a DES model. For example, increasing customer satisfaction and cutting process cost can be translated into increasing process effectiveness, enhancing delivery speed, and reducing excess inventory, for example. That is, these time-based CTQs after translation are lean measures or metrics that can be assessed easily using simulation. In addition, these lean measures (as dependent responses) are mapped to key process variables such as buffer sizes and locations, production lot size, quality control measures, process elements, and allocated resources.

In the measure phase, the 3S-LSS approach focuses on utilizing the process VSM to visualize process flow, parameters, and estimates of the lean measures defined. The VSM provides the flow time line, inventory sizes and locations, cycle times, takt time (productivity), operations uptime, available production time, and much other useful information that describes the process state while summarizing its performance. However, the VSM is a static representation that provides estimates of lead time and productivity measures at a given point in time, capturing the state when mapping was conducted.

A unique feature of the 3S-LSS approach presented here is its dynamic ability to mimic reality. This allows us to extract as much useful information from a simulation as possible to aid decision making. For example, the 3S-LSS approach utilizes static VSM along with other simulation inputs to develop a DES process model that captures the stochastic nature of the process. Through a feedback loop, static VSM can be converted into dynamic VSM via simulation when simulation process output statistics are fed back into the VSM. This activity transformation provides a form of dynamism to the traditional VSM. Dynamic VSM provides a better and more realistic process representation with a statistical summary on each mapping element, and dynamic VSM will be used as a basis for conducting analysis and making process improvement in the 3S-LSS approach.

Should dynamic VSM be ready for analysis, we proceed to that phase of the DMAIC process. Readiness for analysis implies that the VSM involves essential information, process parameters, and structural elements that affect the lean measures defined. Otherwise, we need to restructure the VSM and DES model and apply some judgments and assumptions.

In the analyze phase, the 3S-LSS approach focuses effort on six-sigma analyses that set process parameters to reduce process variability and achieve a six-sigma performance level. A tool that can be used in this phase is a design of experiment (DOE) where process variables are used as experimental factors and the lean CTQs as responses (e.g., lead time, setup time, flow). A simulation-based experimental design is conducted on the process parameters defined in the VSM. The 3S-LSS approach benefits significantly from the flexibility of the DES model in importing and exporting data, adjusting process parameters, and estimating lean CTQs. Using the simulation model, the experimental scope can be extended and experimental time can be cut significantly. Process performance is reflected in the VSM after completing the analysis phase of the DMAIC process.

The DMAIC improve phase is a key in the 3S-LSS approach, where lean techniques are used to alter process structure and control systems to reduce cost and waste, to increase effectiveness, and to achieve other objectives. Again, the DES model will be the environment for testing lean techniques and assessing their impact and gains. Some lean techniques require changing model logic and routing rules, such as push versus pull and Kanban production systems; others may require changing process structure, such as adding in-station process controls, decoupling assembly operations, and moving to cellular manufacturing. The outcome of the improve phase is a future-state VSM with the new lean CTQ performance levels. Gains in process improvement are quantified using the DES model and compared to objectives to check result satisfaction. If improvements are not sufficient, lean techniques may be enhanced or other techniques may be applied, tested, and assessed. Finally, a model is used to test and verify control schemes.

8.3.4   3S-LSS Procedure

The 3S-LSS approach can also be viewed from a perspective that has more granularity, as shown in Figure 8.5, which conveys the fact that the 3S-LSS project can be executed as a flow of activities in each of the DMAIC phases. The figure shows the overall flow of the 3S-LSS process at each DMAIC phase. Also, the 3S-LSS process depicted in the figure is a synthesis of DMAIC and lean techniques and concepts that eventually make the overall process. The simulation model plays a key role in execution of the 3S-LSS process. The flow shown in the figure is not intended to include every possible activity in the project. Instead, the focus is on presenting 3S-LSS as a flow of interrelated

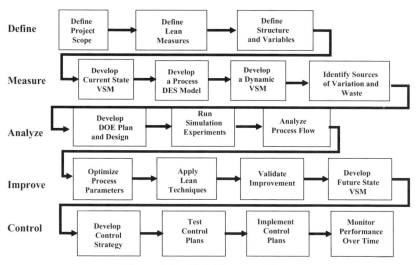

Figure 8.5   3S-LSS process flow.

activities that utilize a wide spectrum of statistical and lean tools over the DMAIC process in the spirit of Chapter 2.

The define phase of the 3S-LSS procedure involves several tools and methods, as shown in Figure 8.6. Phase deliverables include a concise definition of project scope objectives, benefits, and ownership, process structure, and process CTQs. This phase also specifies lean measures to the process CTQs defined and puts together a data and information collection plan. The project team can then start the project with a clear vision and correct understanding.

The measure phase of the 3S-LSS procedure also involves several tools and methods, as shown in Figure 8.7. Phase deliverables include a current-state VSM with lean measures (process elements, flow, and information) and a dynamic process DES model (process logic, animation, and actual

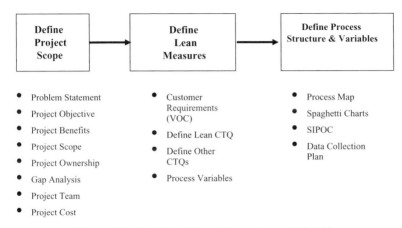

Figure 8.6   Details of the define phase of 3S-LSS.

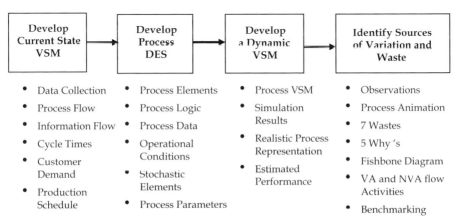

Figure 8.7   Details of the measure phase of 3S-LSS.

performance). This phase also provides a clear definition of process sources of variation and waste to be ready for analyses.

The analyze phase of the 3S-LSS procedure involves several tools and methods, as shown in Figure 8.8. Phase deliverables include a detailed understanding of process critical parameters and flow elements with a prescription for process improvement in terms of structure and parameters.

The improve phase of the 3S-LSS procedure involves several tools and methods, as shown in Figure 8.9. Phase deliverables include a future-state (improved) VSM with tested improvement measures and plan of actions. These measures and actions include changes made to process parameters as a result of DOE optimization and changes made to process after implementing lean techniques. The improved design is presented in the future-state VSM with lean measures.

Finally, the control phase of the 3S-LSS procedure involves several tools and methods, as shown in Figure 8.10. Phase deliverables include firm control

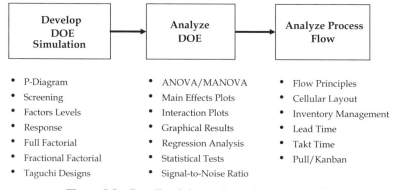

Figure 8.8 Details of the analyze phase of 3S-LSS.

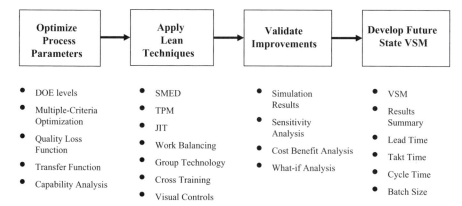

Figure 8.9 Details of the improve phase in 3S-LSS.

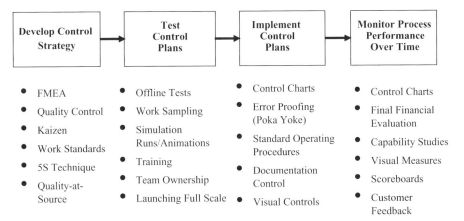

Figure 8.10   Details of the control phase of 3S-LSS.

and monitoring plans that are aimed at sustaining the improved level of performance and making the process robust to future changes. This includes the deployment of inside and outside monitoring plans. A case study of the 3S-LSS process is presented in Chapter 9.

8.4   SIMULATION-BASED DESIGN FOR A SIX-SIGMA ROAD MAP

As discussed in Chapter 3, DFSS is a structured, data-driven approach to design in all aspects of transaction process functions where deployment is launched, to eliminate the defects induced by the design process and to improve customer satisfaction, sales, and revenue. To deliver on these benefits, DFSS applies design methods such as axiomatic design,[2] creativity methods, and statistical techniques to all levels of design decision making in every corner of the business, identifies and optimizes the critical design factors, and validates all design decisions in the use (or surrogate) environment of the end user.

DFSS is not an add-on but represents a cultural change within different functions and organizations where deployment is launched. It provides the means to tackle weak or new processes, driving customer and employee satisfaction. DFSS and six-sigma should be linked to the deploying company's annual objectives, vision, and mission statements. It should not be viewed as another short-lived initiative. It is a vital permanent component to achieve leadership in design, customer satisfaction, and cultural transformation. From marketing and sales, to development, operations, and finance, each business function needs to be headed by a deployment leader or a deployment

---

[2] A perspective design method that employs two design axioms: the independence and information axioms. See Appendix C for more details.

champion. Champions can deliver on their objective through six-sigma operatives called Black belts and green belts, who will execute projects that are aligned with the objectives of the company. Project champions are responsible for scoping projects from within their realm of control and handing project charters (contracts) over to the six-sigma resource. The project champion selects projects consistent with corporate goals and removes barriers. Six-sigma resources complete successful projects using six-sigma methodology and train and mentor the local organization on six-sigma. The deployment leader, the highest initiative operative, sets meaningful goals and objectives for the deployment in his or her function and drives the implementation of six-sigma publicly.

Six-sigma resources are full-time six-sigma operatives, in contrast to green belts, who should be completing smaller projects of their own as well as assisting black belts. They play a key role in raising the competency of the company as they drive the initiative into day-to-day operations. Black belts are the driving force of DFSS deployment. While handling projects, the role of black belts spans several functions, such as learning, mentoring, teaching, and coaching. As a mentor, a black belt cultivates a network of experts in the project on hand, working with the process operators, process owners, and all levels of management. To become self-sustained, the deployment team may need to task their black belts with providing formal training to green belts and team members.

Design for six-sigma is a disciplined methodology that applies the transfer function [FR = f(DP)] to ensure that customer expectations are met, embeds customer expectations into the design, predicts design performance prior to pilot, builds performance measurement systems (scorecards) into the design to ensure effective ongoing process management, and leverages a common language for design within a design tollgate review process described below.

DFSS projects can be categorized as design or redesign of an entity, whether product, process, or service. We use the term *creative design* to indicate new design, design from scratch, and *incremental design* to indicate redesign or design from a datum. In either case, the service DFSS project requires greater emphasis on:

- The voice of the customer collection scheme and addressing customers
- Assessing and mitigating technical failure modes and project risks in their own environment as they are linked to tollgate process reviews
- Project management, with some communication plan to all parties and budget management affected
- A detailed project change management process

8.4.1 ICOV Tollgate Review Process

As discussed in Chapter 3, DFSS has four phases over seven development phases: identify, conceptualize, optimize, and verify. The acronym ICOV is used to denote these four phases.

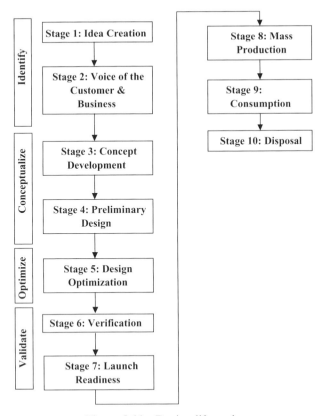

Figure 8.11   Design life cycle.

Due to the fact that DFSS integrates well with a design life-cycle system, it is an event-driven process, especially the development (design) phase. In this phase, milestones occur when the entrance criteria (inputs) are satisfied. At these milestones, the stakeholders, including the project champion, process or design owner, and deployment champion (if necessary), conduct *tollgate reviews*. There are four tollgate reviews, marking the conclusion of each phase of ICOV: one tollgate per phase, as in Figure 8.11. A development phase has a time period, that is, entrance criteria and exit criteria for the bounding toll-gates. The ICOV DFSS phases as well as the seven phases of the design development are depicted in Figures 8.11 and 8.12. The overall DFSS road map is depicted in Figure 8.13. The simulation-based DFSS (3S-DFSS) road map is shown in Figure 8.14 by morphing the simulation process (Figure 6.7) and the DFSS ICOV (Figure 8.13) process.

In tollgate reviews, a decision should be made whether to proceed to the next phase of development, recycle back for further clarification on certain decisions, or cancel the project altogether. Early cancellation of problematic projects is good, as it stops nonconforming projects from progressing further

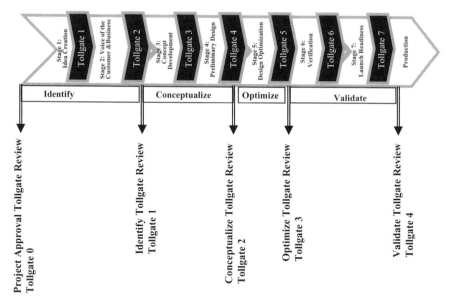

Figure 8.12 ICOV tollgate process.

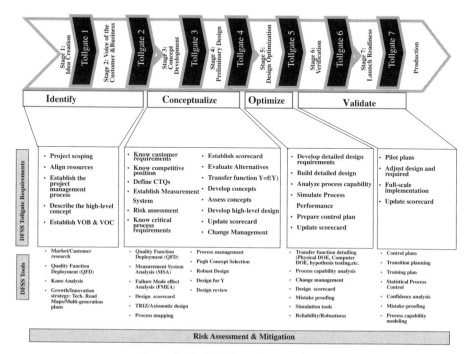

Figure 8.13 DFSS project road map.

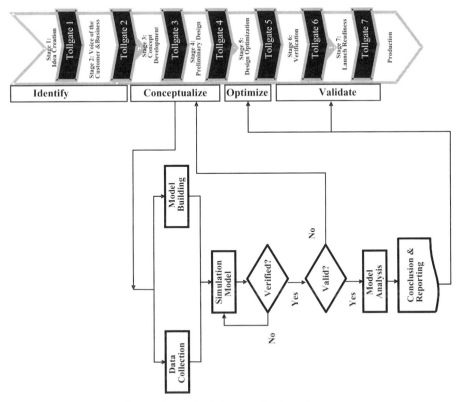

Figure 8.14 3S-DFSS project road map.

while consuming resources and frustrating people. In any case, the black belt should quantify the size of the benefits of the design project in language that will give stakeholders a clear picture, identify major opportunities for improving customer dissatisfaction and associated threats to salability, and stimulate improvements through publication of the DFSS approach.

In tollgate reviews, work proceeds when the exit criteria (required decisions) have been created. As a DFSS deployment side bonus, a standard measure of development progress across the deploying company using a common development terminology is achieved. Consistent exit criteria from each tollgate include both the DFSS deliverables, which arise from the approach itself, and the business- or function-specific deliverables. Detailed entrance and exit criteria by phase are presented in Chapter 5 of El-Haik and Roy (2005).

As depicted in Figure 8.12, tollgates or design milestone events include reviews to assess what has been accomplished in the current developmental stage and preparation of the next stage. The design stakeholders, including the project champion, process owner, and deployment champion, conduct tollgate

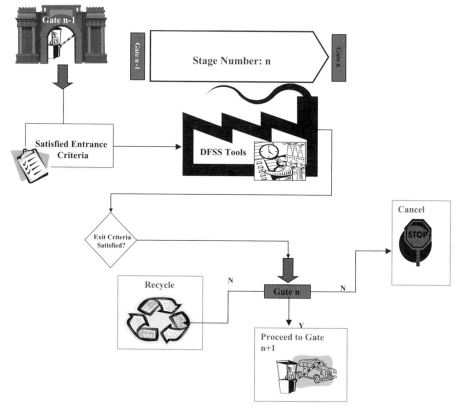

Figure 8.15   DFSS tollgate process.

reviews. In a tollgate review, three options are available to the champion or to his or her delegated tollgate approver (Figure 8.15):

1. Proceed to the next stage of development.
2. Recycle back for further clarification on certain decisions.
3. Cancel the project.

In tollgate reviews, work proceeds when the exit criteria (required decisions) are made. Consistent exit criteria are needed from each tollgate that blend service DFSS deliverables with business unit- or function-specific deliverables.

In Chapter 3 we learned about the ICOV process and the seven developmental stages spaced by bounding tollgates, indicating a formal transition between entrance and exit. In following section we expand on the ICOV DFSS process activities by stage, with comments on the applicable key DFSS tools and methods.

8.4.2 DFSS Phase 1: Identify the Requirements

This phase includes two stages: idea creation (stage 1) and voice of the customer and business (stage 2).

Stage 1: Idea Creation

Stage 1 Entrance Criteria   Entrance criteria may be tailored by the deploying function for the particular program or project provided that the modified entrance criteria, in the opinion of the function, are adequate to support the exit criteria for this stage. They may include:

- A target customer or market
- A market vision with an assessment of marketplace advantages
- An estimate of development cost
- Risk assessment

Tollgate 1: Stage 1 Exit Criteria
- Decision to collect the voice of the customer to define customer needs, wants, and delights
- Verification that adequate funding is available to define customer needs
- Identification of the tollgate keepers[3] leader and the appropriate staff

Stage 2: Customer and Business Requirements Study

Stage 2 Entrance Criteria
- Closure of tollgate 1 (approval of the gatekeeper is obtained).
- A DFSS project charter that includes project objectives, a design statement, and other business levers, metrics, resources, team members, and so on. These are almost the same criteria as those required for DMAIC projects. However, the project duration is usually longer and the initial cost is probably higher. The DFSS team, relative to DMAIC, typically experiences a longer project cycle time. The goal here is either designing or redesigning a different entity, not just patching up the holes in an existing entity. Higher initial cost is due to the fact that the value chain is being energized from process or service development, not from the manufacturing or production arenas. There may be new customer requirements to be satisfied that add more cost to the developmental effort. For DMAIC projects, we may only work on improving a very limited subset of the critical-to-quality characteristics (CTQs).

---

[3] A tollgate keeper is a person or group that will assess the quality of work done by the design team and initiate a decision to approve, reject or cancel, or recycle the project to an earlier gate. Usually, the project champion(s) is tasked with this mission.

- Completion of a market survey to determine customer needs critical to satisfaction/quality [the voice of the customer (VOC)]. In this step, customers are fully identified and their needs collected and analyzed with the help of quality function deployment (QFD) and Kano analysis. Then the most appropriate set of CTQ metrics are determined in order to measure and evaluate the design. Again, with the help of QFD and Kano analysis, the numerical limits and targets for each CTQ are established. In summary, here is the list of tasks in this step.
  - Investigate methods of obtaining customer needs and wants.
  - Determine customer needs and wants and transform them into a VOC list.
  - Finalize requirements.
  - Establish minimum requirement definitions.
  - Identify and fill gaps in customer-provided requirements.
  - Validate application and usage environments.
  - Translate the VOC to critical-to-quality, critical-to-delivery, critical-to-cost, and so on.
  - Quantify CTQs and establish measurement for CTQs.
  - Establish acceptable performance levels and operating windows.
  - Start flow-down of CTQs.
- Assessment of technologies required.
- Project development plan (through tollgate 2).
- Risk assessment.
- Alignment with business objectives: voice of the business relative to growth and innovation strategy.

Tollgate 2: Stage 2 Exit Criteria
- Assessment of market opportunity
- Establishment of a reasonable and affordable price
- Commitment to the development of a conceptual design
- Verification that adequate funding is available to develop a conceptual design
- Identification of the gatekeeper leader (gate approver) and the appropriate staff
- Continuation of flow-down of CTQs to functional requirements

DFSS tools used in this phase include (Figure 8.14):

- Market/customer research
- Quality function deployment (QFD) phase I
- Kano analysis
- Growth/innovation strategy

Even within best-in-class companies, there is a need for and an opportunity to strengthen and accelerate progress. The first step is to establish a set of clear and unambiguous guiding growth principles as a means to characterize company position and focus. For example, growth in emerging markets might be the focus abroad, whereas effectiveness and efficiency of resource use within the context of enterprise productivity and sustainability may be the local position. Growth principles and vision at a high level are adequate to find agreement and focus debate within the zone of interest and exclude or diminish nonrealistic targets. The second key step is to assess the current know-how and solutions of a service portfolio in the context of these growth principles: an inventory of what the senior leadership team knows they have and how it integrates into the set of guiding growth principles. Third, establish a vision of the ultimate state for the company. Finally, develop a *multigeneration plan* to focus the research, product development, and integration efforts in planned steps to move toward that vision. The multigeneration plan is key because it helps the deploying company stage progress in realistic developmental stages, one DFSS project at a time, but always with an eye on the ultimate vision.

In today's business climate, successful companies must be efficient and market-sensitive to supersede their competitors. By focusing on new services, companies can create custom solutions to meet customer needs, enabling customers to keep in step with new service trends and changes that affect them. As the design team engages the customers (surveys, interviews, focus groups, etc.) and processes the QFD, they gather competitive intelligence. This information helps increase design team awareness of competing services or how they stack up competitively with a particular key customer. By doing this homework, the team identifies potential gaps in their development maturity. Several in-house tools to manage the life cycle of each service product from the cradle to the grave needs to be developed to include the multigeneration plan and a customized version of the ICOV DFSS process, if required. The multigeneration plan evaluates market size and trends, service positioning, competition, and technology requirements. This tool provides a simple means to identify gaps in the portfolio while directing the DFSS project road map. The multigeneration plan needs to be supplemented with a decision-analysis tool to determine the financial and strategic value of potential new applications over a medium time horizon. If the project passes this decision-making step, it can be lined up with others in the six-sigma project portfolio for a start schedule.

Research on Customer Activities   This is usually done by the service planning departments (service and process) or market research experts, who should be on the DFSS team. The black belt and his or her team start by brainstorming all possible groups of customers for the product. Use the affinity diagram method to brainstorm potential customer groups. Categories of markets, user types, and service and process applications types will emerge.

From these categories, the DFSS team should work toward a list of clearly defined customer groups from which individuals can be selected.

External customers might be drawn from customer centers, independent sales organizations, regulatory agencies, societies, and special-interest groups. Merchants and, most important, end users should be included. The selection of external customers should include both existing and loyal customers, recently lost customers, and new-conquest customers within the market segments. Internal customers might be drawn from production, functional groups, facilities, finance, employee relations, design groups, distribution organizations, and so on. Internal research might assist in selecting internal customer groups that would be most instrumental in identifying wants and needs in operations and service operations.

The ideal service definition in the eye of the customer may be extracted from customer engagement activities. This will help turn the knowledge gained from continuous monitoring of consumer trends, competitive benchmarking, and customer likes and dislikes into a preliminary definition of an ideal service. In addition, it will help identify areas for further research and dedicated efforts. The design should be described from a customer's viewpoint (external and internal) and should provide the first insight into what a good service should look like. Concept models and design studies using TRIZ and axiomatic design are good sources for evaluating consumer appeal and areas of likes or dislikes.

The array of customer attributes should include all customer and regulatory requirements, and social and environmental expectations. It is necessary to understand requirements and prioritization similarities and differences in order to understand what can be standardized and what needs to be tailored.

8.4.3   DFSS Phase 2: Conceptualize the Design

Phase 2 spans the following two stages: concept development (stage 3) and preliminary design (stage 4).

Stage 3: Concept Development

Stage 3 Entrance Criteria
- Closure of tollgate 2 (approval of the gatekeeper is obtained).
- Defined system technical and operational requirements. Translate customer requirements (CTQs) to service or process functional requirements: CTQs give us ideas about what will make the customer satisfied, but they usually can't be used directly as the requirements for product or process design. We need to translate customer requirements to service and process functional requirements. Another phase of QFD can be used to develop this transformation. Axiomatic design principle will also be very helpful for this step.
- A process or service conceptual design.

- Trade-off of alternative conceptual designs with the following steps:
  - *Generate design alternatives.* After determination of the functional requirements for the new design entity (service or process), we need to characterize (develop) design entities that are able to deliver those functional requirements. In general, there are two possibilities, the first is that the existing technology or known design concept is able to deliver all the requirements satisfactorily; then this step becomes almost a trivial exercise. The second possibility is that the existing technology or known design is not able to deliver all requirements satisfactorily; then a new design concept has to be developed. This new design should be "creative" or "incremental," reflecting the degree of deviation from the baseline design, if any. The TRIZ method (see Chapter 9 of El-Haik and Roy, 2005) and axiomatic design (Appendix C) will be helpful in generating many innovative design concepts in this step.
  - *Evaluate design alternatives.* Several design alternatives might be generated in the last step. We need to evaluate them and make a final determination on which concept will be used. Many methods can be used in design evaluation, which include the Pugh concept selection technique, design reviews, and failure mode effect analysis. After design evaluation, a winning concept will be selected. During the evaluation, many weaknesses of the initial set of design concepts will be exposed and the concepts will be revised and improved. If we are designing a process, process management techniques will also be used as an evaluation tool.
- Functional, performance, and operating requirements allocated to service design components (subprocesses).
- Development of a cost estimate (tollgates 2 through 5).
- Target product or service unit production cost assessment.
- Market:
  - Profitability and growth rate
  - Supply chain assessment
  - Time-to-market assessment
  - Share assessment
- Overall risk assessment.
- Project management plan (tollgates 2 through 5) with a schedule and test plan.
- Team member staffing plan.

Tollgate 3: Stage 3 Exit Criteria
- Assessment that the conceptual development plan and cost will satisfy the customer base
- A decision that the service design represents an economic opportunity (if appropriate)

- Verification that adequate funding will be available to perform the preliminary design
- Identification of the tollgate keeper and the appropriate staff
- Action plan to continue flow-down of the design functional requirements

Stage 4: Preliminary Design

Stage 4 Entrance Criteria
- Closure of tollgate 3 (approval of the gatekeeper is obtained)
- Flow-down of system functional, performance, and operating requirements to subprocesses and steps (components)
- Documented design data package with configuration management[4] at the lowest level of control
- Development-to-production operations transition plan published and in effect
- Subprocesses (steps), functionality, performance, and operating requirements verified
- Development testing objectives complete under nominal operating conditions
- Testing with design parametric variations under critical operating conditions (tests might not utilize the operational production processes intended)
- Design, performance, and operating transfer functions
- Reports documenting the design analyses as appropriate
- Procurement strategy (if applicable)
- Make or buy decision
- Sourcing (if applicable)
- Risk assessment

Tollgate 4: Stage 4 Exit Criteria
- Acceptance of the selected service solution or design
- Agreement that the design is likely to satisfy all design requirements
- Agreement to proceed with the next stage of the service solution or design selected
- Action plan to finish the flow-down of the design's functional requirements to design parameters and process variables

---

[4] A systematic approach to establishing design configurations and managing the change process.

DFSS tools used in this phase:

- QFD
- TRIZ or axiomatic design
- Measurement system analysis
- Failure mode effect analysis
- Design scorecard
- Process mapping
- Process management
- Pugh concept selection
- Robust design
- Design for X
- Design reviews

8.4.4 DFSS Phase 3: Optimize the Design

This phase spans stage 5 only, the design optimization stage.

Stage 5: Design Optimization

Stage 5 Entrance Criteria
- Closure of tollgate 4 (approval of the gatekeeper is obtained).
- Design documentation defined: the design is completed and includes the information specific to the operations processes (in the opinion of the operating functions).
- Design documents are under the highest level of control.
- Formal change configuration is in effect.
- Operations are validated by the operating function to preliminary documentations.
- Demonstration test plan (full-scale testing, load testing) put together that must demonstrate functionality and performance in an operational environment.
- Risk assessment.

Tollgate 5: Stage 5 Exit Criteria
- Agreement that functionality and performance meet customer and business requirements under the operating conditions intended
- Decision to proceed with a verification test of a pilot built to conform to preliminary operational process documentation
- Analyses to document the design optimization to meet or exceed functional, performance, and operating requirements

- *Optimized transfer functions.* DOE (design of experiments) is the back-bone of process design and redesign improvement. It represents the most common approach to quantifying the transfer functions between the set of CTQs and/or requirements and the set of critical factors, the X's, at different levels of design hierarchy. DOE can be conducted by hardware or software (e.g., simulation). From the subset of a few vital X's, experiments are designed to actively manipulate the inputs to determine their effect on the outputs (big Y's or small y's). This phase is characterized by a sequence of experiments, each based on the results of the previous study. Critical variables are identified during this process. Usually, a small number of X's account for most of the variation in the outputs.

The result of this phase is an optimized service entity with all functional requirements released at the six-sigma performance level. As the concept design is finalized, there are still a lot of design parameters that can be adjusted and changed. With the help of computer simulation and/or hardware testing, DOE modeling, Taguchi's robust design methods, and response surface methodology, the optimal parameter settings will be determined. Usually, this parameter optimization phase will be followed by a tolerance optimization step. The objective is to provide a logical and objective basis for setting requirements and process tolerances. If the design parameters are not controllable, we may need to repeat stages 1 to 3 of the service DFSS.

DFSS tools used in this phase:

- Transfer function detailing (physical DOE, computer DOE, hypothesis testing, etc.)
- Process capability analysis
- Design scorecard
- Simulation tools
- Mistake-proofing plan
- Robustness assessment (Taguchi methods: parameter and tolerance design)

8.4.5 DFSS Phase 4: Validate the Design

This phase spans the following two stages: verification (stage 6), which includes DES model validation, as discussed in Chapter 6; and launch readiness (stage 7).

Stage 6: Verification

Stage 6 Entrance Criteria
- Closure of tollgate 5 (approval of the gatekeeper is obtained)
- Risk assessment

Tollgate 6: Stage 6 Exit Criteria   After the parameter and tolerance design are finished, we move to final verification and validation activities, including testing. The key actions are:

- *Pilot test auditing.* The pilot tests are audited for conformance with design and operational process documentation.
- *Pilot testing and refining.* No service should go directly to market without pilot testing and refining. Here we can use design failure mode effect analysis as well as pilot and small-scale implementations to test and evaluate real-life performance.
- *Validation and process control.* In this step we validate the new entity to make sure that the service, as designed, meets the requirements and to establish process controls in operations to ensure that critical characteristics are always produced to specification of the optimize phase.

Stage 7: Launch Readiness

Stage 7 Entrance Criteria
- Tollgate 6 is closed (approval of the gatekeeper is obtained).
- The operational processes have been demonstrated.
- Risk assessment is complete.
- All control plans are in place.
- Final design and operational process documentation has been published.
- The process is achieving or exceeding all operating metrics.
- Operations have demonstrated continuous operations without the support of design development personnel.
- Planned sustaining development personnel are transferred to operations.
  - Optimize, eliminate, automate, and/or control inputs deemed vital in the preceding phase.
  - Document and implement the control plan.
  - Sustain the gains identified.
  - Reestablish and monitor long-term delivered capability.
- A transition plan is in place for the design development personnel.
- Risk has been assessed.

Tollgate 7: Stage 7 Exit Criteria
- The decision is made to reassign the DFSS black belt.
- Full commercial rollout and handover to the new process owner are carried out. As the design entity is validated and process control is established, we launch full-scale commercial rollout. The newly designed service, together with the supporting operation processes, can be handed

over to design and process owners, complete with requirement settings and control and monitoring systems.

- Tollgate 7 is closed. (approval of the gatekeeper is obtained).

DFSS tools used in this phase:

- Process control plan
- Control plans
- Transition planning
- Training plan
- Statistical process control
- Confidence analysis
- Mistake-proofing
- Process capability modeling

8.5 SUMMARY

In this chapter we presented simulation-based lean six-sigma (3S-LSS) and design for six-sigma (3S-DFSS) road maps. Figures 8.4 and 8.14 highlight at a high level the various phases, stages, and activities. For example, 3S-DFSS includes the identify, conceptualize, optimize and validate phases and seven development stages: idea creation, voice of the customer and business, concept development, preliminary design, design optimization, verification, and launch readiness. The road maps also recognize tollgates, milestones where teams update stakeholders on progress and ask for decisions to be made as to whether to approve going into the next stage or phase, recycle back to an earlier stage, or cancel the project altogether. The road maps also highlight most appropriate tools used in the phase. They indicate where tool use is most appropriate. The roadmaps are utilized in the case studies Chapters 9 and 10.

9

SIMULATION-BASED LEAN SIX-SIGMA APPLICATION

9.1 INTRODUCTION

In this chapter we provide the details of 3S-LSS application procedures and discuss the specifics of various elements of the 3S-LSS road map presented in Chapter 8. The focus is on the practical aspects of 3S-LSS use to reengineer real-world processes. These aspects are discussed through an application of the 3S-LSS approach to a manufacturing case study. The case study selected represents an assembly plant of multiple assembly lines separated by work-in-process buffers. Raw materials of assembly components arrive at the plant, are assembled into finished items, and are shipped to customers. Such simple flow is common in assembling many products, such as appliances and electronic devices. The 3S-LSS application procedure presented is not limited to production systems and can be replicated on any process of a transactional nature.

As discussed in Chapter 8, the 3S-LSS process adopted herein is based on combined lean manufacturing techniques and DMAIC six-sigma methodology. Six-sigma's quality focus (e.g., reducing process variability and increasing sigma rating) is combined with lean manufacturing's value-added focus (e.g., reducing process waste and increasing effectiveness). Simulation is utilized as an environment for integrating lean techniques and the DMAIC method. Emphasis is placed on utilizing a simulation-based application of lean

Simulation-Based Lean Six-Sigma and Design for Six-Sigma, by Basem El-Haik and Raid Al-Aomar
Copyright © 2006 John Wiley & Sons, Inc.

techniques and six-sigma analysis to improve a set of process critical-to-quality (CTQ) characteristics. The CTQs are defined in terms of typical lean manufacturing measures (time-based performance metrics) such as productivity, lead time, and work-in-process inventory levels.

Use of the 3S-LSS approach is based on a value-driven understanding of the underlying process steps (through a value stream map), an accurate measurement of process capabilities (through six-sigma analysis), a realistic estimation of process time-based performance (through DES), and an application of six-sigma analysis and lean techniques to improve process performance. Thus, the main steps in this 3S-LSS application involve developing a VSM of the current production system, modeling the production system with discrete event simulation, estimating the process lean measures (CTQs) with simulation, improving the production system in terms of lean measures with experimental design and lean techniques, developing a future-state VSM, and defining actions to control and maintain the improvement achieved.

Focusing the 3S-LSS approach on time-based performance takes advantage of simulation modeling for a system-level application of six-sigma statistical tools of DMAIC and lean techniques. As discussed in Chapter 1, DMAIC is a disciplined and data-driven approach that is used to improve processes by ensuring that current process performance measures meet customer requirements with the least variability. The DMAIC focus in the 3S-LSS approach is on achieving improvement in time-based process performance measures such as throughput, inventory, and lead time. Simultaneously, lean manufacturing techniques are applied to reduce the process waste and increase process effectiveness through faster deliveries and a shorter lead time. As discussed in Chapter 2, these methods include the application of an array of techniques such as a pull production system, Kanbans and visual aids, Kaizen, and work analyses. The main objective is to reduce or eliminate inventory levels, non-value-added activities, and process waste.

In a 3S-LSS context, therefore, both lean techniques and the DMAIC method are aimed at working on solving problems within the current state of a transactional process and on improving process performance by developing a future state with better time-based CTQs. This leads to a powerful problem-solving and continuous improvement approach that can benefit greatly from the synergy of DMAIC, simulation, and lean techniques.

Marrying the accuracy and data-driven nature of the DMAIC method with the work effectiveness mandated by a lean environment often has a tremendous impact on process effectiveness and increases employee awareness of process functionality, variability, and value chain. Other benefits of the 3S-LSS approach include providing a framework for the involvement of all parties and stakeholders of the production or service process and the creation of initiative ownership and effective teamwork dynamics. Ultimately, this is expected to result in multiple process gains.

The material presented herein is intended to give the reader a practical sense for applying the 3S-LSS approach to real-world systems. The focus will be on how six-sigma performance can be achieved in lean manufacturing measures such as productivity, lead time, and work-in-process. The case study highlights the synergy of applying lean manufacturing in conjunction with the DMAIC method and shows how lean techniques can be used to enhance performance by reducing waste in the process flow while maintaining the high level of performance achieved through six-sigma. This case study also highlights the advantages of using simulation as an environment for verifying lean techniques in a DMAIC process. Finally, the case provides the reader with the ability to assess how 3S-LSS could be used in relation to their jobs to solve or enhance transactional processes in production and service systems.

9.2   3S-LSS INTEGRATED APPROACH

As discussed in Chapter 2, lean manufacturing provides a set of methods that companies can apply to any manufacturing, transactional, or service process to reduce waste, eliminate non-value-added actions and cut lead time. Combining lean with six-sigma in the 3S-LSS approach can therefore produce an integrated program for process improvement that brings both short-term results and long-term process change. Today, many companies are turning to a combined lean and six-sigma effort to solve current-state problems and migrate toward a more effective and more profitable future state.

The development of an integrated improvement program that incorporates both lean and six-sigma tools requires more than the application of a few lean techniques in the DMAIC six-sigma methodology. What is really critical is to establish an integrated improvement strategy that is able to benefit from the advantages of the two approaches. The 3S-LSS approach establishes an integrated program for improvement using a simulation-based application of LSS. The numerous advantages that can then be attained from the synergy of successful 3S-LSS implementation are highlighted throughout this chapter.

The 3S-LSS integrated approach is focused first on utilizing DES to the benefit of the DMAIC process when working on improving time-based performance CTQ measures. The DMAIC methodology phases and tools are applied in conjunction with lean techniques based on the process road map presented in Chapter 8. Beginning with the define phase, DES provides a more realistic process map by capturing the complex, stochastic, and dynamic characteristics of the real-world process. Further, and since DMAIC is an iterative process, adjustments made to process structures and logic are implemented easily and quickly with the DES model. In the measure phase, DES is used to verify historical data and provide steady-state process performance measures with various statistics and results. In the analyze phase, the model is used extensively for experimental design and various types of DMAIC analyses. The analysis cost and time are reduced significantly using the DES model. In the improve phase, DES is used to validate improvement plans and actions aimed at fixing problems and improving process performance. This includes

TABLE 9.1 DMAIC–DES Integration

DMAIC Phase	DMAIC–DES Integration
Define (D)	Represent process data, structure, logic, and layout
Measure (M)	Measure process dynamic and stochastic performance
Analyze (A)	Run experimental design and DMAIC analyses
Improve (I)	Validate improvement actions and lean techniques
Control (C)	Test monitoring plans and collecting feedback measures

applying lean manufacturing techniques. Finally, in the control phase, various control schemes and feedback measures are evaluated and tested using the model, especially with the distinctive DES capability to predict future events and scenarios. The integration of DES with the five DMAIC phases is summarized in Table 9.1.

The second focus of the 3S-LSS approach is to integrate the lean curriculum into the DMAIC framework. Within the simulation environment, whereas six-sigma focuses on defects and process variation, lean is linked to process speed, efficiency, and waste. This emphasizes the use of DMAIC as a common problem-solving approach and the use of lean tools to reduce cycle time. Lean techniques in the 3S-LSS approach are applied to eliminate waste in every process activity and to incorporate less effort, less inventory, less time, and less space. As discussed in Chapter 2, main lean principles applied to this end include reduced waiting time, reduced inventory, a pull instead of a push system, small batches or single-unit flow, line balancing, and short process times. Other basic lean manufacturing techniques include the 5Ss (sort, straighten, shine, standardize, and sustain), standardized work, and Kaizen (i.e., continuous improvement). Lean techniques attain such principles by eliminating the seven deadly sources of waste overproduction, inventory, transportation, processing, scrap, motion, and waiting or idle time.

As discussed in Chapter 2, lean integration into the DMAIC approach also consists of define, measure, analyze, improve, and control phases. Major Lean implications regarding the DMAIC process include the following:

1. Define customer and process improvement requirements by prescribing CTQs, design parameters, and corresponding process variables. Lean measures (time-based process performance and cost measures) are expected to be the main CTQs in 3S-LSS studies. Process design variables include parameters affecting such time–based performance and contribute to various process cost elements.

2. Measure the CTQs defined in the current state of the process. To this end, a VSM of the current process state with simulation-based estimates of process CTQs is developed in the 3S-LSS approach. The simulation-based VSM herein is a dynamic process representation that includes close-to-real performance-based assessment of the current process state.

3. Analyze the process current state using a simulation-based VSM by developing transfer functions to estimate lean measures (CTQs) at various process parameters. Typical six-sigma analyses such as DOE and regression models are essential at this phase. From a 3S-LSS perspective, the analysis phase is focused on parameter design rather than structural design. DOE is used to set the process parameters to the levels at which process CTQs (lean measures) are at their best (a six-sigma level is desirable) with the least variation (robust performance).

4. Improve the current state by proposing a better process design (a future state). A future-state VSM is developed to quantify the improvement achieved and to represent the structure and parameters of the future state. From a 3S-LSS perspective, this is the phase where various lean manufacturing techniques are put into action to restructure the process for better performance. The focus here is on the elimination of non-value-added (NVA) elements from within the process by reducing lead time and excessive inventory from within the process so that process effectiveness is increased and process waste is reduced. Various lean techniques (discussed in Chapter 2) are implemented in this phase, including Kanban and pull production systems, batch size, and setup time reduction (SMED).

5. Control the future-state (improved) process by monitoring process variables and behavior and setting controls to maintain the integrated six-sigma lean performance. Sensitivity analysis using a process simulation model can help test monitoring schemes and develop effective plans and measures for sustainable process improvement.

Integration of the main lean manufacturing principles within the five DMAIC phases is summarized in Table 9.2.

Figure 9.1 shows an integrated approach to applying a 3S-lean process to a real-world process based on DMAIC, simulation, and lean techniques. The approach is derived from the 3S-LSS road map presented in Chapter 8. Figure 9.1 demonstrates the progression attained for 3S-LSS from the define phase through the control phase (i.e., implementation of six-sigma improvement to a real world process in terms of lean measures). This is expected to deliver numerous process gains (in the Z-scores) associated with the time-

TABLE 9.2   DMAIC-Lean Integration

DMAIC Phase	DMAIC-Lean Integration
Define	Define lean measures and develop a VSM
Measure	Measure process value-added and non-value-added elements
Analyze	Analyze process flow, inventory level, and production
Improve	Apply lean techniques and principles
Control	Implement Kaizen and 5S for process sustainability

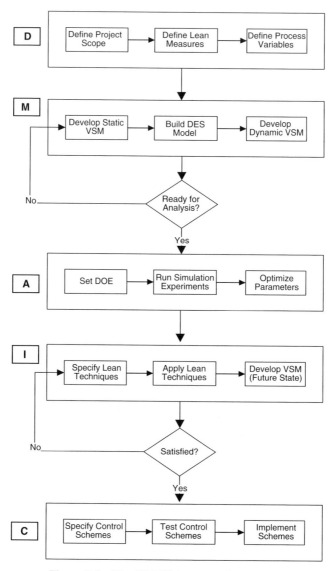

Figure 9.1 The 3S-LSS integrated approach.

based performance metrics selected and in several functional requirements (CTQs).

As shown in Figure 9.1, each DMAIC phase is split into three elements and integrated as a process flow with a simulation-based assessment of process performance. Our focus here will is on addressing the practical considerations that are essential to successful application of the 3S-LSS approach.

In the define phase, the 3S-LSS approach focuses on defining the scope of the improvement project and translating the project objectives into a specified definition of lean measures. This is essential in verifying the applicability of the 3S-LSS approach to the underlying real-world process, where the process performance needs to be expressed in terms of time-based lean measures that can be estimated using a DES model. For example, increasing customer satisfaction can be translated into reducing production cost, increasing delivery speed, and reducing excess inventory. These time-based CTQs are lean measures that can be assessed using simulation. The lean measures defined (as dependent responses) are linked to key process variables such as buffer sizes and locations, production lot size, quality control measures, process elements, allocated resources, and so on. Such parameters shape the lean measures defined.

At the measure phase, the 3S-LSS approach focuses on utilizing the process VSM to visualize process flow, parameters, and estimates of the defined lean measures. As discussed earlier, the VSM provides the flow time line, inventory sizes and locations, cycle times, takt time (productivity), operations uptime, available production time, and much other useful information that describes the current process state and summarizes its performance. However, the VSM is a static process representation that provides spreadsheet-based estimates of lead time and productivity measures.

Thus, the 3S-LSS approach utilizes a static VSM along with other simulation essential data and inputs to develop a DES process model that captures the process complex, dynamic, and stochastic characteristics. The process outcomes and statistics fed into a traditional VSM to transform it into a dynamic VSM provide a better and more realistic process representation with a statistical summary of each VSM element. This dynamic VSM will be used as a basis for conducting analysis and making process improvement in the 3S-LSS approach. Should the dynamic VSM be ready for analyses, we move to the analyze phase of the DMAIC process.

In the analyze phase, the 3S-LSS approach focuses on six-sigma analyses that set process parameters to reduce process variability toward the six-sigma level. This phase mainly utilizes DOE optimization of process parameters using the lean measures defined as a process response to optimize the levels of process parameters. A simulation-based experimental design is conducted on the process parameters defined in the VSM. The 3S-LSS approach benefits greatly from the flexibility of the DES model in importing and exporting data, adjustment of process parameters, and estimating lean measures. With the model, the experimental scope can be extended and experimental time can be cut significantly. Process performance is reflected in the VSM after completing the analysis phase of the DMAIC process.

The improve phase is a key element in the 3S-LSS approach, where lean techniques are used to alter process structure and control systems to reduce cost and waste and increase effectiveness. Again, the DES model will be the effective tool for testing lean techniques and assessing their impacts and gains.

Some lean techniques require changing model logic and routing rules, such as push or pull and Kanban production systems; others may require changing the process structure, such as adding in-station process controls, decoupling assembly operations, moving to cellular manufacturing, and so on. The outcome of the improve phase is a future-state VSM with the new lean measures. The gains of process improvement are quantified using the DES model. These gains are compared to objectives to check satisfaction. If improvements are not sufficient, other lean techniques may be applied, tested, and assessed.

Finally, in the control phase, the simulation model is used to test and verify control schemes intended to maintain the improvement achieved in applying the 3S-LSS approach. Typical six-sigma tools such as failure mode effect analysis, error-proofing, monitoring plans, and quality control charts can be used for this objective. Along with six-sigma tools, a lean 5S approach can be implemented to maintain work organization and workplace order. A lean Kaizen plan for continuous improvement will also serve the purpose of continually reviewing performance and seeking improvement.

9.3   3S-LSS CASE STUDY

A simple yet informative assembly process case study is presented to demonstrate application of the 3S-LSS approach to real-world processes. The case involves assembling a certain product by passing batches of it through four sequential assembly lines. Each assembly line consists of multiple stations of assembly and testing operations that take place on a synchronous conveyor (stopping one station stops the assembly line). All assembly lines are synchronized to transfer products with an indexing time of 60 seconds. Such assembly lines are common in final assembly operations of many products, such as computers, electronic devices, and appliances.

Figure 9.2 shows a high-level process map of the assembly process. Raw materials (assembly components) are received from suppliers at the plant and stored in a raw material inventory warehouse. A certain order size (300 units) of raw materials is shipped daily to the assembly facility based on a fixed schedule. A receiving station at the warehouse is dedicated to store input supplies. Batches of assembly components (100 units each) are prepared and sent to the plant and unloaded at the first assembly line using forklifts. A special unit load (pallet) is designed for easy material handling. Batches of the product are moved through assembly operations using a container. At each assembly operation, parts and modules of the product are integrated together through various assembly tools to form the final product.

The first assembly line consists of 25 sequential and synchronized assembly stations. A buffer (buffer 1) is located between the first and second assembly lines for work-in-process items. The second assembly line consists of 20 sequential and synchronized assembly stations. A sequencing area is located

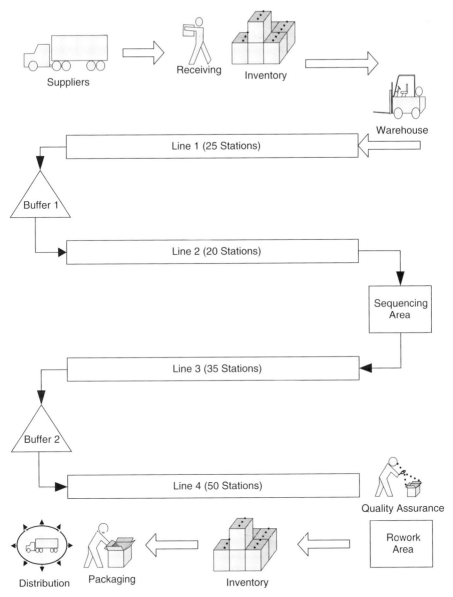

Figure 9.2 Assembly process case study.

right after the second assembly line to resequence out-of-sequence units. The third assembly line consists of 35 sequential and synchronized assembly stations. A buffer (buffer 2) is located between the third and fourth assembly lines. The fourth assembly line consists of 50 sequential and synchronized assembly stations.

A quality control station is located at the end of the fourth assembly line to check product functionality and mark product nonfunctioning problems. Marked items are processed in a rework area where actions are prescribed to fix assembly problems by reworking marked products through corresponding assembly lines. A finished-item storage area is used to store finished products prior to shipping. Finally, a shipping station is used to consolidate products into batches (250 units per lot) of ready-to-ship items and package them for shipping to customers.

Next, we apply the five phases of the 3S-LSS approach to the assembly process case study shown in Figure 9.2. The 3S-LSS application is based on the integrated approach shown in Figure 9.1. At each phase of the applied approach, the emphasis will be on the DMAIC integration with lean techniques and simulation modeling.

9.3.1 Define (Project Scope, Lean Measures, and Process Variables)

In this phase we specify the objectives and scope of work in the underlying case study. This typically includes solving a specific problem or improving certain performance levels. In the assembly process case study, the scope of the work is focused on improving the performance of the assembly process in terms of two major critical-to-quality (CTQs): product price and product delivery reliability. Production cost is a major contributor to product cost, and delivery reliability is a function of order lead time and delivery shortages. After analyzing a voice-of-customer survey, these two major CTQs are sought to have high impact on product market share and overall company competitiveness.

After a thorough analysis of the drivers behind product price and delivery reliability, it was found that three process measures can significantly improve the defined process CTQs: productivity, lead time, and work-in-process inventory. Hence, the company likes to focus this 3S-LSS study on achieving the following objectives: reducing delivery shortages by improving productivity and increasing delivery speed by reducing lead time. The company also aims at reducing process waste and assembly process cost to reduce production cost. This ultimately allows the company to better compete based on price and thus to increase market share. As noticed, these measures are typical time-based process-level lean measures in many production systems. Product quality characteristics are the focus of typical six-sigma projects. The following is a definition of the three lean measures:

1. Process productivity is defined in terms of the throughput of the assembly process measured in the units jobs per hour (JPH). JPH is a common measure of productivity in production system since it quantifies the system's capability in delivering product quantity in the required time frame to avoid delivery shortages. Analyzing the JPH value also pinpoints many process limitations and ineffectiveness such as bottlenecks and delays.

2. Process delivery speed is the elapsed time between placing an order and delivering the order to a customer. Although other process steps may delay order processing such as paperwork, purchases, approval, and so on, the focus in this case is on manufacturing lead time (MLT), the time line that begins with the arrival of raw materials and ends with the shipping of finished items to the receiving warehouse. This includes processing time, material-handling time, and time delay in inventory and buffers. Manufacturing time in this case is compared to processing time in transactional processes. It is often asserted that the time that products spend within a production system incurs extra cost to the company.

3. Process size is defined in terms of the total number of units within the assembly process. An average work-in-progress (WIP) inventory level within the process is used to quantify the process size. This includes items on assembly lines and buffers. It is commonly known that all types of inventory within a production system incur cost. High WIP levels also indicate process ineffectiveness and excess safety stocks (waste).

Many process variables can affect the three lean measures (production rate, MLT, and WIP). After investigation, it was found that the main process variables affecting the three lean measures in this case include the following:

- Cycle times (lines speeds) at various assembly operations
- Downtime elements at various assembly operations
- Arrival rate of raw materials from suppliers
- Shipping speed of finished goods to customers
- Percent of process rework
- Percent of units that are out of sequence (delayed for resequencing)
- Size of buffers and storage areas

In this case example, we analyze five process parameters: order size, the size of buffer 1, the size of the sequencer, the size of buffer 2, and the percentage of process rework. Process rework (15% now) is controlled by investing in in-station process quality controls. The size of the sequencer area (100 units of size) is affected by an 80% in-sequence production. The area is used to resequence units into the process flow. An order size of 300 units of the assembly components is transferred from the receiving station to the ready-for-assembly inventory area. The two buffers within assembly lines (buffers 1 and 2) are sized at 400 and 500 units.

9.3.2 Measure (Static VSM, DES Model, and Dynamic VSM)

In the measure phase of DMAIC, the 3S-LSS emphasizes the role of VSM in presenting current-state performance and defining the opportunities for improvement. Thus, the first step in this phase is to develop a current-state static VSM of the underlying assembly process. By *static* we mean a VSM that captures a snapshot of the process time line at certain defined conditions (mostly deterministic).

A typical VSM provides such process information, as discussed in Chapter 2. The current-state VSM along with process layout, structure, and logic provides inputs to the analyst for building a current-state process DES model that captures the stochastic process behavior as it evolves over time. The outcomes of the DES model (collected statistics) are integrated into the current-state VSM to develop a dynamic VSM that will better represent the actual process time line and performance measures. Thus, in this 3S-LSS phase, we develop the static VSM, the DES model, and the dynamic VSM to the assembly process case study shown in Figure 9.3.

Based on the VSM development procedure discussed in Chapter 2, a current-state VSM of the case study assembly process is developed as shown in Figure 9.4. The VSM presents the time line of material and product flow within the different stages of the assembly process. The production control of the assembly process receives product orders from customers. These orders are translated into orders made to the supplier of raw materials and into production orders sent to the production supervisors of various assembly departments. Production control often specifies order quantities and delivery dates based on the bill of materials using material requirement planning. The VSM depicts the time required to receive and store components and raw materials required for the product assembly (8 days in this case).

As production begins, the VSM records useful information about each process step, including work-in-process (WIP) storage. As shown in Figure 9.3, a typical VSM includes cycle time at each process operation (index time on an assembly conveyor), changeover time between product lot sizes of product models and colors, lot size (100 units), available time per shift of production, line uptime percentage, and number of assembly stations at each line. Assembly operations represent the only value-added (VA) time in the VSM. In this case, this time represents the total assembly time at all stations of the assembly line.

For the WIP inventory, the VSM shows the maximum inventory capacity and maximum time in the buffer measured in days of demand. In this case the demand average is 200 units per day; hence each 200 units in a WIP inventory results in a day in the VSM time line. The VSM summarizes the overall process lead time (15 days on NVA inventory in raw material, WIP, and finished units, and 2 hours and 10 minutes of VA time in assembly operations). The lead time is focused on manufacturing lead time from the arrival of raw material to the assembly plant to the ready for shipment of finished items.

In the 3S-LSS approach, the second step in the DMAIC measure phase is to develop a DES model for the assembly process described in Figure 9.3. The process DES model has a pivotal role in the 3S-LSS approach since it integrates application of six-sigma analysis and lean techniques. As discussed in Chapter 6, DES model building and validation for a current-state process

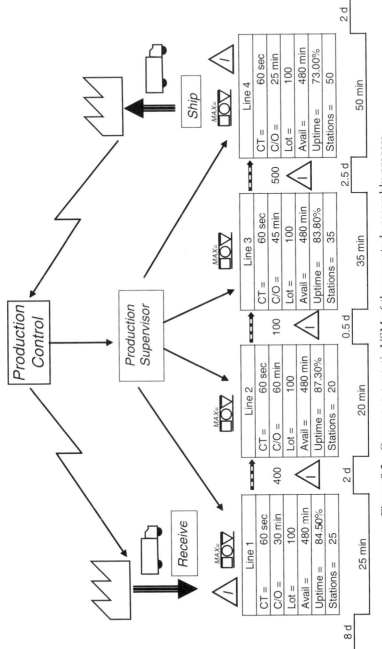

Figure 9.3 Current-state static VSM of the case study assembly process.

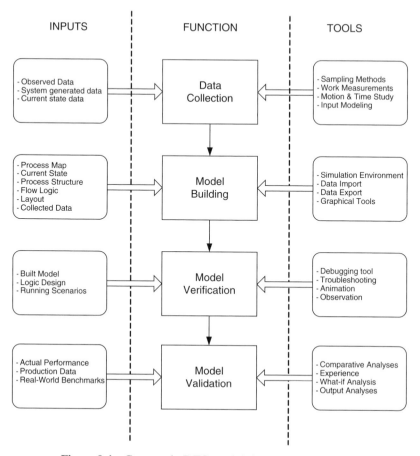

Figure 9.4   Case study DES model development process.

involves four major steps: data collection, model building, model verification, and model validation. Figure 9.4 depicts the inputs, function, and tools used at each step in DES model building in the 3S-LSS approach.

At the data collection stage of building a DES model, historical data are collected on the performance of the four assembly lines. The factory information system can provide useful information regarding production data, maintenance reports, and operating patterns. Other data can be generated from the direct observations of operators and subject matter production supervisors.

A DES model of the assembly process is then built using plant layout, process map, flow logic, and collected production data. WITNESS simulation software was used to develop the model. The model logic was validated using the methods described in Chapter 6. To make sure that the WITNESS model reflects the actual current state of the assembly process, the model is validated using last year's production data. Observed model performance at various

TABLE 9.3   Details of the DES Model Development Process

DES Step	Inputs	Outputs
1. Data collection	Data observed Layout Process map Logical design Any other useful data	Simulation data Cycle times Downtime data Operating pattern Lot size
2. Model building	Current-state VSM Flow logic Simulation data Process structure	Base model of the current state Animated flow Consolidation
3. Model verification	Base model Logical design Input data Output results	Base model verified Correct logic Correct data Correct outcomes
4. Model validation	Behavior observed Production data Benchmark performance	Model validated (ready for analyses) Correct behavior Correct statistics

running scenarios was compared to actual production patterns. Production managers and supervisors can often provide a better validation of the model performance since they are closer to the assembly lines.

Table 9.3 shows how the four steps used in developing the DES model are applied to the assembly process case by showing the inputs and outputs at each DES development step. At the data collection step, the inputs to this step are the data observed at various process elements. Other inputs include assembly process layout, process map, and logical design. From this step we obtain the necessary simulation data for modeling each assembly line (cycle time, changeover, downtime, lot size, operating pattern, etc.). For the model building step, the inputs include the current-state VSM, flow logic, input simulation data, and process structure. This step results in a base DES model of the current process with animated flow and a process consolidated model.

To verify the assembly process base model, we check the model logic and structure compared to the intended design, debug model logic, and verify the correct modeling of inputted data and model outcomes. This step should not be passed until we obtain a verified base model in terms of code, flow, inputs, and outputs. The model verified is then validated by comparing the model behavior, inputs, and outputs to the actual behavior, inputs, and outputs of the assembly process. The outcome of this stage is a validated representative model that can be used for the DMAIC analyze phase with high confidence.

The last step in the measure phase of DMAIC is to merge the static VSM with the process DES model into a dynamic VSM, which will be the platform for six-sigma analyses and lean techniques. The key statistics obtained from

the assembly process DES model are integrated into the static VSM structure
to provide a dynamic and more realistic representation of the process time
line and performance.

The verified and validated base process DES model is run at the current-
state process parameters, and the results are fed into the process VSM. The
assembly process model is run for 8 hours of warm-up and 2000 hours of production. Selected statistics on assembly lines performance are shown in Figure
9.5. The following process parameters (five selected variables at 80% in-
sequence production) are used to generate the dynamic DES-based VSM:

- Order size = 300 units
- Buffer size (WIP1) = 400 units
- Sequencer capacity = 100 units
- Buffer size (WIP2) = 500
- % Rework = 15%

By tracking components flow from raw material delivery buffer to finished
items shipping, the dynamic VSM provides better estimates of the assembly
process performance. The simulation dynamics and stochastic changes are
incorporated into the VSM to estimate the maximum, minimum, and average
level of inventory at each buffer or product holding area. The average level of
inventory is necessary to estimate the overall WIP inventory level in the
assembly plant. For example, the delivery buffer of input components accumulates most of the inventory in the current state. Only about 202 units on
average exist in the shipping inventory. The two WIP buffers hold on average
378 and 152 units, respectively. Including buffer statistics in the dynamic VSM
also provides guidance for determining flow bottlenecks and delays.

The dynamic VSM also provides the average time spent in each buffer or
storage area, which is part of the overall manufacturing lead time (MLT). For
example, raw materials spend about 11 days on average in the delivery process
and buffer before being shipped to the first assembly line. On the other hand,
finished items spend about 2 days in packaging and inventory before being
shipped to customers. Other major elements of MLT can be observed in the
two WIP buffers, the sequencer, and the rework area. Parts spend about 3
hours of value-added time in assembly operations on the four assembly lines.
Unlike the deterministic processing time used in static VSM, processing time
in dynamic VSM consists of assembly time, downtime, setup time (assembly
line cleaning and calibration), and blockage time. The total manufacturing lead
time adds up to 17.44 days. This estimate is more realistic than the 15-day MLT
in the static VSM.

The dynamic VSM provides statistics essential to describe the actual performance of the four assembly lines. Key line statistics include % idle
(starved), % setup, % busy, % down, and % blocked. As shown in Figure 9.6,
a time-in-state chart can be developed for each assembly line using such data.

Line 1

CT	60 sec
C/O	30 min
Lot	100
Avail time	480 min
% Busy	48.71
% Down	19.00
% Blocked	30.50
% Starved	0.34
% Set up	1.45
SAT	47.73
Stations	25
Avg Time	37.82

Line 2

CT	60 sec
C/O	60 min
Lot	100
Avail time	480 min
% Busy	48.37
% Down	15.78
% Blocked	32.85
% Starved	0.10
% Set up	2.90
SAT	48.79
Stations	20
Avg Time	30.33

Sequencer

Max	58
Min	0
Avg Units	23.50
Avg Time	745.07
Rework %	15%

Delivery Buffer

Max	10000
Min	0
Avg Units	6862.48
Avg Time	5311.29

WIP-Buffer 1

Max	400
Min	0
Avg Units	378.98
Avg Time	778.09

WIP-Buffer 2

Max	500
Min	0
Avg Units	152.38
Avg Time	266.84

Shipping Buffer

Max	1175
Min	0
Avg Units	202.90
Rework %	15%
Avg Time	899.77

Line 3

CT	60 sec
C/O	45 min
Lot	100
Avail time	480 min
% Busy	56.39
% Down	17.74
% Blocked	0.97
% Starved	21.76
% Set up	2.55
SAT	47.83
Stations	35
Avg Time	50.06

Line 4

CT	60 sec
C/O	25 min
Lot	100
Avail time	480 min
% Busy	56.98
% Down	36.24
% Blocked	0.00
% Starved	5.36
% Set up	1.42
SAT	37.40
Stations	50
Avg Time	71.51

Shipping

CT	60 min
C/O	25 min
Lot	250
Avail time	480 min
% Busy	14.45
% Down	46.64
% Blocked	0.00
% Filling	38.91
% Set up	0.00
SAT	32.02
Stations	1
Avg Time	185 min

Figure 9.5 Dynamic VSM for the current-state assembly process.

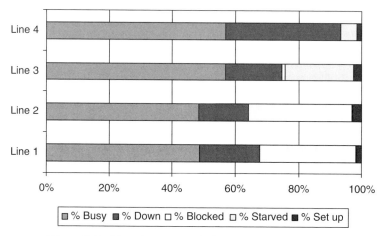

Figure 9.6 Time-in-state statistics in the dynamic VSM.

TABLE 9.4 Values (Averages) of Current-State Lean Measures

Lean Measure	Simulation Run					Grand Average
	1	2	3	4	6	
Productivity (JPH)	28.89	28.66	28.69	27.97	28.06	28.45
MLT (min)	8115.64	7812.81	8284.41	8657.74	9008.24	8375.77
WIP (units)	7948.89	7806.45	7938.09	8143.28	8297.28	8026.79

This data also determine each line's stand-alone throughput (SAT). SAT describes the maximum productivity that is expected from the assembly line with no starving and blocking (i.e., the line operates independent of other lines). For example, the line with the slowest line is line 4, where SAT is the lowest (37.4 UPH on average). Along with buffer statistics, time-in-state statistics provide indications on flow impending and potential improvement actions.

The simulation model of the current-state VSM is also used to provide steady-state estimates of the assembly plant performance in terms of the three defined lean measures (production rate, MLT, and WIP). These time-based performance metrics change dynamically over run time while encompassing stochastic variability due to random process elements such as downtime and shortages. Hence, five independent simulation replications of different random streams (using independent replication method) are run at the three performance metrics.

The results of the five simulation runs, each of 40 hours of warm-up and 2000 hours of run time, are shown in Table 9.4. Model run controls are selected carefully using pilot simulation runs to reflect the steady-state behavior based on the procedure discussed in Chapter 7.

As shown in Table 9.4, the grand averages of the three lean measures are obtained from the dynamic VSM are as follows:

- Production rate = 28.45 JPH (228 units/day)
- MLT = 8375.77 minutes (17.44 days)
- WIP = 8026.79 units (about 40 days of demand based on a demand of 200 units/day)

It is clearly noticeable from the results in Table 9.4 that higher buffer capacities and large storage areas result in a relatively high production rate throughput. Based on a demand of 200 units per day, the process is required to produce 25 UPH while the actual production average is 28.45 UPH. This is what is expected from a typical push production system where units are made to stock or are shipped directly to customers (assuming that there is a demand). Such current-state conditions often result in long MLT and an excessive WIP level, as seen from the results in Table 9.4. Over 8000 units of WIP on average are in the assembly process and they spend on average more than 17 days before being shipped to customers. This process waste is later targeted by lean techniques to reduce both MLT and WIP.

In line with lean thinking, six-sigma process capability is a key objective in 3S-LSS application. Hence, the simulation results will be used to calculate various six-sigma measures. This requires defining specifications for the three lean measures (production rate, MLT, and WIP). The definition of defects and opportunities of defects is different from the case in standard six-sigma calculations. Combining the results of five simulation runs, each of 2000 hours, results in data for an accumulated 10,000 hours. Within the 10,000 hours, products produced are checked for quality status (accepted or rejected) using preset quality specifications. The focus here is on time-based performance of the assembly process rather than on the product specifications. The focus is on recording and tracking changes in the production rate, manufacturing lead time, and work-in-process at each production hour.

To estimate six-sigma process measures, certain specifications for the three time-based performance metrics are set. For a production rate, it is desirable to produce at least 25 JPH. For MLT it is desirable to complete products within 2 weeks [10 working days or 80 hours (4800 minutes)]. For WIP it is desirable to operate the assembly process with up to 5000 units in WIP. A defect in the production rate is reported when obtaining a productivity of less than 25 JPH. A lower production rate is observed to cause shortages in deliveries made to customers. Similarly, a defect in MLT is reported when the average MLT in a production hour exceeds 80 hours, and a defect in WIP is reported when the average WIP level in a production hour exceeds 5000 units. Based on the simulation results in the current-state dynamic VSM, six-sigma measures were determined as shown in Table 9.5.

It is clear from the results in Table 9.5 that the assembly process encompasses high variability (as seen from the high standard deviation) and a large

TABLE 9.5 Summary of Current-State Six-Sigma Calculations

Lean Measure	Production Rate	MLT	WIP
Defect specification	<25 JPH	>4800 min	>5000 units
Mean	28.45 JPH	8375.77 min	8026.79 units
Standard deviation	18.66 JPH	4111.16 min	3456.12 units
Units produced (U)	10,000 hours	10,000 hours	10,000 hours
Opportunities (O)	1	1	1
Total opportunities (TOP)	10,000	10,000	10,000
Defects (D)	4216	7526	7598
Defect %	42.16%	75.26%	75.98%
Defect per unit (DPU)	0.4216	0.7526	0.7598
DPMO	421,600	752,600	759,800
Process yield (Y)	57.84%	24.74%	24.02%
Z-score	0.19	−0.683	−0.706
Sigma quality level[a]	1.69	0.817	0.794

[a] Short term with 1.5σ shift.

number of defects and defect per million opportunity (DPMO). This bad combination of high variability and defects often has a tremendous negative impact on any process performance. This is clearly evident from the resulting low process capability index and the low process six-sigma quality level.

The current-state assembly process reports weak performance in terms of the three lean measures (production rate, MLT, and WIP) for many operational causes. The process accumulates production in inventory and takes an extremely long time to ship products to customers. Although the average production rate is above the limit specified, many production shortages are reported, due to high process variability. Many real-world processes operate such that the impact of variability is ignored and lean measures are not observed. A substantial process reengineering effort is therefore essential to report better six-sigma measures. To this end, the 3S–LSS approach is applied to reduce process variability with the six-sigma method and to improve the time-based performance with lean techniques.

The following definitions and formulas arc used in the six-sigma calculations in Table 9.5:

m = number of operation steps

U = units produced per year

D = defects (production hours at which productivity is below 25 JPH, MLT is above 4800 minutes, and WIP is above 5000 units)

O = opportunities for defects = 1 (system-level view of the time-based measures)

Total opportunities = TOP = $U \times O$

Defect per unit = DPU = D/U

Defect per unit opportunity = DPU/O = $D/(U \times O)$

Defect per million opportunity = DPMO = DPO $\times 10^6$

Y = yield = percent of defect-free units (production) [determined based on DPU as Poisson probability of zero defects (i.e., $Y = e^{-DPU}$) where DPU = $-\ln(Y)$]

% Defect = 100% − yield

Rolled throughput yield = YRT = product of m operation steps Y_{TP}

Z-score = the corresponding Z to the process yield (from normal tables)

Process sigma quality level = $Z_{short\ term}$ = $Z_{long\ term}$ + 1.5 (with 1.5σ process shift assumption, simulation is used to determine the long-term Z-score, and the short-term Z-scores is reported by adding the 1.5σ shift to the value obtained from standard normal tables)

9.3.3 Analyze (DOE, ANOVA, and Regression)

The analysis stage of the 3S-LSS approach is focused mainly on using DOE, ANOVA, and regression to analyze the impact of defined process control factors (order size, size of Buffer 1, sequencer size, size of Buffer 2, and % rework) on the three lean performance measures (production rate, MLT, and WIP). The focus is placed on achieving maximum improvement in process capability and sigma rating. The objective is to provide optimal or nearly optimal settings to process control factors so that process performance in terms of production rate reaches its best possible level. Many process noise factors are impeded from entering the process simulation model by subjecting process elements (assembly lines) to random failures. Process performance in terms of MLT and WIP will be the focus of lean techniques in a later section. Many other statistical and flow analysis in lean and six-sigma can also be applied at this phase. The objective is to focus this case on DOE as a primary six-sigma tool.

Screening DOE is used to filter potential process control factors (design parameters). DOE is then used to determine the impact effect of the five design parameters selected on the process production rate and to provide the best level settings of these parameters so that the assembly line performance is at its best. Focusing the analysis on three levels of each parameter leads to $(3)^5$ or 243 full factorial experiments. Being simulation-based, the 3S-LSS approach has the advantage of avoiding the potentially high costs and tremendous efforts of physical experimentation.

The marginal cost of simulation-based experiments is driven primarily by computer run time and simulation analyses. This facilitates conducting full factorial experimental design in a 3S-LSS application. In this case it is decided to rely on fractional factorial design to make the analysis comparable to typical DOE studies. A fractional factorial design 3^{k-p} with $k = 5$ and $p = 2$ is developed based on an L_{27} Taguchi design.[1] The three levels are selected carefully to cover the performance expected for each design parameter (DP) at various running conditions. Pilot simulation runs are used to select factor levels. Instead of using an order size of 300 units, pilot runs recommended three reduced levels for the order size to be used in DOE and similarly, three reduced levels of the two WIP buffers, the sequencer, and the rework rate. This results in the following DOE levels of control factors:

- A = DP_1: order size (150, 200, 250)
- B = DP_2: buffer 1 (100, 200, 300)
- C = DP_3: sequencer (10, 25, 50)
- D = DP_4: buffer 2 (50, 100, 150)
- E = DP_5: rework (5%, 10%, 15%)

The assembly process performance in terms of the three lean measures at each L_{27} design is assessed using the DES model. To model process noise (stochastic variability), five simulation replications of varying random streams are used to estimate the three lean measures at each design combination. Table 9.6 presents the simulation results in terms of the process production rate using five replications. The production rate is given in terms of productivity mean and signal-to-noise (S/N) ratio[2] [with a nominal-is-best (NB) criterion]. The NB for an S/N ratio is selected since our goal is to reduce variability, and NB is inversely related to variance. Selecting a larger-the-better form of S/N ratio will push toward a higher JPH value mean shift.

DOE analyses in this section are focused on enhancing the assembly process to meet the desired 25 JPH level with less variability. Because of interrelationships, adjusting process parameters to enhance the production rate is expected to result in improving MLT and WIP as well. Tables 9.7 and 9.8 present the simulation results of L_{27} design for the MLT and WIP measures, respectively.

It is noted from the results in Tables 9.6, 9.7, and 9.8 that values of the three lean measures (production rate, MLT, and WIP) vary significantly from one parameter design to another (indicating large main effect values of control factors). At each parameter design, the amount of variability within the five simulation runs reflects the stochastic nature of simulation results. The stan-

---

[1] The authors realize that L_{18} can be used as well. L_{18} is a saturated array alternative.

[2] Signal-to-noise ratio is a metric used to optimize a system's robustness (insensitivity) to noise, independent of putting the performance on target. The S/N ratio measures relative quality, because it is intended to be used for comparative purposes (comparison of design parameters effects). Optimization (in the context of robustness) means that we must seek to identify the best expression of a design (product and process) that is the lowest total cost solution to customer wants. "Good enough" or "within specification" are not concepts that are compatible with optimization activity using robust design. To analyze a system's robustness and adjustability, the unique S/N ratio metric is available, which make it possible for us to use powerful experimental methods to optimize a product's insensitivity to sources of variability (called *noise*).

TABLE 9.6 L₂₇ DOE: Production Rate Results of Five Simulation Replications

| | Control Factor | | | | | Simulation Run | | | | | Production Rate | |
No.	A	B	C	D	E	1	2	3	4	5	S/N	Mean
1	1	1	1	1	1	18.68	18.66	18.65	18.65	18.66	63.66	18.66
2	1	1	1	1	2	18.59	18.67	18.67	18.60	18.58	52.46	18.62
3	1	1	1	1	3	18.62	18.62	18.63	18.58	18.66	56.26	18.62
4	1	2	2	2	1	18.67	18.68	18.65	18.65	18.62	58.17	18.65
5	1	2	2	2	2	18.67	18.67	18.66	18.66	18.63	61.10	18.66
6	1	2	2	2	3	18.67	18.65	18.64	18.62	18.66	59.73	18.65
7	1	3	3	3	1	18.68	18.67	18.67	18.63	18.67	59.62	18.66
8	1	3	3	3	2	18.67	18.65	18.66	18.67	18.65	65.42	18.66
9	1	3	3	3	3	18.65	18.64	18.65	18.66	18.66	66.96	18.65
10	2	1	2	3	1	24.86	24.89	24.87	24.85	24.84	62.23	24.86
11	2	1	2	3	2	24.80	24.85	24.79	24.84	24.86	58.03	24.83
12	2	1	2	3	3	24.84	24.88	24.84	24.78	24.86	56.44	24.84
13	2	2	3	1	1	22.35	22.04	22.28	22.82	22.87	35.91	22.47
14	2	2	3	1	2	22.21	22.16	21.59	22.82	22.33	34.08	22.22
15	2	2	3	1	3	21.68	21.52	21.78	22.30	21.82	37.46	21.82
16	2	3	1	2	1	24.82	24.86	24.80	24.81	24.76	56.75	24.81
17	2	3	1	2	2	24.82	24.89	24.76	24.78	24.74	52.42	24.80
18	2	3	1	2	3	24.66	24.64	24.63	24.47	24.53	49.53	24.59
19	3	1	3	2	1	25.15	26.13	24.75	25.49	25.55	33.92	25.41
20	3	1	3	2	2	25.21	25.31	25.28	25.68	25.25	42.49	25.35
21	3	1	3	2	3	25.14	24.67	24.70	24.72	25.17	39.89	24.88
22	3	2	1	3	1	27.76	26.86	28.40	28.04	27.68	33.73	27.75
23	3	2	1	3	2	27.71	27.66	28.00	28.17	27.71	41.90	27.85
24	3	2	1	3	3	27.80	27.60	27.29	28.46	27.98	36.07	27.83
25	3	3	2	1	1	22.11	22.10	22.50	22.93	22.56	36.21	22.44
26	3	3	2	1	2	22.18	22.42	22.37	22.25	22.72	40.62	22.39
27	3	3	2	1	3	22.27	21.66	22.02	22.72	21.56	33.39	22.05

dard deviation of both MLT and WIP also varies significantly from one parameter design to another.

Since current-state productivity is already above the desired production rate demand, the focus of DOE will be on reducing production rate variability while maintaining an average JPH value above or equal to a demand rate of 25 JPH. Instead of maximizing throughput mean, therefore, the DOE analysis is focused on using the production rate nominal-is-best S/N ratio as a design response. We show later how lean techniques can be complementary to six-sigma by focusing on achieving improvement in terms of MLT and WIP.

To focus the DOE analysis on the production rate, we first need to check the normality of JPH data. The normality assumption is an essential element in six-sigma calculations. If the data are not normal, plotting the natural

TABLE 9.7 L$_{27}$ DOE: MLT Results of Five Simulation Replications

No.	Control Factor					Simulation Run					MLT	
	A	B	C	D	E	1	2	3	4	5	S.D.	Mean
1	1	1	1	1	1	1,871.08	1,858.90	2,026.60	2,060.83	1,892.00	94.51	1,941.90
2	1	1	1	1	2	2,098.47	2,133.20	2,084.80	2,008.48	1,958.30	71.42	2,056.70
3	1	1	1	1	3	2,149.66	2,139.10	2,184.30	2,024.15	2,079.10	63.51	2,115.30
4	1	2	2	2	1	1,814.77	1,826.40	1,988.30	1,944.37	1,817.60	81.98	1,878.30
5	1	2	2	2	2	2,029.18	1,966.90	1,929.80	2,058.57	1,887.60	70.12	1,974.40
6	1	2	2	2	3	1,971.77	2,008.00	2,040.40	2,008.98	1,942.50	37.81	1,994.30
7	1	3	3	3	1	1,827.33	1,796.40	1,862.80	1,971.58	1,769.70	78.55	1,845.60
8	1	3	3	3	2	2,065.24	2,000.60	2,044.30	2,131.63	1,797.50	126.73	2,007.90
9	1	3	3	3	3	2,020.34	2,174.50	2,039.90	2,116.80	1,908.80	101.07	2,052.10
10	2	1	2	3	1	1,705.57	1,653.90	1,735.10	1,745.59	1,619.40	53.94	1,691.90
11	2	1	2	3	2	2,137.19	1,794.40	2,043.10	1,878.21	1,616.10	205.38	1,893.80
12	2	1	2	3	3	2,040.41	1,824.00	2,427.00	1,805.57	1,911.80	255.18	2,001.80
13	2	2	3	1	1	4,029.41	5,568.10	6,104.60	3,388.36	4,030.80	1,152.79	4,624.30
14	2	2	3	1	2	4,234.20	5,891.90	6,634.40	3,190.96	3,837.50	1,447.75	4,757.80
15	2	2	3	1	3	4,995.73	6,833.60	6,998.60	4,646.51	5,770.10	1,057.28	5,848.90
16	2	3	1	2	1	1,973.35	1,745.40	1,754.10	1,857.14	2,070.60	141.02	1,880.10
17	2	3	1	2	2	1,880.34	2,486.90	1,864.70	1,754.15	1,725.30	311.84	1,942.30
18	2	3	1	2	3	2,179.96	2,780.20	2,297.10	2,492.92	2,571.50	234.98	2,464.30
19	3	1	3	2	1	6,723.56	7,393.80	8,698.10	6,305.91	6,934.70	918.96	7,211.20
20	3	1	3	2	2	6,679.45	7,818.60	8,153.80	6,853.03	7,328.40	624.78	7,366.70
21	3	1	3	2	3	6,794.25	8,128.50	7,583.40	6,542.01	7,003.90	641.20	7,210.40
22	3	2	1	3	1	4,687.73	6,134.70	4,892.60	4,340.38	4,374.60	734.49	4,886.00
23	3	2	1	3	2	4,497.08	4,811.40	5,621.00	4,503.46	4,994.10	462.36	4,885.40
24	3	2	1	3	3	4,549.38	5,733.20	6,191.20	4,097.50	5,133.90	850.38	5,141.00
25	3	3	2	1	1	9,512.91	10,322.20	10,533.90	8,845.92	9,317.00	705.96	9,706.40
26	3	3	2	1	2	9,345.61	10,935.90	10,124.00	9,032.32	10,197.50	753.04	9,927.10
27	3	3	2	1	3	9,347.48	10,263.40	10,983.90	8,870.83	11,149.50	997.97	10,123.00

TABLE 9.8 L₂₇ DOE: WIP Results of Five Simulation Replications

No.	Control Factor					Simulation Run					WIP	
	A	B	C	D	E	1	2	3	4	5	S.D.	Mean
1	1	1	1	1	1	515.66	497.04	539.57	550.42	523.73	20.79	525.28
2	1	1	1	1	2	562.28	553.01	561.48	540.76	545.24	9.58	552.55
3	1	1	1	1	3	595.94	565.66	529.71	563.07	563.77	23.45	563.63
4	1	2	2	2	1	493.65	476.42	519.55	510.48	475.72	19.75	495.16
5	1	2	2	2	2	550.82	536.58	501.02	530.90	491.86	24.85	522.24
6	1	2	2	2	3	530.30	536.18	549.03	510.68	518.55	14.99	528.95
7	1	3	3	3	1	540.96	503.31	508.69	520.94	479.21	22.75	510.62
8	1	3	3	3	2	547.04	521.44	519.55	537.48	479.11	26.01	520.92
9	1	3	3	3	3	535.68	537.18	540.76	538.57	523.73	6.67	535.18
10	2	1	2	3	1	707.87	660.86	704.28	649.10	715.94	30.37	687.61
11	2	1	2	3	2	790.64	714.14	820.32	716.93	660.86	64.18	740.58
12	2	1	2	3	3	770.62	760.86	866.93	726.59	752.19	53.70	775.44
13	2	2	3	1	1	2883.67	3926.49	3894.12	2212.95	2467.63	797.59	3076.97
14	2	2	3	1	2	3309.66	3808.07	4569.22	2275.80	2710.26	902.77	3334.60
15	2	2	3	1	3	3545.72	4579.78	4467.53	3082.37	3853.09	628.54	3905.70
16	2	3	1	2	1	798.41	749.90	736.26	718.82	853.29	54.54	771.34
17	2	3	1	2	2	808.37	973.80	772.11	808.57	679.58	106.38	808.49
18	2	3	1	2	3	1043.53	1220.82	982.77	1017.33	1103.09	93.36	1073.51
19	3	1	3	2	1	6234.16	6374.85	7284.20	5853.21	6137.55	542.01	6376.79
20	3	1	3	2	2	6260.65	6795.46	6883.95	5943.93	6564.69	388.87	6489.74
21	3	1	3	2	3	6212.99	7030.34	6898.99	6295.30	6329.00	380.71	6553.32
22	3	2	1	3	1	3951.37	5446.79	3957.84	3431.15	3868.50	766.85	4131.13
23	3	2	1	3	2	4108.04	4347.49	4465.61	3589.12	4165.31	336.93	4135.11
24	3	2	1	3	3	3972.59	4829.06	5046.39	3350.97	3988.82	693.22	4237.57
25	3	3	2	1	1	7708.65	7945.03	8043.03	7388.45	7653.30	257.71	7747.69
26	3	3	2	1	2	7655.21	8124.79	7912.90	7471.08	7882.29	251.86	7809.25
27	3	3	2	1	3	7603.09	8031.24	8128.31	7430.62	8202.85	341.83	7879.22

logarithmic of data points can be used as a normalization technique. Normality is an assumption in inferential statistics. Other statistical methods can be used to treat nonnormal data. The normality assumption is made to conduct six-sigma calculations of productivity. A normal distribution fitting curve of individual JPH values is shown in Figure 9.7, and a residuals normality plot is shown in Figure 9.8.

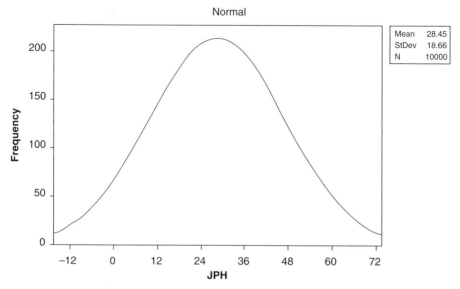

Figure 9.7   Histogram of individual JPH values.

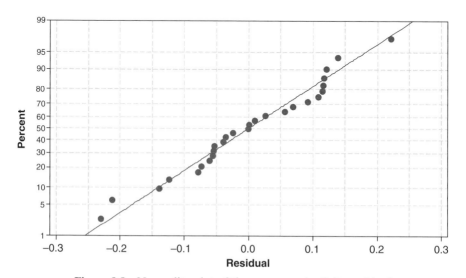

Figure 9.8   Normality plot of the mean productivity residuals.

The measure phase in the 3S-LSS approach strives to determine the process quality level based on productivity as a time-based performance measure. Hence, the normality assumption of JPH data justifies the applicability of many six-sigma calculations and statistical inferences.

It is important first to check the interactions among the five design parameters before analyzing their impacts on process mean and S/N ratio. As shown in Figure 9.9, strong interactions are evident among control factors since impact lines intersect and most lines are not parallel. Factor interaction often results in masking the main effects of control factors. Thus, the main effects of factors will be used to select best factor levels as a starting parameter design activity. This design will be enhanced later to include the effects of factor interactions.

It is also typical in experimental design to analyze the interactions between control factors and process internal and external noise factors (the outer array in Taguchi's experimental design). The DES model provides an advantage of modeling potential stochastic variability at various model elements. The impact of model noise factors is included in random generation from probability distributions and through simulation replications. All L_{27} designs are obtained under the same noise factors of downtime, arrival rates, and other sources of variability. Sources of variability that are typically related to operating conditions such as climate and external noise factors are not modeled. Since better designs result in higher S/N ratio values, the DOE analysis will be focused on selecting the best levels of design parameters at which the JPH S/N ratio is maximized.

The ANOVA and multiple linear regression models for both productivity means and S/N ratio are shown in Figure 9.10. The analysis shows that except for factor E (% rework), all other design factors have a significant impact on the productivity mean and S/N ratio with less than perfect linear model when interactions are not modeled.

To determine best levels of design factors based on main effects, we can use the response table for the production rate means and S/N ratio. Design factors are ranked based on their impact as shown in Table 9.9. The table shows that factor A (order size) has the most significant effect on both the productivity mean and S/N ratio. This is expected since the order size is used to feed the assembly line with assembly components, where larger orders create an accumulation of supplies that prevents shortages and maintains high productivity. Pushing more parts into the process results in a longer MLT and higher WIP level. Factor D (the size of buffer 2) has the next greatest effect on the productivity mean and S/N ratio. The factor with almost negligible impact on productivity is factor E (% rework).

Analyzing the impact of the five design parameters on productivity shows that factor A, order size, is the factor with highest impact on both the productivity mean and S/N ratio. However, this impact is not similar on both responses since high order size increases mean productivity and reduces the S/N ratio. Other factors vary slightly in their impacts, such as C and E. Factors B and D have a similar impact on both responses. This is also illustrated in the

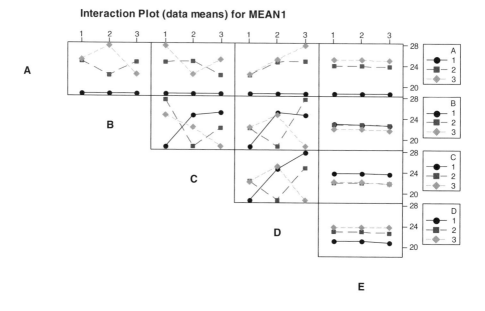

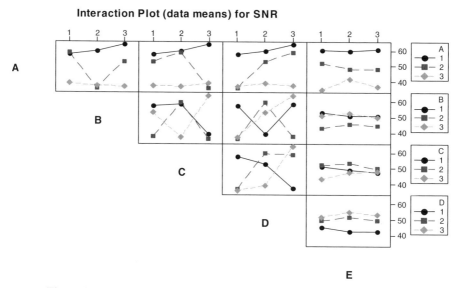

Figure 9.9   Interactions plots for the productivity mean and *S/N* ratio.

main effects plots on the productivity mean and *S/N* ratio, as shown in Figure 9.11.

Based on the plots of main effects on the productivity mean, the following level design of parameters are selected: (A: high, B: low, C: low, D: high, E: medium). This results in an average productivity value of 27.32 JPH and in an

Regression and Analysis of Variance for JPH Means:

```
Source            DF    Seq SS   Adj SS   Adj MS        F      P
A                  2   212.461  212.461  106.230  5431.47  0.000
B                  2     5.925    5.925    2.962   151.46  0.000
C                  2    18.466   18.466    9.233   472.07  0.000
D                  2    35.020   35.020   17.510   895.27  0.000
E                  2     0.203    0.203    0.102     5.19  0.018
Residual Error    16     0.313    0.313    0.020
Total             26   272.387
```

The regression equation is
MEAN1 = 16.3 + 3.23 A - 0.502 B - 0.855 C + 1.37 D - 0.100 E

```
Predictor     Coef  SE Coef      T      P
Constant    16.277    1.349  12.07  0.000
A            3.2277   0.2967  10.88  0.000
B           -0.5017   0.2967  -1.69  0.106
C           -0.8551   0.2967  -2.88  0.009
D            1.3688   0.2967   4.61  0.000
E           -0.1002   0.2967  -0.34  0.739
```

S = 1.25868   R-Sq = 87.8%   R-Sq(adj) = 84.9%

Regression Analysis of Variance

```
Source            DF       SS      MS      F      P
Regression         5  239.118  47.824  30.19  0.000
Residual Error    21   33.270   1.584
Total             26  272.387
```

Regression and Analysis of Variance for JPH S/N ratio:

```
Source  DF    Seq SS   Adj SS   Adj MS      F      P
A        2   2338.68  2338.68  1169.34  77.83  0.000
B        2    313.93   313.93   156.96  10.45  0.001
C        2    140.11   140.11    70.06   4.66  0.025
D        2    479.91   479.91   239.96  15.97  0.000
E        2      9.33     9.33     4.66   0.31  0.737
Error   16    240.38   240.38    15.02
Total   26   3522.34
```

S = 3.87603   R-Sq = 93.18%   R-Sq(adj) = 88.91%

The regression equation is
SNR = 65.8 - 11.4 A - 0.25 B - 1.50 C + 5.02 D - 0.25 E

```
Predictor      Coef  SE Coef      T      P
Constant     65.802    6.130  10.73  0.000
A           -11.398    1.349  -8.45  0.000
B            -0.247    1.349  -0.18  0.856
C            -1.502    1.349  -1.11  0.278
D             5.021    1.349   3.72  0.001
E            -0.247    1.349  -0.18  0.856
```

S = 5.72123   R-Sq = 80.5%   R-Sq(adj) = 75.8%

Figure 9.10   Results of Productivity Rate ANOVA and regression analysis.

TABLE 9.9   Response Tables for Productivity Means and S/N Ratio

	Response for Means				
Level	A	B	C	D	E
1	18.65	22.90	23.72	21.03	22.64
2	23.92	22.88	21.93	22.87	22.60
3	25.10	21.89	22.01	23.77	22.44
Delta	6.46	1.00	1.80	2.74	0.20
Rank	1	4	3	2	5

	Response for Signal-to-Noise Ratios[a]				
Level	A	B	C	D	E
1	60.38	51.71	49.20	43.34	48.91
2	49.21	44.24	51.77	50.45	49.84
3	37.58	51.21	46.19	53.38	48.42
Delta	22.80	7.47	5.57	10.04	1.42
Rank	1	3	4	2	5

[a] Nominal is best $[10\text{Log}(\bar{Y}^2/s^2)]$.

S/N ratio of 49.33. Based on the plots of the main effects on the productivity S/N ratio, the following level design of parameters is selected: (A: low, B: low, C: medium, D: high, E: medium). This results in low productivity (an average of only 18.66 JPH) and a high S/N ratio of 68.76. The latter corresponds to a tremendous production shortage and cannot be accepted practically. We can resolve the contradiction between the productivity mean and S/N ratio by running further experiments to include interactions and compromise the design selected (a design that leads an average JPH value of at least 25 with less variability). Further experiments resulted in arriving at the following design: (A: high, B: low, C: medium, D: high, E: medium). This compromised design results in an average value of 26.75 JPH and an S/N ratio of 46.21. This design is not one of L_{27}. Table 9.10 shows the results of Taguchi predictions and simulation confirmation runs at the three parameter designs.

9.3.4   Improve (DOE Optimized and Lean Applied)

In the improve phase of 3S-LSS, the focus is on optimizing the DOE parameters and then applying relevant lean techniques to enhance the time-based performance in terms of MLT and WIP. Process parameters are set based on their optimal levels obtained from DOE analysis in Section 9.3.3. The process simulation model is used to verify the DOE-improved design. Process structural changes are then made through the application of lean techniques. Similarly, the simulation model is used as a testing and verification platform for various lean techniques. Results of process parametric and structural improve-

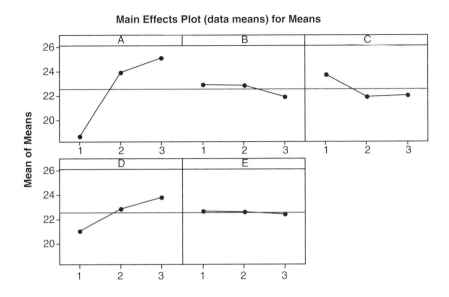

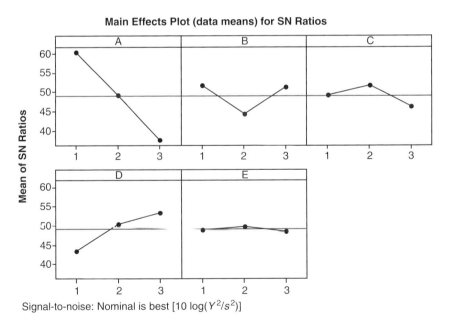

Figure 9.11 Main effects of control factors on productivity mean and *S/N* ratio.

TABLE 9.10   Results of Taguchi Predictions and Confirmation Runs

Factor levels for predictions			
A	3	1	3
B	1	1	1
C	1	2	2
D	3	3	3
E	1	2	2
Taguchi's predicted values			
S/N ratio	48.96	70.8506	48.06
Mean	27.91	19.6176	26.07
Simulation confirmation run			
S/N ratio	49.33	68.76	46.21
Mean	27.32	18.66	26.75

ments are shown by updating the dynamic VSM to develop a future state of the reengineered assembly process.

We start from the best DOE Taguchi design reached using the productivity rate as a process response. Using Taguchi's DOE to optimize MLT and WIP will result in a design that contradicts that reached for the productivity rate. For example, better throughput is achieved by increasing the order size factor and pumping more units into the process inventory and buffers. This, on the other hand, increases the average WIP level within the process. Higher WIP levels also correspond to longer MLT. MANOVA analysis can be used to develop a compromized parameter design based on the three lean measures. Instead of getting into this dilemma, therefore, we work on the best DOE design reached based on the productivity rate to improve the process performance in terms of WIP and MLT. The DOE-improved design entails the following settings of process variables:

- Order size = 250 units
- WIP1 buffer = 100 units
- Sequencer = 25 units
- WIP2 buffer = 150 units
- % Rework = 10%

The averages of the three lean measures at this DOE-optimal design are as follows:

- Production rate = 26.75 JPH (214 units/day)
- MLT = 4621.95 minutes (11.55 days of production)
- WIP = 3999.40 units (20 days of demand based on a demand of 200 units/day)

TABLE 9.11 Summary of DOE-Improved Simulation Results

Lean Measure	Simulation Run					Mean Value
	1	2	3	4	5	
Productivity (JPH)	25.94	26.10	28.03	26.36	27.34	26.75
MLT (min)	4052.63	3832.73	5284.56	4663.40	5276.41	4621.95
WIP (units)	3910.13	3353.26	3882.74	4341.71	4509.14	3999.40

TABLE 9.12 Summary of DOE-Improved Six-Sigma Calculations

Lean Measure	Production Rate	MLT	WIP
Defect specification	<25 JPH	>4800 min	>5000 units
Mean	26.75 JPH	4621.95 min	3999.40 units
Standard deviation	8.72 JPH	1885.07 min	451.69 units
Units produced (U)	10,000 hours	10,000 hours	10,000 hours
Opportunities (O)	1	1	1
Total opportunities (TOP)	10,000	10,000	10,000
Defects (D)	2479	5031	3518
Defect %	24.79%	50.31%	35.18%
Defect per unit (DPU)	0.2479	0.5031	0.3518
DPMO	247,900	503,100	351,800
Process yield (Y)	75.21%	49.69%	64.82%
Z-score	0.681	−0.008	0.381
Six-sigma quality level[a]	2.181	1.492	1.881

[a] Short term with 1.5σ shift.

Table 9.11 shows a summary of the DOE-improved simulation results at the three lean measures with five simulation replications. Compared to results of simulation runs at the current state, results indicate less stochastic variability.

Using the simulation results, the six-sigma measures can be recalculated for the DOE-improved design as shown in Table 9.12. The six-sigma results in the table show clearly the substantial improvement achieved through enhancing the process parameter design and reducing process variability. Process mean productivity is adjusted slightly toward the 25 JPH target. Significant reduction in MLT and WIP means is achieved (44.82% and 50.17%, respectively). This is also combined with 53.27%, 54.15%, and 86.93% reduction in the standard deviation of the production rate, MLT, and WIP, respectively. As a result, the number of defects is reduced significantly and the process six-sigma level is improved substantially. The updated dynamic VSM at the DOE-improved design is shown in Figure 9.12.

Comparing the new VSM to the current-state VSM in Figure 9.5 reveals significant improvement in MLT and WIP at the productivity-based DOE. These two time-based measures are the focus of lean techniques since they

Line 1

CT	60 sec
C/O	30 min
Lot	100
Avail time	480 min
% Busy	45.76
% Down	17.67
% Blocked	34.84
% Starved	0.35
% Set up	1.38
SAT	48.57
Stations	25
Avg Time	38.56

Delivery Buffer

Max	7833
Min	0
Avg Units	3219.44
Avg Time	2215.65

Line 2

CT	60 sec
C/O	60 min
Lot	100
Avail time	480 min
% Busy	50.25
% Down	15.80
% Blocked	30.53
% Starved	0.67
% Set up	2.75
SAT	48.87
Stations	20
Avg Time	29.95

Sequencer

Max	10
Min	0
Avg Units	8.42
Avg Time	747.28
Rework %	15%

WIP-Buffer 1

Max	100
Min	0
Avg Units	88.35
Avg Time	193.06

Line 3

CT	60 sec
C/O	45 min
Lot	100
Avail time	480 min
% Busy	48.17
% Down	18.92
% Blocked	6.87
% Starved	23.90
% Set up	2.14
SAT	47.36
Stations	35
Avg Time	53.14

WIP-Buffer 2

Max	150
Min	0
Avg Units	72.26
Avg Time	149.63

Line 4

CT	60 sec
C/O	25 min
Lot	100
Avail time	480 min
% Busy	53.11
% Down	36.99
% Blocked	0.00
% Starved	8.71
% Set up	1.19
SAT	37.09
Stations	50
Avg Time	73.45

Shipping Buffer

Max	1453
Min	0
Avg Units	208.72
Rework %	15%
Avg Time	936.23

Shipping

CT	60 min
C/O	25 min
Lot	250
Avail time	480 min
% Busy	13.60
% Down	47.26
% Blocked	0.00
% Filling	39.14
% Set up	0.00
SAT	31.65
Stations	1
Avg Time	185 min

Figure 9.12 DOE-improved dynamic VSM.

provide indications on process speed, waste, and effectiveness. Thus, further improvement can still be achieved in MLT and WIP through lean techniques. Lean techniques will also target the relatively high variability in productivity that is caused by various sources of variability in the assembly process [i.e., downtime (MTBF and MTTR), % rework, and % out of sequence].

We begin the lean techniques application by analyzing the assembly process based on the seven types of wastes. Clearly, our objective is to identify and eliminate or reduce waste. These wastes contribute to the majority of the non-value-added elements in the process VSM. In our case, these wastes include the following:

1. *Overproduction.* An excess of about 1.75 JPH is produced over the customer demand required (25 JPH). Some of the production is even produced before it is actually needed. In the long run, overproduction results in accumulating finished items inventory, which increases the inventory holding cost, WIP level, and MLT.

2. *Transportation.* Material flow in the raw material warehouse can be improved to reduce material withdrawal time and cost. Some production lines are long (e.g., line 4 has 50 stations) with multiple idle transfer stations. Eliminating these stations will reduce the flow product time. Also, some of the assembly supplies (bolts, nuts, screws, etc.) need to be located closer to their point of use.

3. *Motion.* The macro-level nature of VSM should not make us ignore the micro-movements performed by workers before, during, and after processing. A study of work methods and procedures through a detailed motion and time study is recommended at manual operations to increase motion economics and reduce cycle time and waste.

4. *Waiting.* Some input components wait for days in the raw material warehouse before being used at the assembly lines. This is caused primarily by large order quantity, high safety stock, supply fluctuations, and poor production scheduling. There is also a long delay for finished items before they are shipped to customers. This is usually caused by the push production system, demand fluctuations, high safety stock, and poor forecasting. Within production, parts spend a long time in WIP inventory, due to congestion, rework, and frequent downtimes. The resequencing process is long and the delay in the rework area is also high. The measures taken to reduce waiting time include the reduction of inventory order size and production batch size, the switch to a pull production system driven by steady demand, at-source process quality measures to reduce rework, and effective maintenance through TPM to increase availability and reduce downtime. Better methods of item resequencing and rework routing will also reduce the waiting times of parts in the flow.

5. *Processing.* To eliminate non-value-added processing, a thorough study of each assembly station in the four assembly lines is required to identify areas of excessive use of materials, tools, and equipment. Work balancing and

standardization will also identify these opportunities for waste reduction. Documentation coupled with the development of work standards is a key lean measure to check the viability of work elements, adjust process sequence, and eliminate or combine extra steps in the work procedure.

6. *Inventory.* The process still suffers from the accumulation of raw materials, WIP, and finished items. As mentioned earlier, a large inventory of raw materials is caused mostly by large order quantity, high safety stock, and poor production scheduling. Accumulation of finished items is usually caused by the push production system, demand fluctuations, high safety stock, and poor forecasting. Finally, WIP accumulation is due to rework and frequent downtimes. Hence, the same measures taken to reduce waiting time will reduce inventory.

7. *Defects.* The process suffers from a high potential for making defective products (reworked or scrapped). The 5 to 15% rework and the long downtime of assembly lines (especially line 4) represent an opportunity to further reduce defects and rework. In-station-process controls can be utilized to reduce scrap and defects from within the assembly lines, and TPM can be used to reduce downtime and increase the availability of assembly lines.

In summary, the analysis of the seven types of process wastes has led to recommending use of the following lean techniques to improve MLT and WIP while keeping productivity high enough to prevent shortages in delivery:

- Determining the takt time based on the customer demand
- Balancing and standardizing assembly lines to achieve a takt-paced production and to increase utilization
- Changing the push production system to a JIT/pull production system using Kanbans of standard small lot size
- Reducing inventory level by order size and minimizing WIP flow
- Applying SMED for setup reduction of production lines
- Reconfiguring layout to reduce travel time
- Applying in-station quality control to reduce rework and sequence interruptions
- Applying total productive maintenance to reduce downtime and increase the availability of assembly lines, especially line 4.
- Reducing rework percentage by using at-source quality controls
- Applying the 5Ss to clean and organize the workplace
- Applying a Kaizen plan for continuous improvement

Takt time, which is the elapsed time between units produced, is determined and set based on customer demand (i.e., 200 units/day). This translates into 25 JPH or 2.4 minutes of takt time. Thus, the first lean action to apply is to set the assembly process so that a unit is assembled every 2.4 minutes. We should, however, keep in mind that obtaining a unit every 2.4 minutes on average represents a net productivity of the assembly process under setup, downtime, rework, and other sources of process variability. The four assembly lines run at the same gross throughput of 60 JPH (i.e., they are capable of producing a unit every minute under ideal conditions). Due to setup, downtime, starving, blocking, and rework, the process produces an average of 26.75 JPH. This rate is above what is actually needed by customers (i.e., the 25 JPH or 200 units/day). It is, therefore, necessary for the assembly process to eliminate the 1.75 JPH overproduction in order to become demand-driven. This is achieved through balancing and standardizing the four assembly lines to achieve a takt-paced production and increase utilization.

Instead of slowing assembly lines, we can balance work by first controlling the material flow. There are three sizes of material movement within the assembly process. For raw materials, DOE resulted in an order size of 250 units of assembly components. For WIP items, production is moved on assembly lines using batch sizes of 100. Finally, the process ships finished items to customers in packaged batches of 200 units each. This imbalance in the material flow is a major cause of overproduction, inventory accumulation, and delays. If we simply slow down production to force the process to generate 200 units every 8 hours without adjusting material flow, we are likely to end up with a large amount of inventory accumulated before assembly lines and in a raw material warehouse. This is likely to result in higher values of MLT and WIP.

A better lean technique is to return to the process and reduce the order size to 200 in order to match a demand rate of 200 units/day. This action results in the following averages of lean measures: 24.93 JPH, 1648.50 MLT, and 1800.41 WIP. This is a perfect case since it almost matches the daily requirements and drastically reduces the MLT and WIP levels. Of course, the impacts of the action proposed will be verified through simulation to make sure that shortages in deliveries are minimized, if not eliminated. This solution was not preferred by DOE since the focus was on maximizing the mean productivity. As discussed in the DOE section, reducing the supply to a lower level (i.e., 150 units/day) results in low productivity (18.7 JPH on average), which creates tremendous shortages in the amount shipped to customers.

Changing the push production system to a JIT pull production system using Kanbans of standard small lot size is another major lean technique that is typically applied to control material flow and eliminate or reduce overproduction and excessive inventory. As discussed in Chapter 2, the pull philosophy is aimed at making sure that only the assembly components needed are pulled, only the WIP needed is advanced, and only the finished items needed are shipped. This philosophy is materialized by using Kanbans to control material flow and prevent excessive inventory.

The process simulation model can be utilized to test several scenarios for implementing a JIT/Kanban lean technique. For example, we can test an ideal pull system with a one-unit flow (lot size = 1) and without using buffers. Running a process simulation under these conditions results in a significant drop in an average throughput as low as 11.4 JPH. Clearly, such a throughput cannot meet the demand requirements. Due to minimizing material flow, low MLT (670.4 minutes) and WIP (480.6 units) averages are obtained. The drop in productivity is likely to occur when eliminating buffers under elements of randomness and production interruptions caused by downtime, rework, setup, and so on. Thus, a compromise is needed to implement the pull strategy with a certain lot size and with a certain level of buffers.

JIT implementation in the assembly process requires using a Kanban (card) system to signal parts withdrawal and authorize product assembly. As discussed in Chapter 2, a Kanban is a card that is attached to a storage and transport container. It identifies the part number and container capacity, along with other relevant information. To attain the benefits of a pull production system, therefore, Kanban containers are used to better control material flow and limit the WIP level in the assembly process. Each Kanban container represents the minimum production lot size.

Kanbans are used to pull from WIP buffers and to pull from assembly lines. A transport Kanban (T-Kanban) authorizes the withdrawal and movement of a container of material. This will be the case when line 1 withdraws assembly components from input storage. A production Kanban (P-Kanban) authorizes the assembly of a container of assemblies. This will be the case when shipping pulls from the last assembly line (line 4). If there is an intermediate buffer, a two-card system is used. This will be the case when one assembly line pulls from another line upstream. For example, as shown in Figure 9.13, one card (T-Kanban) circulates between the user (line 2) and the WIP buffer, and the other card (P-Kanban) circulates between the WIP buffer and the producer (line 1). Figure 9.13 demonstrates the idea of using two Kanbans (cards) to control material flow within the assembly process.

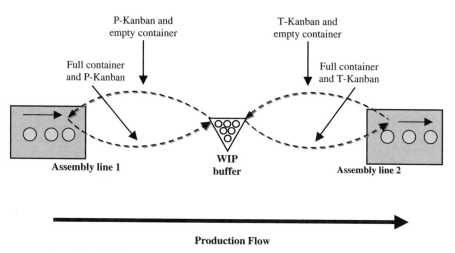

Figure 9.13   Using a two-card Kanban system in the assembly process.

Kanbans are used in the assembly process to guarantee material replenishment with the lowest possible inventory. The maximum WIP level is determined by the number of cards (containers) that are in circulation in the assembly process. The number of Kanbans between a user and a producer is determined as follows:

$$\frac{\text{demand rate} \times \text{lead time for the container cycle}}{\text{container capacity}}$$

The lead time for the container cycle includes the elapsed time for container filling and retrieval multiplied by a safety factor. For example, in the assembly process, products are shipped based on a demand of 200 units/day. This demand rate corresponds to 8 hours of lead time (assuming that line 4 produces 25 JPH). If we decide to use a container capacity of 100 units, the lead time for the container cycle will be (100 units × 2.4 min/unit = 240 minutes).

Container capacity is typically determined based on the economic lot size. Sometimes, the capacity is decided based on container weight, size, and the material-handling system used. A key requirement in this regard is to use a standardized container that protects its content with least waste and cost. For example, a container of order size capacity (200 units) is used to move raw materials to the start of assembly process. A container of lot size capacity (50 units) is used to move WIP units between assembly lines. Finally, a container of material-handling unit load capacity (100 units) is used to move finished items to packaging and shipping.

In the balanced assembly process, since demand rate is aligned with production rate, one container (size = 200) is needed to pull materials from inventory, and another (size = 100) is needed to move finished items to shipping. Within assembly lines and because of intermediate buffers, two containers (size = 50) are needed, one for transporting WIP units from the WIP buffer to the next line and another to move WIP units from the line to the WIP buffer. A single card is used for withdrawing raw material and finished items and a two-card system is used for pulling WIP units. Assembly lines will produce the desired components only when they receive a card and an empty container indicating the demand for more parts.

Other requirements for implementing an effective Kanban pull production system in the assembly process include the following:

- Creating a balanced workload on all assembly lines through a uniform and stable daily production
- Reducing setup and changeover time
- Reducing the order size to 200 units/order
- Implementing a single-unit flow at assembly lines
- Implementing better planning of production resources to meet the needs of the production process

- Reducing production and supply lead times by reconfiguring the production process so that travel and movements are reduced, receiving and shipping points are closer, and workstations are closer (this improves the coordination between successive processes)
- Eliminating the sequencer area by taking measures to prevent out-of-sequence issues
- Reducing rework to 1%

Along with balanced flow in a pull production system, standardized work procedures are applied to make sure that assembly tasks are organized in the best sequence and to ensure that products are being assembled using the best method every time. Standardized work procedures bring consistency (and as a result, better quality) to the workplace, and therefore they should be documented and shared with task-relevant personnel and used in training, team building, and benchmarking.

The SMED lean technique is also applied by observing changeover frequency and analyzing current model changeover procedures and practices. SMED effort is focused on changing the way in which changeovers are performed in order to reduce waste in changeover materials, effort, and time. SMED is applied as follows:

- Reduce the setup frequency by reducing the model changes and increasing the back-to-back flow of each product type.
- Reduce the setup time (preferably below 10 minutes) by analyzing setup components and eliminating unneeded elements.

The assembly process tackles three product models. Frequent changeovers therefore take place to meet the product mix demand requirements. The objective of SMED application to the assembly process is to increase the back-to-back production of each product model to 200 units to cover the one-shift demand. SMED application also recommends performing setup and cleaning and calibration prior to the start of each shift.

To reduce the setup time below 10 minutes at the four assembly lines, the setup time is analyzed and reduced using the SMED technique. Table 9.13

TABLE 9.13   Results of SMED Application to Line 4

Activity	Current Time	SMED Action	New Time
Setup preparation	5 min	Checklist	1 min
Removal or installation	20 min	Standardized tools	5 min
Setting or calibration	15 min	Visual controls	3 min
Validation trials	5 min	Elimination	0 min
Avg. total setup time	45 min		9 min

shows the results of SMED application to line 4. In a similar manner, SMED is applied to the other three assembly lines to reduce the setup time to 9 minutes.

Several other SMED means, actions, and tool kits can also be used to help reduce the setup time in the assembly process and to maintain low setup frequency and duration. This may include the following:

- Visual controls
- Checklist
- Specially designed setup cart
- Overhang tools
- Quick fasteners and clamping tools
- Standardized dies
- Stoppers and locating pins

Maintaining a certain level of raw material inventory in storage is aimed at making assembly components instantly available at their point of use. Similarly, finished items inventory is aimed at making them available instantly as at their point of delivery. The goal is to reduce out-of-stock items, shortages, delay, search time, travel, and material handling. Making products ready for delivery per customer order reduces delivery time, allows taking new orders (unplanned), and increases customer satisfaction.

An effective inventory control systems is therefore essential to be coupled with the pull production system in an assembly process to reduce the overall process inventory while meeting demand requirements. Effective inventory control measures need to be taken at the three types of inventory in the assembly process (inputs, WIP, and finished items) to establish an effective value chain (supply chain, production flow, and shipping). The following provides guidelines for better inventory controls:

1. For input inventory, we need to select the inventory control system that results in lowest overall inventory levels and least annual inventory cost. Mathematical inventory models are available to determine the inventory parameters at which total inventory cost is reduced. For example, a continuous review Q-system can be used to determine the economic order quantity and the reorder point so that a total holding and ordering cost is minimized. A periodic-review P-system can also be used to determine best time between orders and inventory level at which total cost is minimized. The inventory level often includes the quantity needed to cover expected demand during lead time plus a certain safety stock that is determined based on the demand history (distribution). Simulation can be used to model inventory situations where assumptions required for applying mathematical models do not hold.

2. For WIP inventory, we need to track the flow of components within the assembly process and apply a Kanban system and visual controls to

streamline the flow and avoid excess inventory. Simulation results in the dynamic VSM play a major role in identifying areas of excessive inventory (bottlenecks) and impending flow. Theory-of-constraints techniques, which can be used to resolve system bottlenecks, suggest exhausting the effort in removing intangible constraints such as logical design and work procedures before investing capital in more resources to increase capacity.

3. For finished-items inventory, we need to enhance the demand forecasting system, apply the pull production system, and avoid the make-to-stock policy. Reducing the lead time for order fulfillment can also help reduce safety stocks of finished items. This may require developing an effective distribution channel to deliver products to customers. Along with that we should maintain a continuous effort to identify obsolete and defective inventory and eliminate situations that cause future buildup of inventory.

Production interruptions due to failures in assembly stations is a major source of process variability as seen from the results of production rate, MLT, and WIP. A key lean objective in the 3S-LSS approach is to reduce response variability and increase performance consistency. To this, end, total productive maintenance (TPM) is a lean technique that is often applied to reduce downtime, increase the reliability of assembly lines, and increase the overall system availability. Implementing TPM requires the development of both preventive and corrective maintenance plans with a focus on predicting and preventing potential failures. In cases where failures occur, TPM focuses on methods and tools to reduce repair time and minimize downtime impact on system performance.

Lean application to the assembly process is also focused at reducing rework percentage to a maximum of 1% to increase consistency and reduce flow time and process waste. One measure to reduce rework and scrap is to apply in-process quality controls or quality-at-source. This requires providing employees with time, tools, and means to check the quality of assemblies and subassemblies at each assembly station before sending them to the next assembly station. Examples of means necessary for in-process quality control include gauges, tools, and other inspection equipment. Other examples include providing workers with sample parts, mating parts, pictures, and other visual items to check the specifications of their products before sending them over.

In summary, applying to the case example the lean techniques discussed has led to making the following changes to the assembly. These changes are implemented into the process simulation model to develop a future-state dynamic VSM:

1. Pull system: apply a demand-driven assembly based on 200 units/day. Use Kanbans to control and minimize material flow.
2. Inventory control: reduce order size to 200 and receive input supplies daily.
3. Apply a single-unit flow at each assembly line.

4. Increase the in-sequence percentage to 100% through a strict mechanism to substitute out-of-sequence items.

5. SMED: reduce setup time at all assembly lines to 9 minutes.

6. In-process quality control: apply at-source quality measures to reduce rework to 1%.

7. TPM: apply TPM to reduce downtime at assembly lines by 50%.

8. No sequencing area: eliminate out-of-sequence units flow and delay.

After introducing the foregoing changes to the assembly process DES model, five simulation replications were run to estimate the three lean measures (production rate, MLT, and WIP). Many signs of drastic improvement in the three lean measures were observed from process animation (smooth entity flow, low inventory levels, less variability in productivity levels, fewer shortages in shipments, etc.). Simulation statistics show that the average productivity value is further adjusted to the demand-based level of 25 JPH with less variability. The new productivity mean is 24.93 JPH. Production consistency is improved mainly due to the balanced workflow and the demand-driven pull/Kanban system. Variability in production levels is reduced as a result of reducing setup time and applying TPM and in-process quality controls. MLT and WIP are substantially reduced as a result of reducing order size, inventory levels, and using single-unit flow at assembly lines. The new mean for MLT is 1209.23 minutes (about half week) and the new mean WIP level is 506.02 units (about a half week of demand). Lean-improved simulation results are shown in Table 9.14.

Using simulation results for the lean-improved process, six-sigma measures are recalculated at the three lean measures (production rate, MLT, and WIP). Table 9.15 summarizes the results of six-sigma calculations for the lean-improved process. By maintaining the 25 JPH as a lower specifications limit on productivity, only 673 production hours report shortages in the lean-improved process. This is compared to 2479 defects in the DOE-improved process. This results in upgrading the process six-sigma quality level from 2.181 JPH to about 3σ.

When using the initial specifications for MLT and WIP (4800 minutes and 5000 units, respectively), there were no defects to report due to the drastic drop in MLT and WIP that was achieved after the application of lean

TABLE 9.14 Summary of Lean-Improved Simulation Results

Lean Measure	Simulation Run					Mean Value
	1	2	3	4	5	
Productivity (JPH)	24.93	24.94	24.92	24.94	24.94	24.93
MLT (min)	1158.53	1153.50	1318.94	1156.44	1258.74	1209.23
WIP (units)	506.37	512.45	479.78	511.45	520.02	506.02

TABLE 9.15   Results of Lean-Improved Six-Sigma Calculations

Lean Measure	JPH	MLT	WIP
Defect specification	<25 JPH	>1200 min	>600 units
Mean	24.93 JPH	1209.23 min	506.02 units
Standard deviation	4.93 JPH	124.44 min	221.20 units
Units produced (U)	10,000 hours	10,000 hours	10,000 hours
Opportunities (O)	1	1	1
Total opportunities (TOP)	10,000	10,000	10,000
Defects (D)	673	334	365
Defect %	6.73%	3.34%	3.65%
Defect per unit (DPU)	0.0673	0.0334	0.0365
DPMO	67,300	33,400	36,500
Process yield (Y)	93.27%	96.66%	96.35%
Z-score	1.496	1.833	1.793
Sigma quality level[a]	2.996	3.333	3.293

[a] Short term with 1.5σ shift.

techniques. Both measures, therefore, report a better than six-sigma quality level. Hence, it was decided to further improve the initial specifications of MLT and WIP by reducing their upper specification limits to 1200 minutes and 600 units, respectively. This can be viewed as a paradigm shift in the process toward excellence and lean performance. At these new specifications, the process reports 334 MLT defects and 365 WIP defects. This corresponds to 3.33σ and 3.29σ levels, respectively.

It is essential to assert the fact that reaching a 3σ quality level in time-based performance measures such as production rate, MLT, and WIP is in fact a substantial improvement over the current state of the assembly process. After applying 3S-LSS, the 3σ quality level is achieved in system-level measures of transactional processes. This should not be compared to achieving a 3σ quality level in tangible products specifications, where 3σ may not be satisfactory. In practice, 3σ processes are exceptional, since they set high business standards and achieve world-class quality in time-based metrics, which are often difficult to measure and control. This drastic process change will definitely materialize into huge cost savings, increased agility, and stronger competitiveness.

Finally, the lean-improved assembly process is used to develop a future-state dynamic VSM. Simulation statistics at the final process configuration are used to update the dynamic VSM. The future-state VSM is shown in Figure 9.14.

9.3.5   Control (DMAIC Controls and Lean 5S and Kaizen)

In this final phase of the 3S-LSS approach, we specify and test effective monitoring plans and control schemes that are aimed at maintaining the effectiveness achieved and the high sigma rating. Without control and monitoring,

Line 1

CT	60 sec
C/O	30 min
Lot	1
Avail time	480 min
% Busy	41.67
% Down	9.00
% Blocked	7.96
% Starved	41.00
% Set up	0.38
SAT	54.37
Stations	25
Avg Time	39.58

Line 2

CT	60 sec
C/O	60 min
Lot	1
Avail time	480 m in
% Busy	0.00
% Down	7.34
% Blocked	46.98
% Starved	45.30
% Set up	0.38
SAT	55.37
Stations	20
Avg Time	40.00

Sequencer

Max	10
Min	0
Avg Units	2.79
Avg Time	62.79
Rework %	15%

Shipping

CT	60 min
C/O	25 min
Lot	250
Avail time	480 min
% Busy	12.30
% Down	41.48
% Blocked	0.00
% Filling	46.22
% Set up	0.00
SAT	35.11
Stations	1
Avg Time	185 min

Delivery Buffer

Max	202
Min	0
Avg Units	62.06
Avg Time	378.17

WIP-Buffer 1

Max	50
Min	0
Avg Units	22.82
Avg Time	54.76

Shipping Buffer

Max	870
Min	0
Avg Units	107.5 6
Rework %	15%
Avg Time	258.1 6

Line 4

CT	60 sec
C/O	25 min
Lot	1
Avail time	480 min
% Busy	0.00
% Down	18.04
% Blocked	0.00
% Starved	81.58
% Set up	0.38
SAT	48.95
Stations	50
Avg Time	100.00

WIP-Buffer 2

Max	50
Min	0
Avg Units	14.9 5
Avg Time	35.5 0

Line 3

CT	60 sec
C/O	45 min
Lot	1
Avail time	480 min
% Busy	42.11
% Down	10.06
% Blocked	2.92
% Starved	44.54
% Set up	0.38
SAT	53.74
Stations	35
Avg Time	55.26

Figure 9.14 Lean-improved future-state dynamic VSM.

the process may start to lose some of the enormous gains obtained in the three lean measures (production rate, MLT, and WIP). The objective is to build on the momentum gained from six-sigma and lean applications and to set the process attitude toward excellence and continuous improvement. The process simulation model and the future-state VSM are excellent tools to verify the impact of process controls and monitoring plans.

Several six-sigma control tools can be used to maintain high sigma ratings of the assembly process. Examples include the following:

- Error-proofing measures to prevent passing defects and creating out-of-sequence units. Various in-station process controls and visual aids can be used to eliminate defects early in the process and guide production toward achieving a 100% built-to-sequence orders.
- Process failure mode and effect analyses to reduce the failure rates and increase the availability of assembly lines.
- Statistical process control (SPC) to monitor process performance in terms of production rate, MLT, and WIP. Simulation can be used to generate data needed to develop prediction control charts. These charts can be compared to SPC charts that are developed based on actual production data.

Similarly, several lean techniques can be applied for sustaining the leanness of the assembly process and maintaining its effectiveness. This can be achieved primarily by implementing the 5S method periodically and establishing Kaizen plans for continuous improvement. This should be accompanied with work standardization and proper documentation. Hence, controlling the assembly process with lean techniques includes applying the following:

- *Work documentation.* Documenting best work procedures and best practices for maintenance, quality control, and safety, and training workers on continuous use of these documents. These documents should be made available to workers and not kept in manuals for display and audits.
- *Work standardization.* Developing work standards for each workstation of the four assembly lines based on simulation-generated data, work sampling, direct observations, and motion and time studies.
- *Visual controls.* Visual aids are an essential element in the lean assembly process to support JIT/Kanban production, to schedule and sequence production, and to coordinate maintenance and quality checks.
- *5S application.* The 5S method is applied at each workstation in the assembly process to establish a lean infrastructure and facilitate application of various lean techniques.
- *Kaizen.* Continuous improvement plans are set to review the assembly process performance periodically, identify potential problems, and implement improvement action.

9.4   SUMMARY

In this chapter we presented an application of the 3S-LSS approach to an assembly process case study. The application provided details on the 3S-LSS road map presented in Chapter 8 with a focus on the practical aspects of the 3S-LSS approach. The principal aspects discussed include project scoping and process abstraction, defining the process time-based measures, checking the applicability of the 3S-LSS approach, static and dynamic value stream mapping, building a process discrete event simulation model, conducting a simulation-based statistical analyses (DOE, ANOVA, and regression) to improve the process parametric design, applying simulation-based lean techniques to improve the process flow and structure, and measuring and recording progress with six-sigma metrics and VSM.

The 3S-LSS application to the case study demonstrated many process features and benefits of high practical value. It was shown that a significant increase in the six-sigma quality level can be achieved in the typically difficult to measure and hard to control time-based process metrics. The case study was focused on measuring and improving process productivity, lead time, and inventory level. Results showed that a simulation-based dynamic VSM is a useful 3S-LSS tool for conducting six-sigma analyses and for implementing lean techniques. As a data-driven representation of process structure and parameters, the tool facilitates the 3S-LSS application to reengineer transactional processes in both manufacturing and services.

10

SIMULATION-BASED DESIGN FOR SIX-SIGMA APPLICATION

10.1 INTRODUCTION

In this chapter we cover the DFSS road map specific elements as adopted from El-Haik and Roy (2005), which led to the successful deployment of a simulation-based six-sigma redesign in a medical clinic legacy environment. The selection of this case study favors the service industry; however, the process can be replicated on any process of transactional basis, including batch and lean manufacturing. This redesign represented a change in the structure of a dental clinic with greater insight into its management. The clinic cited here is one of many owned by a multistate chain health provider in the eastern United States.

As discussed in Chapter 3, the DFSS combines design analysis (e.g., requirement cascading) with design synthesis (e.g., process engineering) within a framework of a DFSS development plan within the chain. Emphasis is placed on the identification, optimization, and verification of functional requirements (FRs) using the transfer function and scorecard vehicles. A transfer function in its simplest form is a mathematical relationship between the FRs and their mapped-to significant factors (called X's or design parameter). It is useful to predict FR performance by monitoring and recording their mean shifts and variability performance at the current datum (baseline design) and new future redesign. In this case study, no distinction is made between critical-to-

Simulation-Based Lean Six-Sigma and Design for Six-Sigma, by Basem El-Haik and Raid Al-Aomar
Copyright © 2006 John Wiley & Sons, Inc.

satisfaction or critical-to-quality (CTS/CTQ) and FRs for simplicity (see Chapter 3 and El-Haik and Roy, 2005).

DFSS is a disciplined and rigorous approach to service, process, and product design by ensuring that new design entities meet customer requirements at launch. It is a design approach that ensures complete understanding of process steps, capabilities, and performance measurements by using, in addition to transfer functions, *tollgate reviews* (see Section 8.4) to ensure accountability of all the design team members, belts (green, black, and master), project champions, and the remainder of the stakeholders.

The service DFSS road map (Figure 10.1) has four phases (identify, conceptualize, optimize, and validate, denoted ICOV) in seven developmental stages. Stages are separated by milestones called tollgates. A design stage constitutes a collection of design activities and can be bounded by entrance and exit tollgates. A tollgate represents a milestone in the service design cycle and has some formal meaning defined by the business and recognized by management. The ICOV stages are an average of the authors' studies of several deployments. It need not be adopted blindly but customized to reflect the deployment interest. For example, industry type, service production cycle, and volume are factors that can contribute to the shrinkage or elongation of some of the phases. Generally, the life cycle of a service or a process starts with some form of idea generation, whether in free invention format or using a more disciplined format such as multigeneration service planning and growth strategy.

The DFSS process objective is to attack the design vulnerabilities both conceptually and operationally by employing tools and methods for their elimination or reduction. DFSS is used to design or redesign a product or service. The process sigma level expected for a DFSS product or a service functional requirement is at least 4.5 (with normal distribution assumption) over the long term, but can be 6σ or higher, depending on the designed entity. The production of such a low defect level from product or service launch means that customer expectations and needs must be understood completely before a design can be operationalized. That is, quality is defined by the customer.

The material presented herein is intended to give the reader a high-level understanding of simulation-based DFSS and it uses and benefits, with concentration on simulation, basically the 3S process in a DFSS project environment. Following this chapter, readers should be able to assess how it could be used in relation to their jobs and to identify their needs for further learning.

In this chapter the DFSS tools are laid on top of four phases, as detailed in El-Haik and Roy (2005). The service DFSS approach can be phased into the following:

- Identify customer and design requirements (prescribe the functional requirements, design parameters, and corresponding process variables).
- Conceptualize (characterize) the concepts, specifications, and technical project risks.

- Optimize the design transfer functions and mitigate risks.
- Verify that the optimized design meets the intention (customer, regulatory, and business).

In this chapter, the ICOV and DFSS acronyms are used interchangeably.

10.2   3S-DFSS PROCESS

Figure 10.1 (which we saw earlier as Figure 8.13) provides the road map that we follow in this case study (see El-Haik and Roy, 2005). In the context of Figure 10.1, we demonstrate the progression from stage 1, idea creation, through all stages, with more emphasis placed on the first seven stages [i.e., transformation of the case study clinic via simulation-based DFSS to higher Z-scores (sigma values) in several functional requirements].

DFSS project scoping within the 3S approach is depicted in Figure 10.2. A simulation project flowchart is illustrated in Figure 10.3. By integrating the DFSS road map (Chapter 8) and the simulation procedure, we generate the simulation-based DFSS project road map that is used in this case study (Figure 10.4). Needless to say, deployment of the road map in the clinic case study was conducted by a black belt six-sigma team.

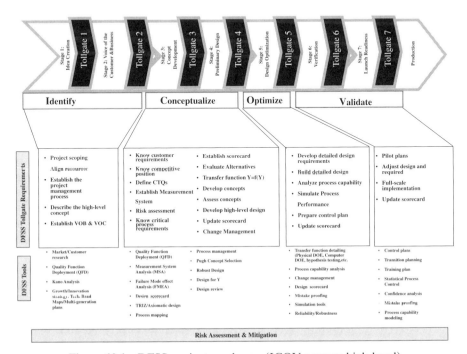

Figure 10.1   DFSS project road map (ICOV process high level).

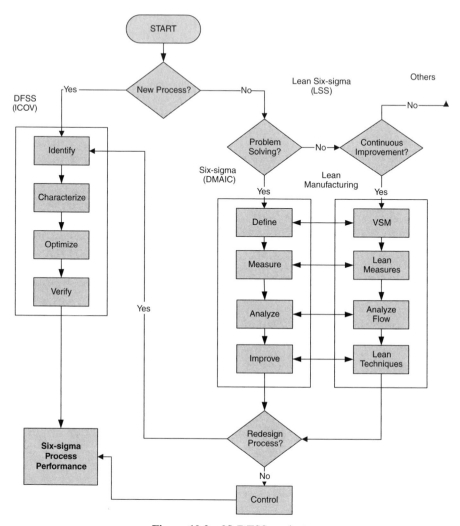

Figure 10.2   3S-DFSS project.

10.3   3S-DFSS CASE STUDY: DENTAL CLINIC REDESIGN

10.3.1   Stage 1: Idea Creation

The subject clinic understudy has a layout that is intended to provide primary care and preventive dental services in an outpatient environment. The medical equipment utilizes current technologies such as dental examination and operatory, general radiographic services with a dedicated dark room, and reception and medical records. The layout is shown in Figure 10.5. All equipment in the facility is patient-ready and from the same manufacturer, a policy managed by the chain of clinics. However, extra resources may be acquired

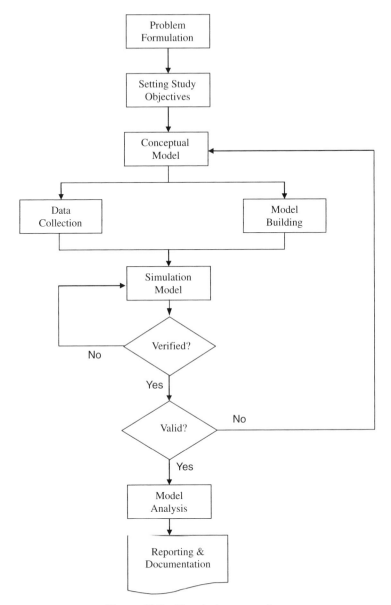

Figure 10.3 Simulation procedure.

for some fixed cost based on the results of this case study. For example, additional capacity can be gained by increased machine and human resources utilization, and enhanced up-time via reconditioned or remanufactured equipment. Employee training can be acquired, at an incremental cost, to administrative and medical staff, including clinical services, operations, finance and administration, general health care, and dental specialties at all levels. These skills and resources can be acquired to train the users to utilize the

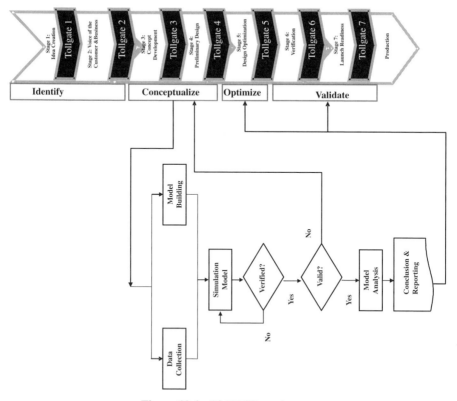

Figure 10.4 3S-DFSS road map.

optimum clinic redesign (following the DFSS project) and to transfer critical knowledge to the in-house professionals to realize the full benefit of the investment in clinic facilities and services.

An existing mother chain model of business for which our subject clinic belongs needs to be entertained. The chain management is in continuous demand to lower operation cost while growing the business. The clinic local management wanted to instill the same rigor in their businesses to leverage patient satisfaction while controlling the running cost. The latter would include rent, wages, and office space as well as the *quality loss*[1] of their services.

---

[1] Quality loss is the loss experienced by customers and society and is a function of how far performance deviates from target. The quality loss function (QLF) relates quality to cost and is considered a better evaluation system than the traditional binary treatment of quality (i.e., within/outside specifications). The QLF of a nominally-the-best FR has two components: mean (μ_{FR}) deviation from targeted performance value (T) and variability (σ_{FR}^2). It can be approximated by a quadratic polynomial of the functional requirement. Other forms of QLF depend on the direction of optimization, such as smaller-the-better (minimization) and larger-the-better (maximization). See Appendix D.

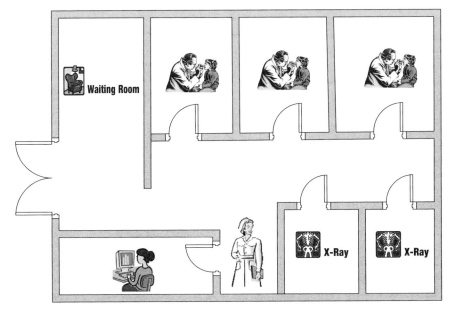

Figure 10.5   Clinic high-level layout.

10.3.2   Stage 2: Voice of the Customer and the Business

The case study black belt team realized that the actual customer base would include different demographics (adult and child patients as well as chain and local management) looking for results, a local management that wanted no increased operational cost on internal operations, space, and services. Through interviews laced with a dose of common sense, the following customer wants and needs were derived:

- Patient voices
  - Expedited processing
  - Less waiting time
  - Less time to get an appointment
- Clinic management voice
  - Maximum number of scheduled patients processed per day
  - Maximum number of walk-in patients processed per day
  - Reduced operating cost

Figure 10.6 shows the affinity diagram that was used to determine these high-level needs. You will notice that there are several comments under each high-level heading, representing the various voices of the business and patients of the process. These are entered in the voice-of-customer room in the house

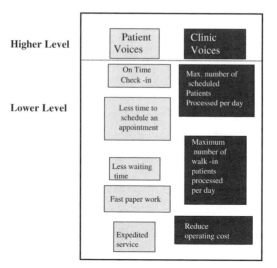

Figure 10.6   Clinic affinity diagram.

of quality. These voices were also processed through a Kano analysis in Figure 10.7 to determine their priority. Notice that the voice "less waiting time" is a delighter (a "wow" factor in this case study) and of medium importance, reflecting patients' lack of confidence of current performance or future accomplishments. The current management did not satisfy most of these voices, and this would be improved by the redesigned 3S-DFSS process.

These voices need to be translated to process metrics using the quality function deployment (QFD) process. QFD is a planning tool that allows the flow-down of high-level customer needs and wants through to design parameters and then to process variables critical to fulfilling the high-level needs. By following the QFD methodology (phase 1), relationships are explored between quality characteristics expressed by customers and substitute quality requirements expressed in engineering terms (Cohen, 1988, 1995). In the context of DFSS, we call these requirements *critical-to characteristics*, representing the customer perspectives in engineering measures. In the QFD methodology, customers define their wants and needs using their own expressions, which rarely carry any actionable technical terminology. The voice of the customer can be affinitized into a list of needs and wants that can be used as input to a relationship matrix, which is called QFD's *house of quality* (HOQ). Figure 10.8 depicts the generic HOQ.

Going room by room we see that the input is entered into room 1, in which we answer the question "what". The "whats" are either the results of VOC synthesis for HOQ 1 or a rotation of the "hows" from room 3 into the following HOQs. QFD can be sequenced into four phases,[2] as shown in Figure 10.9. The

---

[2] Only the first two QFD phases are used in the case study.

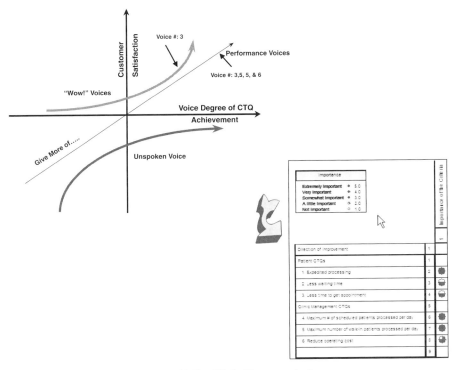

Figure 10.7   Clinic Kano analysis.

"whats" are rated in terms of their overall importance and placed in the importance column.

Next we move to room 2 and compare our performance and our competitors' performance against the "whats" in the eyes of the customer. This is
usually a subjective measure and is generally scaled from 1 to 5. A different
symbol is assigned to the various providers so that a graphical representation
is depicted in room 2. Next we must populate room 3 with the "hows." For
each "what" in room 1 we ask "how" we can fulfill this. We also indicate the
direction in which the improvement is required to satisfy the "what": maximize, minimize, or target, indicating the optimization direction desired. In the
first HOQ (HOQ 1), these become "how" the customer measures the "what."
Usually, there are four house of quality phases in a DFSS project. However,
in the clinic case study, we merged HOQ 1 and HOQ 2 for simplicity, which
enables us to treat the "hows" as the FRs. In this case study, in the following
house (HOQ 3), the "hows" become design parameters (DPs). A word of
caution: Teams involved in designing new services or processes often jump
right to specific solutions in HOQ 1. It is a challenge to stay solution-free until
the DPs. There are some rare circumstances where the VOC is a specific function that flows unchanged straight through each house.

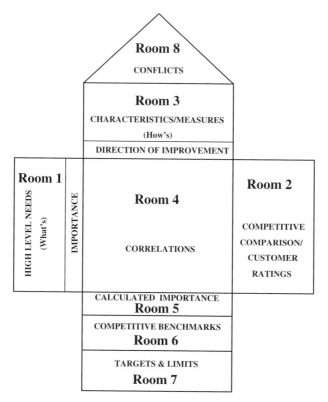

Figure 10.8 QFD house of quality.

Within room 4 we assign the weight of the relationship between each "what" and each "how" using 9 for strong, 3 for moderate, and 1 for weak. In the actual HOQ these weightings are depicted with graphical symbols, the most common being the solid circle for strong, an open circle for moderate, and a triangle for weak (Figure 10.10).

Once the relationship assignment is completed, by evaluating the relationship of every "what" to every "how," the importance can be derived by multiplying the weight of the relationship and the importance of the "what," then summing for each "how." This is the number in room 5. For each of the "hows" we can also derive quantifiable benchmark measures of the competition and the project; in the eyes of industry experts, this is what goes in room 6. In room 7 we can state the targets and limits of each of the "hows." Comparing the competitive assessments can help us check and confirm our thinking early in the DFSS process. The selection of the best target values needed to *win* is the reason we do QFD. This step should be approached with care, considering all of the information we have learned so far. Ultimately, the target values should reflect what is needed for a project to contribute to business objectives. Target setting of the FRs is a balancing activity among customer satisfaction, per-

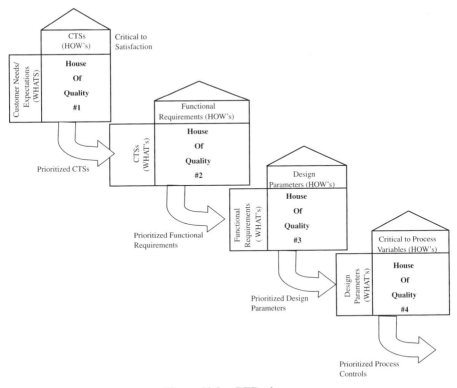

Figure 10.9   QFD phases.

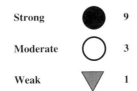

Figure 10.10   Rating values for affinities.

formance of competitor clinics in the area, and current operations capability. A high-level target-setting process is depicted in Figure 10.11, which implies the fact that to gain more satisfaction the efficiency of all processes contributing to "time per visit" should be improved. Notice the alignment between the room 2 and room 6 assessments. This can be achieved by redesigning the involved process via activities such as eliminating non-value-added steps and improving resource utilization.

Finally, in room 8, often called the *roof*, we assess the interrelationship of the "hows." If we were to maximize one of the "hows," what happens to the others? If it were also to improve in measure, we classify it as a *synergy*,

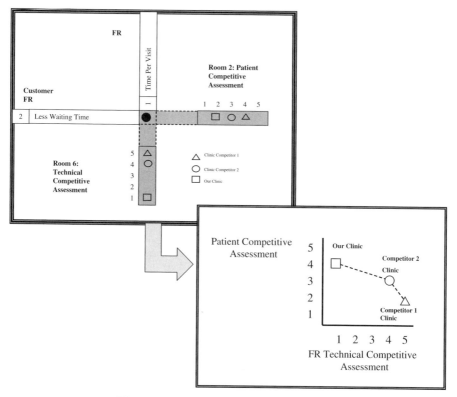

Figure 10.11 QFD target-setting process.

whereas if it were to move away from the direction of improvement, it would be classified as a *compromise*. Wherever a relationship does not exist, it is just left blank. For example, if we wanted to improve the time per visit by increasing the number of hygienists and other resources, the operating cost may degrade. This is clearly a compromise. That is, the clinic management may acquire experienced talent but at the cost of affordability. Although it would be ideal to have correlation and regression values for these relationships, often they are just based on common sense or business law. This completes each of the eight rooms in the HOQ. The next step is to sort based on the importance in room 1 and in room 5 and then to evaluate the HOQ for completeness and balance. We take the remaining elements and begin to build our HOQ 1, as shown in Figure 10.12.

At this stage, the black belt needs to think through what risks would be associated with transformation to a 3S-DFSS solution. In addition to the typical nontechnical risks of schedule and resources that are typically needed, risks in the categories of unstable playing field due to competition in the area, changing strategy, and change in patient base should be considered (e.g., migration to suburbs versus downtown). We suggest the employment of a

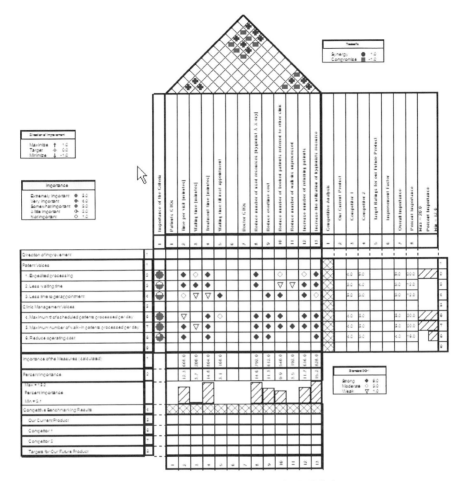

Figure 10.12 Clinic case study: HOQ 1.

design failure mode and effect analysis (DFMEA). The complete set of risks is shown in the failure mode effect analysis (FMEA) of Table 10.1. FMEA is a disciplined procedure that recognizes and evaluates the potential failure of a product or process and the effects of a failure and identifies actions that would reduce the chance of potential failure. FMEA helps the DFSS team improve its design and delivery processes by asking "what can go wrong?" and "where can variation come from?" Service design and production, delivery, and other processes are then revised to prevent the occurrence of failure modes and to reduce variation.

Following the normal method of treating all RPN values greater than 125 and all SEV or OCC values of 9 or higher, the following became the initial focal point for taking actions within the scope:

TABLE 10.1 Clinic Case Study FMEA

Risk Category	Potential Failure Mode	Potential Failure Effects	SEV	Potential Cause	OCC	Current Controls	DET	RPN
Management	Unstable environment	Lack of profit	7	Corporate directions	6	Strong linkage to chain	2	84
	Unstable environment	Lack of resources	9	Loss of customers	2	Monthly operations review	2	36
	Unstable environment	Lack of cost reduction	7	Expansion/improvement	8	Patient feedback reviews	2	112
	Unstable environment	Lack of resources	9	Recruitment process	8	Patient feedback reviews	2	144
Patients	Lack of appreciation	Continued dissatisfaction	7	Unwilling to give up feedback	8	Monthly report	4	224
	Unsatisfied	Negative support	9	Poor employee performance	8	None	8	576
Employees	Lack of qualified employees	Unsatisfied patients	7	Unable to qualify employees	3	None	4	84
	Lack of qualified employees	Diminished performance	6	Too small at new clinics	4	Detailed analysis of employees	2	48
	Moving costs	Diluted net savings	5	Long-term agreements	5	Employee notification	6	150
Technology	Availability	Lack of measurements	5	X-ray resources	5	X-ray project	3	75
	Affordable	Diluted net savings	5	Conversion/implementation	2	Appropriation process	2	20
Project	Lack of qualified resources	Delayed benefits	7	Unwilling to release employees	6	Personnel review	3	126
	Lack of qualified resources	Continued legacy operations	8	Unable to find qualified employees	3	Recruiters	2	48

[a] SEV stands for failure mode severity, and OCC stands for occurrence, a subjective measure for probability of failures.

- Unsatisfied patients due to poor service performance, resulting in dissatisfaction
- Lack of appreciation from patients, indicated by their unwillingness to give up feedback
- High switching costs to change employees due to long-term contracts or unique highly paid skills, which result in diluted savings: for example, the lack of critical resources (e.g., hygienist), which deflects assigned physician resources from leveraging their face-to-face time with patients
- Lack of qualified resources for a redesigned clinic model in the area of operations

The black belt needed to ensure that any new items that lowered the clinic's cost of operation maintained the level of quality performance to which the business was accustomed (Figure 10.12). The other items were deemed as acquiring human and nonhuman resources but no further action at this ICOV stage.

10.3.3 Stage 3: Concept Design

Quality function deployment is accomplished using a series of matrixes, called *houses*, to deploy critical customer needs through the phases of ICOV development. The QFD methodology is deployed through several phases (see El-Haik and Roy, 2005). We processed the first phase in Section 10.3.2. Here we present the second QFD house (Figure 10.13), cascading into the design parameters. It is interesting to note that the QFD is linked to voice of the customer (VOC) tools at the front end and then to design scorecards and customer satisfaction measures throughout the design effort. These linkages with adequate analysis provide the feed forward (requirements flow-down) and feed backward (capability flow-up) signals, which allow for the synthesis of design concepts. Each of these QFD phases deploys the house of quality, with the only content variation occurring in rooms 1 and 3 (Figure 10.9).

The FR measures and their targets and limits were determined from HOQ 1 to be using a process similar to that depicted in Figure 10.11[3]:

- FR1: time per visit, the total time inside per visit (lead time, minutes)
  - Measured: tracked in the clinic computer system forms
  - USL: 2 hours
  - Current average: 1.5 hours
  - Target: 1 hour
- FR2: waiting time, time in the waiting room prior to treatment (minutes)
  - USL: 30 minutes
  - Current average: 25 minutes
  - Target: 10 minutes

---

[3] Note that FR5 and FR6 are outside the scope of this case study.

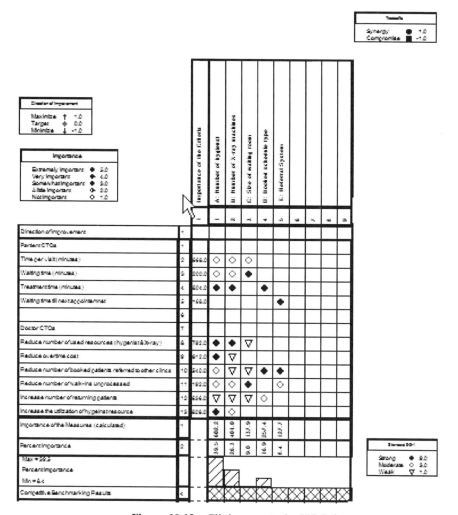

Figure 10.13   Clinic case study: HOQ 2.

- FR3: treatment time, time in the treatment rooms and x-ray (minutes)
  - USL: 45 minutes
  - Current average: 35 minutes
  - Target: 25 minutes
  - PVs: time work studies, SOPs, efficient equipment
- FR4: waiting time until the next appointment (days)
  - USL: 1 week
  - Current average: 2 weeks
  - Target: 4 days

- FR5: number of resources used (hygienist and x-ray)
  - Current: 3 hygienists
  - X-ray machines: 2
- FR6: overtime cost
  - USL: 2 hours per day
- FR7: number of patients booked who are referred to other clinics, with the following rule: If the number of patients scheduled (returning and new) exceeds 100,
  - USL: 100 patients/year
  - Current average: 250 patients/year
  - Target: 0 patients/year
- FR8: number of walk-ins unprocessed with the following rule: Walk-in patients are one step lower in priority than booked patients (admitted according to their availability to fill gaps and increase utilization and returns)
  - Current average: 150 patients/year
  - Target: 0 patients/year
- FR9: number of returning (satisfied) patients who schedule a second appointment once told that they need additional treatment)
  - Current average: 4500 patients/year
  - Target: 6000 patients/year
- FR 10: utilization of hygienist resource (three hygienists in the baseline)
  - LSL: 80% per shift per hygienist
  - USL: 93% per shift per hygienist

From the list of FRs, concepts for the fulfillment of the requirements can be developed. We also have trade-off compromises for several of the FRs, such as the fact that the number of resources is a trade-off with the redesign cost. Through benchmarking it was determined that the current concept, hopefully after improvement, would provide the best solution for achieving the service delivery, quality, and cost reduction objectives while staying close to the needs of the patients and being able to leverage the current patient base competitively. For example, new business value streams, such as internal medicine, pharmacy, and labs and their hybrids, can be used as different concepts, including internal operations variants, to share current and newly acquired resources (reception, accounting, follow-up, etc.). These can be processed using the Pugh controlled convergence method for optimum selection. For the purpose of the clinic case study, the current concept is well suited to deliver on customer voices (evidenced by other outperforming clinics in the chain) – after realizing improvements using the 3S-DFSS process.

Once the conceptual design is determined, the next phase entails detailed design to the level of design parameters.

10.3.4 Stage 4: Preliminary Design

For a design project, we generally use a DFSS engine such as four-phase QFD (Figure 10.9) or a prescriptive method such as axiomatic design (Appendix C). The four-phase QFD process is customarily followed and usually leads to the production of an as-it-should-be process map (Appendix E). Like most DFSS tools, process mapping requires a cross-functional team effort with involvement from process owners, process members, customers, and suppliers. Brainstorming, operation manuals, specifications, operator experience, and process walk are very critical inputs to the mapping activity. A detailed process map provides input to other tools, such as FMEA, transfer function DOEs, capability studies, and control plans.

However, by utilizing the two phases of QFD executed so far and projecting the findings on the current clinic map, we can execute this preliminary design stage. Don't forget: This case study is scoped as a redesign project, and the current concept was adopted in the previous ICOV stage.

The QFD HOQ 1 can be further detailed by mirroring the clinic process map. Process mapping can be used when designing or improving a process or part of a process within the DFSS project scope. The current clinic process map had resources basically doing transactional work with ad hoc operation capability. This structure was replicated in almost all chain clinics. When the black belt team completed the activity of process mapping, the output depicted in Figure 10.14 was obtained.

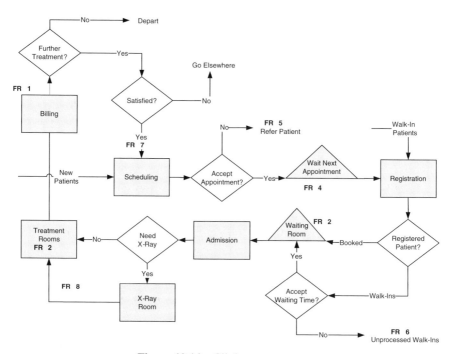

Figure 10.14 Clinic process map.

The design parameters from HOQ 2 in Figure 10.13 were developed as follows:

- DP1 = "number of hygienists"
- DP2 = "number of x-ray machines"
- DP3 = "size of waiting room"
- DP4 = "schedule type booked"

  The booked schedule can have two capacities: 1 week (about 100 patients) and 2 weeks (about 200 patients). A short schedule limits the system capacity of patients but allows customers to have faster succeeding appointments.

- DP5 = "referral system"

  The following rule is used for referring patients to other clinics within the networked chain as well as other partners: If the patients require further treatment and the booked schedule is full, the patient is referred to other clinics with whom this clinic has an agreement. So the number of customers referred is, as a functional requirement, dependent on scheduling capacity.

Also notice that what we would like to see in the relationship room (room 4 in the HOQ) matrix is as close as possible to a diagonal of strong relationships. This will facilitate reducing the design coupling vulnerability in HOQ 2 (Appendix C). At this point we are highlighting the following uncoupled subset of room 4 (Figure 10.13):

- FR2 = "waiting time (minutes)"
- FR3 = "treatment time (minutes)"
- FR4 = "waiting time until next appointment"
- DP3 = "size of waiting room"
- DP4 = "schedule type booked"
- DP5 = "referral system"

$$\begin{Bmatrix} FR2 \\ FR3 \\ FR4 \end{Bmatrix} = \begin{bmatrix} X & 0 & 0 \\ 0 & X & 0 \\ 0 & 0 & X \end{bmatrix} \begin{Bmatrix} DP3 \\ DP4 \\ DP5 \end{Bmatrix}$$

That is, the design team is free as to which way they utilize the design parameters to satisfy the functional requirements. There is no design sequence to follow under the assumption that only these three DPs will be used for the three FRs mentioned above.

10.3.5 Stage 5: Design Optimization

The 3S-DFSS method employed is applied to redesign a dental clinic operated by two dentists. Currently, the expansion budget allows for up to four treatment rooms, five hygienists, two x-ray machines, and one receptionist for admission, billing, and appointment booking. The objective is, however, to develop a six-sigma clinic design in the most economical and effective way possible.

The QFD of the clinic shows the FRs along with their interrelationships. After investigation, it was found that clinic performance in terms of FRs is dependent on four control (design) parameters: number of hygienists, number of x-ray machines, size of waiting room, and type of booking schedule. QFD also shows the relationship of design parameters on their mapped-to FRs.

Using the process map and data collected (benchmarked), a DES model of the dental clinic is developed using WITNESS simulation package. The model is set to run 1 year of clinic operations while importing different settings of control factors, incorporating the impacts of random factors into model performance and estimating values of clinic FRs. Model run time is set to 2000 hours (50 weeks/year, 5 days/week, and 8 hours/day). Random factors are modeled through sampling from theoretical probability distributions available in the simulation package based on random number generators. Output statistics are set to track clinic performance over time and to provide an end-of-run estimate of the FRs specified.

The clinic simulation model serves as a model for clinic performance in terms of FRs based on clinic design (structure, design parameters, random elements, and flow logic). The flexibility of simulation facilitates the initial assessment of various designs in order to disqualify infeasible and poor designs early in the design process. Also, using the model as a flexible experimental platform, a screening DOE can be executed to drop noncritical design factors (those with no statistical significance). Furthermore, animating the clinic operation helped in understanding and debugging the functionality and checking service conformance to medical, legal, and city regulations.

In this section we present the application of the simulation procedure to the clinic 3S-DFSS case study. As discussed earlier, the identify phase of ICOV considers the simulation modeling of the clinic when defining the initial clinic structure and logical design. Also, FRs are defined taking into consideration system performance being measured in time-based measures that can be tracked and assessed with simulation. The focus on clinic design with simulation takes place primarily in the characterize phase of ICOV. The clinic simulation model will be utilized intensively during the optimize phase of ICOV. Simulation will be also helpful when verifying the clinic design and testing the reality of the clinic performance and the clinic operational details. Simulation application to the characterize phase involves four steps (as shown in Figure 10.4).

In general, every step used in an optimization DOE process allows for the adjustment of various factors, affecting the FRs of the clinic service produced. Experimentation allows us to adjust the settings in a systematic manner and to learn which factors have the greatest impact on the resulting functional requirement. Using this information, the settings can be improved continuously until optimum quality is obtained.

A key aspect of DFSS philosophy is that during the design stage, inexpensive parameters can be identified and studied, and can be combined in a way that will result in performance that is insensitive to uncontrollable sources of variation. The team's task is to determine the combined best settings (parameter targets) for each of the design parameters, which have been judged by the design team to have potential to improve the output(s) of interest. The selection of factors will be done in a manner that will enable target values to be varied during experimentation with no major impact on service cost. The greater the number of potential control factors that are identified, the greater the opportunity for optimization of the functional output in the presence of *noise factors*.[4] In a product design, noise factors are factors that cannot be specified by the designer, such as environmental factors, manufacturing variation (piece to piece), and degradation over time. In a 3S DFSS project, noise factors have a stochastic and randomness nature, reflecting event occurrence at times that follow empirical or mathematical probability density distributions.

Once the basic design parameters were decided upon, it became a matter of finding the optimum solution that delivers all the FRs. In this stage, the design documentation is defined, that is, the design includes all the information specific to the operations processes with the design documents under the highest level of control. The ultimate deliverable is an agreement that functionality and performance meet the customer and business requirements under the operating conditions intended. Optimized transfer functions through vehicles such as DOE (design of experiments) are the backbone of the DFSS process. It represent the most common approach to quantifying the transfer functions between the set of FRs and the set of critical factors, the DPs. In this case study, DOE can be conducted only by simulation. From the subset of vital few DPs, experiments are designed to manipulate the inputs actively to determine their effect on the FR outputs. Usually, this phase is characterized by a sequence of experiments, each based on the results of the previous study. Critical design parameters are identified during this process. Usually, a small number of DPs account for most of the variation in the FRs. The 3S-DFSS tools that can be used are process capability analysis, design scorecard, and DES simulation tools.

In an optimization DOE, the primary objective is usually to extract the maximum amount of *unbiased* information regarding the parameters affecting a process or product from as few observations as possible, due to cost. In any case, this is not a trivial task. The development of an optimization DOE can be carried over the following steps.

---

[4] Noise factors are factors that can't be controlled or can be controlled at unaffordable cost, such as randomness and variability factors in this case study (patient arrival, processing time variation, billing time variation, treatment time variation, etc.).

1. *Define the problem and set the objectives*: to optimize the FRs of the clinic, that is, to minimize variability and shift the mean of the FRs to their targets of the clinic case study be treated as a multiobjective optimization study at this ICOV stage. This entails developing an overall objective function such as the *quality loss function*.

2. *Select the responses*: the set of FRs in room 1 of HOQ 2.

3. *Select the design parameters and their levels.* In a DOE, factors can be classified as control or noise. Control factors are design parameters or process variables, which are freely specified by the design team, consulting their knowledge about the process design for the purpose of empirical optimization in this case. These were identified in HOQ 3 in room 3. The design parameters used in the clinic case optimization study are repeated below with their levels:

- DP1 (factor A) = "number of hygienists" at two levels (low: 3, high: 5)
- DP2 (factor B) = "number of x-ray machines" at two levels (Low: 1, high: 2)
- DP3 (factor C) = "size of waiting room" at two levels [low: small (5 seats), high: large (10 seats)]
- DP4 (factor D) = "booked schedule" at two levels [low: short (1 week: 100 patients), high: long (2 weeks: 200 patients)]

    From a continuity standpoint, there are two types of factors: continuous and discrete. A *continuous factor* can be expressed over a defined real number interval with one continuous motion of the pencil: for example, the size of the clinic waiting room (DP3). A *discrete factor* is also called a *categorical variable* or an *attribute variable*: for example, the number of hygienists or the number of x-ray machines (DP1).

4. *Identify the noise variables.* The random (noise) factors included in the optimization-simulation DOE to represent stochastic behavior are:

- *Patient arrival process* (exponentially distributed with mean = 12 minutes)
- *X-ray cycle time*: uniform (10, 15) minutes
- *Treatment type (length) in rooms*: normal (40, 7) minutes
- *Billing time*: uniform (5, 10) minutes

5. *Select the DOE design.* The general representation of the number of experiments is l^k, where l is the number of levels of decision variables and k is the number of factors. Using only two levels of each control factor (low, high) often results in 2^k factorial designs. *Factorial design* looks at the combined effect of multiple factors on system performance. Fractional and full factorial DOE are the two types of factorial design. A 2^4 full factorial design is used in this case study.

The type of experimental design to be selected will depend on the number of factors, the number of levels in each factor, and the total number of experimental runs that can be afforded. If the number of factors and levels are given, full factorial experiment will need more experimental runs, thus be more costly, but it also provides more information about the design under study. The fractional factorial will need fewer runs and thus will be less costly, but it will also provide less information about the design.

6. *Plan the experiment in terms of resources, supplies, schedule, sample size, and risk assessment.*

- Understand the current state of a DFSS project by understanding the technical domains, which are active in the process being optimized. A review of the process logic, systems behaviors, and underlying assumptions is appropriate at this stage.
- Develop a shared vision for optimizing service and process mapping. The black belt is advised to conduct a step-by-step review of the DOE methodology to understand and reinforce the importance of each strategic activity associated with experimentation and to facilitate consensus on the criteria for completion of each activity.
- The DFSS team will need an appreciation for what elements of experimental plan development will require the greatest time commitment. The team will also need to discuss the potential impact associated with compromise of the key DOE principles, if any.
- Plan time for measurement system(s) verification.
- Allocate a budget figure for performing the DOE simulation plan. Estimate cost and seek resource approval to conduct the DOE test plan.

One simulation replicate is used at each run of the 2^4 DOE. The clinic case study simulation model was run for a period of one year (250 days, 8 hours/day). The data collected are listed in Table 10.2.

7. *Analyze the results.* As you might expect, the high-order interactions were pooled as an estimate for error. The analysis was conducted using MINITAB (Figure 10.15).

The significant effects can be identified from the Pareto chart (Figure 10.16) as well as the estimated coefficients for FR1. That is, the main effects A (DP1), B (DP2), C (DP3), and interactions AC and BC are significant. The transfer function can be depicted as[5]

$$FR1 = 137.200 - 49.8350(DP1) + 4.67500(DP2) + 20.2425(DP3)$$
$$- 5.84500(DP1)(DP2) - 15.3525(DP1)(DP3)$$

The rest of the functional requirements analysis is listed in Figures 10.17 to 10.30 using the ANOVA table–Pareto chart template.

8. *Interpret and conclude.* A preferred method in a multiresponse optimization case study such as the clinic case study is to combine them in one objective function. Here, we adopted Taguchi's loss function (see Appendix D). There are some tactical differences about the approach here and traditional application of the quality loss function. Nevertheless, there are many

---

[5] DP1 and DP2 are treated as continuous variables for simplification. They are discrete factors.

TABLE 10.2 Clinic DOE Data

Std. Order	Run Order	DP1	DP2	DP3	DP4	FR1	FR2	FR3	FR4	FR7	FR8	FR9	FR10
14	1	1	−1	1	1	94.76	28.86	65.91	6.1	1285	37	5880	75.03
1	2	−1	−1	−1	−1	142.44	51.16	91.28	3.57	790	446	3914	99.17
12	3	1	1	−1	1	82.64	15.79	66.86	6.18	1249	96	5958	76.9
9	4	−1	−1	−1	1	142.44	51.16	91.28	7.07	786	446	3774	99.17
4	5	1	1	−1	−1	82.64	15.79	66.86	3.11	1280	96	5986	76.9
5	6	−1	−1	1	−1	210.59	118.58	92.01	3.54	844	436	1237	99.64
13	7	−1	−1	1	1	210.59	118.58	92.01	7.03	828	436	1103	99.64
15	8	−1	1	1	1	234.67	132.23	102.44	7.29	786	497	285	99.85
6	9	1	−1	1	−1	94.76	28.86	65.91	3.07	1276	37	5879	75.03
2	10	1	−1	−1	−1	82.31	16.64	65.67	3.09	1247	115	6078	74.01
3	11	−1	1	−1	−1	160.44	58.71	101.73	3.68	723	532	2884	99.51
8	12	1	1	1	−1	89.75	22.81	66.94	3.08	1342	20	5864	78.09
10	13	1	−1	−1	1	82.31	16.64	65.67	6.13	1238	115	6046	74.01
11	14	−1	1	−1	1	160.44	58.71	101.73	7.29	727	532	2492	99.51
7	15	−1	1	1	−1	234.67	132.23	102.44	3.68	802	497	204	99.85
16	16	1	1	1	1	89.75	22.81	66.94	6.12	1317	20	5869	78.09

```
Factorial Fit: FR1 versus DP1, DP2, DP3, DP4

Estimated Effects and Coefficients for FR1 (coded units)

Term Effect     Coef         SE    Coef          T       P
Constant       -397.9    832.341   -0.48      0.653
DP1            -347.5    -173.7    95.886     -1.81   0.130
DP2             -18.0      -9.0    47.958     -0.19   0.858
DP3            -205.1    -102.6   239.501     -0.43   0.686
DP4              -0.0      -0.0   277.354     -0.00   1.000
DP1*DP2          -5.8      -2.9     0.319     -9.16   0.000
DP1*DP3         -76.8     -38.4     1.596    -24.05   0.000
DP1*DP4          -0.0      -0.0    31.920     -0.00   1.000
DP2*DP3           0.2       0.1     0.798      0.14   0.890
DP2*DP4          -0.0      -0.0    15.960     -0.00   1.000
DP3*DP4          -0.0      -0.0    79.800     -0.00   1.000


S = 2.55359 R-Sq = 99.94% R-Sq(adj) = 99.81%


Analysis of Variance for FR1 (coded units)

Source              DF    Seq SS    Adj SS   Adj MS        F       P
Main Effects         4   46642.3    650.33  162.583    24.93   0.002
2-Way Interactions   6    4317.9   4317.95  719.658   110.36   0.000
Residual Error       5      32.6     32.60    6.521
Total               15   50992.8


Estimated Coefficients for FR1 using data in uncoded units

Term              Coef
Constant        137.200
DP1             -49.8350
DP2               4.67500
DP3              20.2425
DP4              -0.000000
DP1*DP2          -5.84500
DP1*DP3         -15.3525
DP1*DP4          -0.000000
DP2*DP3           0.092500
DP2*DP4          -0.000000
DP3*DP4          -0.000000
```

Figure 10.15 FR1 ANOVA Analysis.

merits for such deviation, as in the subject case study. This will enable optimizing the total loss as the FRs deviate from their ideal or set targets. As a first step, the FR responses are summed together according to the direction of optimization. The smaller-the-better (STB) responses are FR1, FR2, FR4, FR7, and FR9; the larger-the-better (LTB) responses are FR8 and FR9, and FR3 is a nominal-the-best (NTB) response. Second, these responses are converted to a quality loss using the appropriate quality loss function proportionality constant, k, based on the direction of optimization (e.g., NTB, LTB, or STB). Third, the FR loss (by optimization direction) is summed to obtain a total loss for the clinic. The net result of the two steps above is shown in

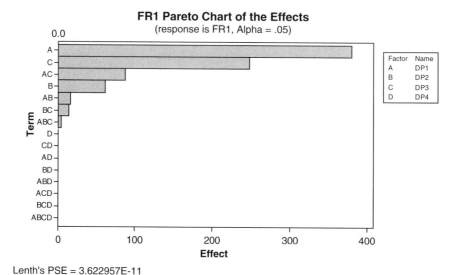

Lenth's PSE = 3.622957E-11

Figure 10.16   FR1 Pareto chart.

```
Factorial Fit: FR2 versus DP1, DP2, DP3, DP4
Estimated Effects and Coefficients for FR2 (coded units)
Term Effect    Coef       SE    Coef       T       P
Constant     -406.0   823.595          -0.49   0.643
DP1          -307.9   -153.9   94.879  -1.62   0.166
DP2           -11.4     -5.7   47.454  -0.12   0.909
DP3          -203.3   -101.6  236.985  -0.43   0.686
DP4           -0.0     -0.0   274.440  -0.00   1.000
DP1*DP2        -3.5     -1.8    0.316  -5.56   0.003
DP1*DP3       -76.1    -38.0    1.579 -24.08   0.000
DP1*DP4        -0.0     -0.0   31.584  -0.00   1.000
DP2*DP3         0.3      0.1    0.790   0.18   0.866
DP2*DP4        -0.0     -0.0   15.792  -0.00   1.000
DP3*DP4        -0.0     -0.0   78.961  -0.00   1.000

S = 2.52676 R-Sq = 99.89% R-Sq(adj) = 99.68%

Analysis of Variance for FR2 (coded units)

Source              DF   Seq SS   Adj SS   Adj MS       F       P
Main Effects         4  25589.7   538.66  134.665   21.09   0.002
2-Way Interactions   6   3900.3  3900.33  650.055  101.82   0.000
Residual Error       5     31.9    31.92    6.384
Total               15  29521.9

Estimated Coefficients for FR2 using data in uncoded units

Term         Coef
Constant    55.5975
DP1        -34.5725
DP2          1.78750
DP3         20.0225
DP4         -0.000000
DP1*DP2     -3.51250
DP1*DP3    -15.2125
DP1*DP4     -0.000000
DP2*DP3      0.112500
DP2*DP4     -0.000000
DP3*DP4     -0.000000
```

Figure 10.17   FR2 ANOVA.

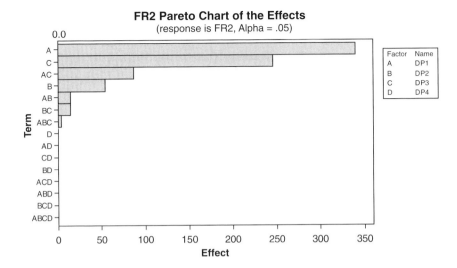

Lenth's PSE = 2.779070E-11

Figure 10.18   FR2 Pareto chart.

```
Factorial Fit: FR3 versus DP1, DP2, DP3, DP4
Estimated Effects and Coefficients for FR3 (coded units)
Term Effect    Coef       SE    Coef          T       P
Constant       8.10   10.2038  0.79       0.463
DP1          -39.62  -19.81   1.1755    -16.85   0.000
DP2           -6.61   -3.31   0.5879     -5.62   0.002
DP3           -1.87   -0.93   2.9361     -0.32   0.763
DP4           -0.00   -0.00   3.4001     -0.00   1.000
DP1*DP2       -2.33   -1.17   0.0039   -298.04   0.000
DP1*DP3       -0.70   -0.35   0.0196    -17.89   0.000
DP1*DP4        0.00    0.00   0.3913      0.00   1.000
DP2*DP3       -0.06   -0.03   0.0098     -2.87   0.035
DP2*DP4       -0.00   -0.00   0.1957     -0.00   1.000
DP3*DP4       -0.00   -0.00   0.9783     -0.00   1.000

S = 0.0313050 R-Sq = 100.00% R-Sq(adj) = 100.00%

Analysis of Variance for FR3 (coded units)
Source              DF   Seq SS   Adj SS   Adj MS         F       P
Main Effects         4  3860.06   5.3029   1.3257   1352.79   0.000
2-Way Interactions   6    87.37  87.3706  14.5618  14858.95   0.000
Residual Error       5     0.00   0.0049   0.0010
Total               15  3947.43

Estimated Coefficients for FR3 using data in uncoded units
Term                Coef
Constant        81.6050
DP1            -15.2600
DP2              2.88750
DP3              0.220000
DP4             -0.00000000
DP1*DP2         -2.33250
DP1*DP3         -0.140000
DP1*DP4          0.00000000
DP2*DP3         -0.0225000
DP2*DP4         -0.00000000
DP3*DP4         -0.00000000
```

Figure 10.19   FR3 ANOVA.

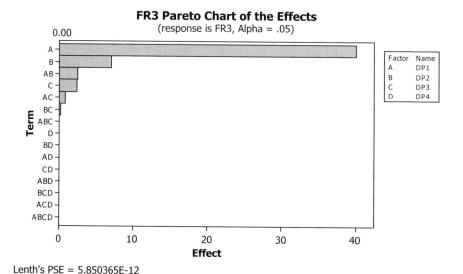

Figure 10.20 FR3 Pareto chart.

```
Factorial Fit: FR4 versus DP1, DP2, DP3, DP4
Estimated Effects and Coefficients for FR4 (coded units)
Term Effect    Coef      SE        Coef       T        P
Constant      174.63    8.15688   21.41     0.000
DP1           -39.03  -19.52      0.93968  -20.77    0.000
DP2             2.44    1.22      0.46998    2.60    0.049
DP3            -2.48   -1.24      2.34710   -0.53    0.619
DP4           114.37   57.19      2.71805   21.04    0.000
DP1*DP2        -0.04   -0.02      0.00313   -6.29    0.001
DP1*DP3        -0.02   -0.01      0.01564   -0.70    0.516
DP1*DP4       -12.69   -6.34      0.31281  -20.28    0.000
DP2*DP3         0.00    0.00      0.00782    0.30    0.776
DP2*DP4         0.84    0.42      0.15641    2.70    0.043
DP3*DP4        -0.78   -0.39      0.78203   -0.50    0.639

S = 0.0250250 R-Sq = 99.99% R-Sq(adj) = 99.98%
Analysis of Variance for FR4 (coded units)

Source               DF    Seq SS    Adj SS    Adj MS        F        P
Main Effects          4   46.0299   37.3654   9.34136  14916.34   0.000
2-Way Interactions    6    0.2874    0.2874   0.04791     76.50   0.000
Residual Error        5    0.0031    0.0031   0.00063
Total                15   46.3204

Estimated Coefficients for FR4 using data in uncoded units

Term           Coef
Constant     5.00188
DP1         -0.391875
DP2          0.0518750
DP3         -0.0131250
DP4          1.64938
DP1*DP2     -0.0393750
DP1*DP3     -0.00437500
DP1*DP4     -0.126875
DP2*DP3      0.00187500
DP2*DP4      0.0168750
DP3*DP4     -0.00312500
```

Figure 10.21 FR4 ANOVA.

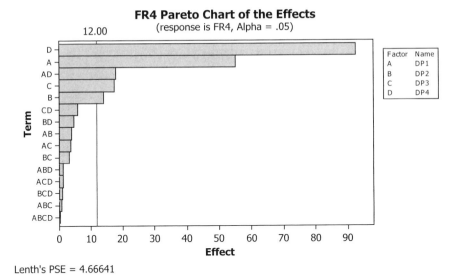

Lenth's PSE = 4.66641

Figure 10.22 FR4 Pareto chart.

```
Factorial Fit: FR7 versus DP1, DP2, DP3, DP4
Estimated Effects and Coefficients for FR7 (coded units)
Term Effect     Coef       SE     Coef        T      P
Constant       -596.1  3216.84    -0.19    0.860
DP1              83.3     41.6   370.58     0.11  0.915
DP2            -321.3   -160.6   185.35    -0.87  0.426
DP3            -227.5   -113.8   925.63    -0.12  0.907
DP4           -1975.0   -987.5  1071.92    -0.92  0.399
DP1*DP2          22.0     11.0     1.23     8.92  0.000
DP1*DP3          -8.7     -4.4     6.17    -0.71  0.510
DP1*DP4        -150.0    -75.0   123.36    -0.61  0.570
DP2*DP3          15.0      7.5     3.08     2.43  0.059
DP2*DP4        -150.0    -75.0    61.68    -1.22  0.278
DP3*DP4        -125.0    -62.5   308.41    -0.20  0.847

S = 9.86914 R-Sq = 99.95% R-Sq(adj) = 99.85%

Analysis of Variance for FR7 (coded units)

Source              DF  Seq SS  Adj SS  Adj MS      F      P
Main Effects         4  987042  3256.8  814.21   8.36  0.019
2-Way Interactions   6    8553  8553.0 1425.50  14.64  0.005
Residual Error       5     487   487.0   97.40
Total               15  996082

Estimated Coefficients for FR7 using data in uncoded units

Term           Coef
Constant    1032.50
DP1          246.750
DP2           -4.25000
DP3           27.5000
DP4           -5.50000
DP1*DP2       22.0000
DP1*DP3       -1.75000
DP1*DP4       -1.50000
DP2*DP3        6.00000
DP2*DP4       -3.00000
DP3*DP4       -0.50000
```

Figure 10.23 FR7 ANOVA.

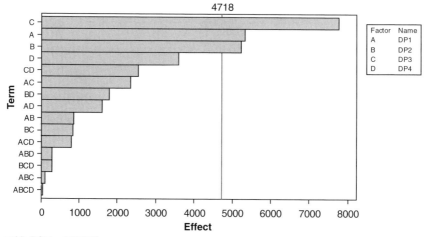

Lenth's PSE = 1835.53

Figure 10.24   FR7 Pareto chart.

```
Factorial Fit: FR8 versus DP1, DP2, DP3, DP4

Estimated Effects and Coefficients for FR8 (coded units)

Term Effect   Coef       SE     Coef        T       P
Constant      -1293   1967.88   -0.66     0.540
DP1            -684    -342     226.70    -1.51    0.192
DP2             -99     -50     113.39    -0.44    0.680
DP3            -418    -209     566.25    -0.37    0.727
DP4              0       0      655.74     0.00    1.000
DP1*DP2         -23     -11       0.75   -15.16    0.000
DP1*DP3         -68     -34       3.77    -9.03    0.000
DP1*DP4          0       0       75.47     0.00    1.000
DP2*DP3          -7      -4       1.89    -1.90    0.115
DP2*DP4          0       0       37.73     0.00    1.000
DP3*DP4          0       0      188.67     0.00    1.000

S = 6.03738 R-Sq = 99.97% R-Sq(adj) = 99.92%

Analysis of Variance for FR8 (coded units)

Source               DF   Seq SS   Adj SS   Adj MS     F       P
Main Effects          4   687843   2814.6   703.66   19.30   0.003
2-Way Interactions    6    11475  11474.8  1912.46   52.47   0.000
Residual Error        5      182    182.3    36.45
Total                15   699500

Estimated Coefficients for FR8 using data in uncoded units

Term         Coef
Constant    272.375
DP1        -205.375
DP2          13.8750
DP3         -24.8750
DP4           0.00000
DP1*DP2     -22.8750
DP1*DP3     -13.6250
DP1*DP4       0.00000
DP2*DP3      -2.87500
DP2*DP4       0.00000
DP3*DP4       0.00000
```

Figure 10.25   FR8 ANOVA.

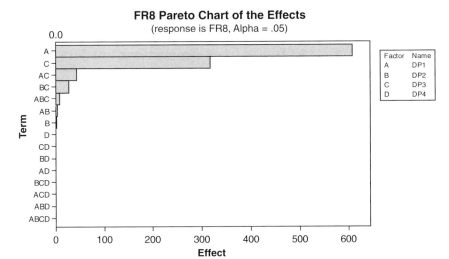

Lenth's PSE = 1.109329E-10

Figure 10.26   FR8 Pareto chart.

```
Factorial Fit: FR9 versus DP1, DP2, DP3, DP4
Estimated Effects and Coefficients for FR9 (coded units)
Term Effect    Coef        SE      Coef        T       P
Constant      78266.6   30555.5     2.56    0.051
DP1           23711.9   11855.9    3520.0    3.37    0.020
DP2             732.6     366.3    1760.5    0.21    0.843
DP3           34530.5   17265.2    8792.2    1.96    0.107
DP4           34556.2   17278.1   10181.8    1.70    0.150
DP1*DP2         247.3     123.7      11.7   10.55    0.000
DP1*DP3        3018.4    1509.2      58.6   25.76    0.000
DP1*DP4        3318.7    1659.4    1171.8    1.42    0.216
DP2*DP3          96.1      48.0      29.3    1.64    0.162
DP2*DP4         -90.6     -45.3     585.9   -0.08    0.941
DP3*DP4        8515.6    4257.8    2929.5    1.45    0.206
S = 93.7431 R-Sq = 99.94% R-Sq(adj) = 99.83%
Analysis of Variance for FR9 (coded units)
Source               DF    Seq SS    Adj SS    Adj MS      F       P
Main Effects          4  71198318   1517052    379263   43.16   0.000
2-Way Interactions    6   6869511   6869511   1144919  130.29   0.000
Residual Error        5     43939     43939      8788
Total                15  78111768
Unusual Observations for FR9
Obs StdOrder FR9 Fit SE Fit Residual St Resid
14 11 2492.00 2598.19 77.73 -106.19 -2.03R
R denotes an observation with a large standardized residual.
Estimated Coefficients for FR9 using data in uncoded units
Term            Coef
Constant     3965.81
DP1          1979.19
DP2          -273.063
DP3           675.688
DP4           -39.9375
DP1*DP2       247.313
DP1*DP3       603.687
DP1*DP4        33.1875
DP2*DP3        38.4375
DP2*DP4        -1.8125
DP3*DP4        34.0625
```

Figure 10.27   FR9 ANOVA.

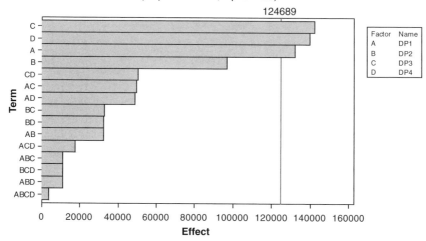

Lenth's PSE = 48506.0

Figure 10.28   FR9 Pareto chart.

Factorial Fit: FR10 versus DP1, DP2, DP3, DP4

Estimated Effects and Coefficients for FR10 (coded units)

Term	Effect	Coef	SE Coef	T	P
Constant		54.111	21.8653	2.47	0.056
DP1	-18.885	-9.442	2.5189	-3.75	0.013
DP2	3.550	1.775	1.2598	1.41	0.218
DP3	5.425	2.713	6.2916	0.43	0.684
DP4	0.000	0.000	7.2860	0.00	1.000
DP1*DP2	0.675	0.337	0.0084	40.25	0.000
DP1*DP3	0.875	0.438	0.0419	10.43	0.000
DP1*DP4	0.000	0.000	0.8385	0.00	1.000
DP2*DP3	0.013	0.006	0.0210	0.30	0.778
DP2*DP4	0.000	0.000	0.4193	0.00	1.000
DP3*DP4	0.000	0.000	2.0963	0.00	1.000

S = 0.0670820 R-Sq = 100.00% R-Sq(adj) = 100.00%

Analysis of Variance for FR10 (coded units)

Source	DF	Seq SS	Adj SS	Adj MS	F	P
Main Effects	4	2228.43	0.42133	0.10533	23.41	0.002
2-Way Interactions	6	7.78	7.78040	1.29673	288.16	0.000
Residual Error	5	0.02	0.02250	0.00450		
Total	15	2236.23				

Estimated Coefficients for FR10 using data in uncoded units

Term	Coef
Constant	87.7750
DP1	-11.7675
DP2	0.812500
DP3	0.377500
DP4	0.0000000
DP1*DP2	0.675000
DP1*DP3	0.175000
DP1*DP4	0.0000000
DP2*DP3	0.0050000
DP2*DP4	0.0000000
DP3*DP4	0.0000000

Figure 10.29   FR10 ANOVA.

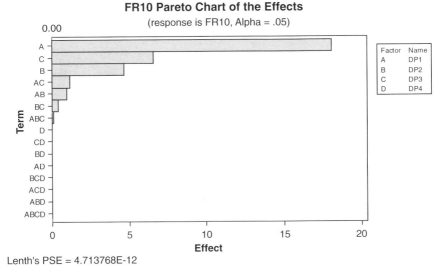

Lenth's PSE = 4.713768E-12

Figure 10.30 FR10 Pareto chart.

TABLE 10.3 Quality Loss Function Grouping

Std. Order	Run Order	DP1	DP2	DP3	DP4	NTB QLF	STB QLF	LTB QLF	Total QLF
14	1	1	−1	1	1	0.43723	3.62355	0.81928	4.88005
1	2	−1	−1	−1	−1	2.14184	1.59664	0.05587	3.79435
12	3	1	1	−1	1	0.47786	3.70649	0.19306	4.37741
9	4	−1	−1	−1	1	2.14184	1.48838	0.05587	3.68609
4	5	1	1	−1	−1	0.47786	3.74777	0.19306	4.41869
5	6	−1	−1	1	−1	2.20994	0.23009	0.05562	2.49566
13	7	−1	−1	1	1	2.20994	0.19607	0.05562	2.46163
15	8	−1	1	1	1	3.29935	0.07716	0.0542	3.43072
6	9	1	−1	1	−1	0.43723	3.62006	0.81928	4.87657
2	10	1	−1	−1	−1	0.42725	3.85042	0.1669	4.44456
3	11	−1	1	−1	−1	3.21829	0.88694	0.05403	4.15926
8	12	1	1	1	−1	0.48136	3.6196	2.58199	6.68296
10	13	1	−1	−1	1	0.42725	3.80938	0.1669	4.40353
11	14	−1	1	−1	1	3.21829	0.67678	0.05403	3.9491
7	15	−1	1	1	−1	3.29935	0.07574	0.0542	3.42929
16	16	1	1	1	1	0.48136	3.61883	2.58199	6.68218

Table 10.3. As a last step, we regress the total loss against predictors DP1 to DP4 and their interactions using MINITAB. This will provide an overall transfer function of the total loss as in Figure 10.31.

The transfer function of the QLF is given by

```
Regression Analysis: TOTAL QLF versus DP1, DP2, ...

The regression equation is
TOTAL QLF = 4.26 + 0.835 DP1 + 0.380 DP2 + 0.107 DP3 - 0.0269 DP4 + 0.0641
       DP1*DP2 + 0.578 DP1*DP3 + 0.0170 DP1*DP4 + 0.308 DP2*DP3 - 0.0044
       DP2*DP4 + 0.0232 DP3*DP4


Predictor      Coef   SE Coef       T      P
Constant    4.26075   0.06716   63.44  0.000
DP1         0.83499   0.06716   12.43  0.000
DP2         0.38045   0.06716    5.66  0.002
DP3         0.10663   0.06716    1.59  0.173
DP4        -0.02691   0.06716   -0.40  0.705
DP1*DP2     0.06412   0.06716    0.95  0.384
DP1*DP3     0.57807   0.06716    8.61  0.000
DP1*DP4     0.01696   0.06716    0.25  0.811
DP2*DP3     0.30846   0.06716    4.59  0.006
DP2*DP4    -0.00443   0.06716   -0.07  0.950
DP3*DP4     0.02318   0.06716    0.35  0.744


S = 0.268645  R-Sq = 98.3%  R-Sq(adj) = 94.8%

Analysis of Variance

Source            DF       SS      MS      F      P
Regression        10  20.6130  2.0613  28.56  0.001
Residual Error     5   0.3608  0.0722
Total             15  20.9738


Source    DF    Seq SS
DP1        1   11.1554
DP2        1    2.3159
DP3        1    0.1819
DP4        1    0.0116
DP1*DP2    1    0.0658
DP1*DP3    1    5.3466
DP1*DP4    1    0.0046
DP2*DP3    1    1.5223
DP2*DP4    1    0.0003
DP3*DP4    1    0.0086
```

Figure 10.31   Quality loss function ANOVA.

$$
\begin{aligned}
\text{total loss} = {} & 4.26075 + 0.83499(\text{DP1}) + 0.38045(\text{DP2}) + 0.10663(\text{DP3}) \\
& - 0.02691(\text{DP4}) + 0.06412(\text{DP1})(\text{DP2}) \\
& + 0.57807(\text{DP1})(\text{DP3}) + 0.01696(\text{DP1})(\text{DP4}) \\
& + 0.30846(\text{DP2})(\text{DP3}) - 0.00443(\text{DP2})(\text{DP4}) \\
& + 0.02318(\text{DP3})(\text{DP4})
\end{aligned}
$$

It is obvious that DP1, DP1 × DP3 interaction, and DP2 are the most significant effects. From the interaction plots in Figure 10.32 as well as the cube plot in Figure 10.33, we can determine the optimum design parameter coded settings that minimize the total loss:

- DP1 (factor A) = "number of hygienists" at low = 3 (coded level = −1)
- DP2 (factor B) = "number of x-ray machines" at the low level, 1 (coded level = −1)
- DP3 (factor C) = "size of waiting room" at the high level, 10 seats (coded level = 1)

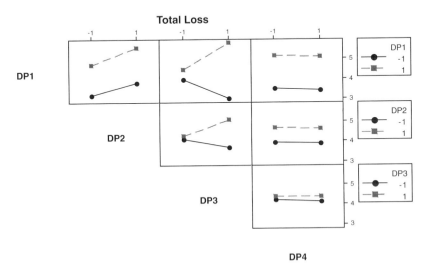

Figure 10.32   QLF interaction matrix.

Cube Plot (data means) for TOTAL QLF

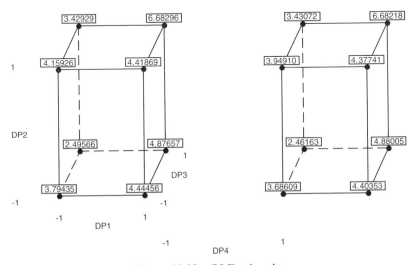

Figure 10.33   QLF cube plot.

- DP4 (factor D) = "booked schedule" at any of the following two levels [low: short (1 week: 100 patients), high: long (2 weeks: 200 patients)]

This conclusion can be confirmed by the three-dimensional contour plots (Figures 10.34 to 10.36) of the loss against all two factorial combinations of significant effects of the set {DP1, DP2, DP3}.

The minimum loss can be found by substituting the coded optimum settings as follows:

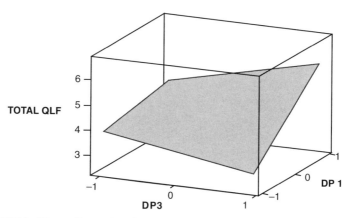

Figure 10.34   Three-dimensional contour plot of total loss against DP1 and DP3.

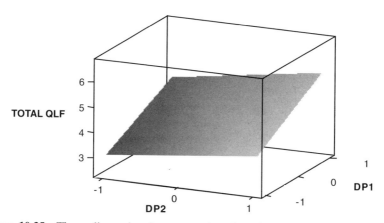

Figure 10.35   Three-dimensional contour plot of total loss against DP1 and DP2.

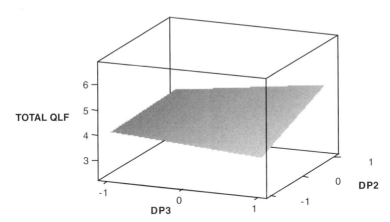

Surface Plot of TOTAL QLF vs DP2, DP3

Figure 10.36   Three-dimensional contour plot of total loss against DP2 and DP3.

$$
\begin{aligned}
\text{total loss} &= 4.26075 + 0.83499(-1) + 0.38045(-1) + 0.10663(1) - 0.02691(1) \\
&\quad + 0.06412(-1)(-1) + 0.57807(-1)(1) + 0.01696(-1)(1) \\
&\quad + 0.30846(-1)(1) - 0.00443(-1)(1) + 0.02318(1)(1) \\
&= 2.3133
\end{aligned}
$$

10.3.6   Stage 6: Verification

This stage of ICOV DFSS methodology is differentiated from the prevalent "launch and learn" method through design validation. Design validation helps identify unintended consequences and effects of design, develop plans, and reduce risk for implementation. That is, we want to verify that the optimum design (DP1 = DP2 = −1, DP3 = 1, DP4 = 1 or −1) performance is capable of achieving the expectations of customers and stakeholders at unprecedented performance levels. Optimum design is what we need to accomplish this assessment in a low-risk, cost-effective manner. In this section we cover the service-relevant aspects of design validation.

Design Validation Steps   In general, design validation can take on many different forms, depending on the magnitude of the design or redesign change and the complexity of the environment in which the design will operate. Typically, it is desirable to understand the performance of the optimum design with the highest confidence achievable.

When we think of product-specific validation, it often means building prototypes and testing them in a *beta environment*, where real patients utilize the optimum design in their actual visit and provide feedback. Writing this book is similar to having people read the draft (prototype) chapters and provide valuable feedback to the authors. For the beta testing to provide useful feedback to the design team, frequent debriefs around the experience and detailed notes around the operating environment are necessary.

As we move into a clinic service validation environment, many of the same concepts are directly applicable from the product validation environment, but often it is just as easy to run the new service or process in parallel to the legacy process. The key elements of validation then are to create a prototype, subject it to a realistic test environment, summarize the performance observed, and feed any required changes back into the process.

Generally, to produce a useful prototype of the optimum design that is sufficient to allow for evaluation, consideration must be given to the completeness and stability of the elements that make up the design. These elements can be classified as the design parameters that are manifested into components and subsystems (hygienists, x-ray machines, etc.) as well as the production process (reception, diagnosis, treatment, checkout, etc.) variables with all steps and subprocesses. We can look at components, subsystems, and system to determine the lack of knowledge of performance and the interaction of pre-existing (carryover) subsystems into the optimum system.

Risk can be assessed from failure mode effect analysis or other sources (e.g., history) and/or DFSS tool use. If any of the elements are of the optimum design high-risk category, they should not be compromised and must be included in the prototype.

Prototypes usually involve some trade-off between the final completed design and some elements that are not available until later. Prototype designs are tested under controlled conditions. Prototypes should be evaluated using objective and robust analytical techniques objectively to demonstrate design performance for a product or service pilot. The testing method is critical in accomplishing objective decision making and obtaining this information in the most cost-effective manner.

When testing prototypes, it is vital to follow a rigorous process to assess the performance and learn from any unusual observances. Unexpected results often occur, and without the correct approach these will have to be written off as unusual observations when in fact they may be significant events that can either differentiate the new design or plague it.

Some of the specific types of testing include pilot tests and preproduction (low-rate service, such as on the weekend or in the middle of the week). A pilot tests prototype design under limited real-world conditions to verify original design assumptions versus actual performance. Preproduction takes a new product or process through concept and pilot evaluation to full production or rollout.

In validation, the DFSS team conducts a progressive sequence of verification tests to quantify the ability of the service to satisfy FRs under conditions that approximate the conditions of actual use. The DFSS team needs to conduct specification and verification tests and make initial adjustments to test and measurement equipment. Verify and record initial conditions according to a planned test plan. The team then begins collecting data for those tests that are designed to analyze performance and record any abnormalities that may occur during testing and may facilitate subsequent concern resolution.

10.3.7   Stage 7: Launch Readiness

Based on successful verification in the clinic production environment, the team will assess the readiness of all of the service process infrastructure and resources. For instance, have all the standard operating procedures been documented and people trained in the procedures? What is the plan for process switchover or ramp-up? What contingencies are in place? What special measures will be in place to ensure rapid discovery? Careful planning and understanding of the behavior desired are paramount to successful transition from the design world into the production environment.

10.3.8   Stage 8: Production

In this stage, if the team has not already begun implementation of the design solution in the service environment, the team should do so now. Validation of some services may involve approval by medical regulatory agencies (e.g., approval of quality systems) in addition to the chain's own rigorous assessment of the capability. Appropriate controls are implemented to ensure that the design continues to perform as expected and that anomalies are addressed as soon as they are identified (update the FMEA) in order to eliminate waste, reduce variation, and further error-proof the design and any associated processes.

10.3.9   Stage 9: Service Consumption

Whether supporting patients with the help of the service processes, which themselves need periodic maintenance (e.g., x-ray machines), the support that will be required and how to provide it are critical in maintaining consistent high-quality performance levels. Understanding the total life cycle of service consumption and support are paramount to planning adequate infrastructure and procedures.

10.3.10   Stage 10: Phase Out

Eventually, all products and services become obsolete, replaced by either new technologies or new methods. Also, the dynamic and cyclical nature of patient attributes dictates continuous improvement to maintain adequate market share. Usually, it is difficult to turn off the switch, as many patients have different dependencies on clinic services and processes. Just look at the bank teller and the ATM machine: One cannot just convert to a single new process. There must be a coordinated effort, and a change in management is often required to provide incentives to customers to shift to the new process.

10.4   SUMMARY

In this chapter we presented a simulation-based design for six-sigma case study using the road map is depicted in Figure 10.4. The road map highlights at a high level the identify, charcaterise, optimize, and validate phases and the seven service development stages: idea creation, voice of the customer and business, concept development, preliminary design, design optimization, verification, launch readiness. The road map also recognizes tollgates, design milestones where DFSS teams update stockholders on developments and ask that decisions be made as to whether to approve going into the next stage, recycling back to an earlier stage, or canceling the project altogether.

The case study also highlights most appropriate DFSS tools by the ICOV phase. It indicates where it is most appropriate for tool use to start, such as transfer function and quality function deployment. Following the DFSS road map helps accelerate the process introduction designed and aligns the benefits for customers and stakeholders. The tools and tollgates (Figure 10.1) allow for risk management, creativity, and a logical documented flow that is superior to the "launch and learn" mode with which many new organizations, processes, or services are deployed. Not all projects will use the complete tool kit of DFSS tools and methodology, and some will use some of the tool kit to a greater extent than will others.

11

PRACTICAL GUIDE TO SUCCESSFUL DEVELOPMENT OF SIMULATION-BASED SIX-SIGMA PROJECTS

11.1 INTRODUCTION

Like any other project, simulation-based projects are selected according to specific requirements and are executed based on a specific procedure. Similar to six-sigma projects, the degree of 3S project success is often a function of correct selection, scoping, planning, and implementation. Although several methods and tools can be followed to manage a 3S project in various applications, a generic framework for successful project management and development can be proposed based on experience. This framework clarifies the ambiguity involved in project selection and specifies the correct method for project development.

In this chapter we provide a practical guide for successful development of 3S projects based on the road maps discussed in Chapter 8. The guide starts by discussing the unique characteristics of 3S projects. 3S approaches are often applied to engineer the processes in systems of a transactional nature whose performance can be measured quantitatively using time-based critical-to-quality characteristics (CTQ) performance metrics. These systems are data-driven service and logically designed applications that can either be designed using the DFSS method or improved using the DMAIC and LSS methods.

Next, we provide a framework for successful development of a 3S project. The first phase in the framework proposed is project definition, which consists

Simulation-Based Lean Six-Sigma and Design for Six-Sigma, by Basem El-Haik and Raid Al-Aomar
Copyright © 2006 John Wiley & Sons, Inc.

of project selection, scoping, and chartering. The second phase is project execution, which includes data collection, model building, and six-sigma application. The final phase is project implementation and maintenance, which includes the implementation of 3S changes or design and measures for sustaining the performance achieved.

This chapter concludes with a section on selecting and comparing software tools used in the application of 3S projects. This includes three software categories: six-sigma software tools, simulation software tools, and simulation-six-sigma software tools. We include examples of software tools in each category, together with links and information on software features and vendors. A list of software vendors is provided in Appendix F.

The main guidelines for successful 3S project development include:

- The characteristics of the 3S application for scoping
- Ingredients for a successful 3S Program
- Correct selection of the 3S project
- Developing a business case for the 3S project
- Developing a charter for the 3S project
- An effective 3S project implementation, following Chapters 1 to 10
- Utilizing the proper software tools

11.2 CHARACTERISTICS OF A 3S APPLICATION

A 3S project can be viewed as an integration element between quality and productivity measures. This is a key distinction of 3S projects where six-sigma quality is coupled with simulation-based productivity improvement methods. This extends traditional quality methods scope to cover a system-level design and improvement in both manufacturing and services: hence, six-sigma and design for six-sigma. Many quality and productivity methods and initiatives were introduced during the past few decades. This includes quality circles, statistical process control, total quality management, enterprise resource planning, and lean manufacturing. Although they have been sound in theory, practical applications of these initiatives have not always delivered their premises over the long term. Typical deployment failure modes include incorrect project selection, unclear definition of scope, and poor execution, culture, and impatience for long-run results. Many of these pitfalls can be avoided by establishing a clear understanding of the deployment infrastructure and application characteristics.

As discussed in previous chapters, the 3S approach is a simulation-based application of six-sigma methods (design for six-sigma and lean six-sigma) to design new processes or to improve existing processes. In Chapter 8 we presented road maps for the successful implementation of the two 3S approaches (3S-DFSS and the 3S-LSS). From previous chapters it is clear that the underlying process in a 3S application needs to be represented by a logically designed transactional service or manufacturing process whose performance is measured quantitatively using one or more time-based (CTQs) performance metrics. The 3S projects are applied to processes and systems whose structure and logic can be represented with discrete event simulation. If the logic is complicated beyond the capability of simulation modeling, the 3S approach may not be applicable. This applies to both manufacturing and service systems.

The following is a brief discussion of 3S application characteristics.

11.2.1   Transactional Nature

The 3S projects are characterized by their application to transactional processes in both manufacturing and service environments. The term *transactional* often refers to some intended transformation that is applied to an entity of interest in a service or manufacturing system. The process is transactional once it consists of a logically organized operations with intended value-added transformations. From a lean manufacturing view, all process operations (transformations) are expected to be value-added. In manufacturing, the value-added transformation is applied to entities that are mainly raw materials and components in order to produce finished goods. Examples of transactional manufacturing processes include various types and configurations of job shops, assembly lines, and flexible manufacturing systems. In services, on the other hand, the value-added transformation can be applied directly to customers or to customer-related entities where some type of service is provided to customers to gain their business. Examples of transactional service processes include banking services, clinics and hospitals, fast-food restaurants, insurance services, mass transportation, and airline services.

The 3S approaches apply to processes of a transactional nature due to their amenability to simulation modeling. As discussed in Chapter 4, process simulation resembles certain a flow logic of entities such as components and customers through a set of operations so that certain value-adding transactions can be carried out on the flowed entity. The majority of real-world production and business systems involve some type of transactional process. We can tell whether the system is transaction-based simply by checking how the product is produced and how the service is provided. The answer is usually expressed using a flowchart, an operation chart, a process map, or a value stream map. By definition, all these process forms and representations indicate a transactional nature.

11.2.2   Time-Based (CTQ) Process Measures

Another key characteristic of 3S projects states that the performance of the underlying system is measured with time-based metrics. Time-based process metrics include those dynamic measures that change discretely or continuously over time. These measures are common in manufacturing and service systems and include plant productivity, order lead time, number of customers in a waiting room, number of daily served customers, and inventory level, among others.

Although initially, the six-sigma method was developed to work on product quality, many recent six-sigma applications are focused on analyzing process or system-level performance. For example, customers may complain about long lead times for their orders rather than complaining about product quality. On the other hand, system-level simulation was developed fundamentally to model the time-based performance of real-world systems and to estimate time-based metrics such as productivity and lead time. Hence, the merge of six-sigma and simulation in the 3S approach provides a great advantage in facilitating the analysis of systems' time-based performance.

11.2.3 Process Variation in a Data-Driven Approach

Six-sigma is a data-driven methodology based mainly on reducing process variation using statistical analysis and experimental design. Simulation is also a data-intensive application of model building and statistical analysis. Hence, the 3S approach is characterized by being data-driven with applications that aim at reducing variability in system-level time-based metrics, the CTQs. Data-driven implies that each time-based selected metric can be estimated quantitatively based on certain model data and conditions. Qualitative criteria are typically harder to measure and may not be suitable for 3S application. Examples include measures such as comfort, trust, loyalty, and appeal.

11.2.4 Simulation-Based Application to Manufacturing Systems

A wide range of manufacturing applications can benefit from the 3S approach. Many six-sigma projects have targeted the effective design and improvement of manufacturing systems. Simulation modeling was used initially in manufacturing systems applications. 3S manufacturing applications target job shops, plants, assembly line, material-handling systems, and many other manufacturing systems. 3S can be used to design new manufacturing systems or to improve existing ones using a defined set of time-based (CTQ) performance measures.

Elements of Manufacturing Systems   In manufacturing systems, value is added to an entity (product, material, component, etc.) by routing entities through a sequence of processing operations or stations and resources. Basic elements that are modeled and tracked in any manufacturing system include:

   1. *System entities*. Material, parts, and components represent the entities in manufacturing systems. The entities are routed within the plant operations and use plant resources before departing the manufacturing systems as finished goods.

2. *System resources.* Manufacturing system resources mainly include machines, tools, equipment, and labor. Capacity and utilization design and enhancement at manufacturing resources are a focus of 3S projects.

3. *Warehouses and inventory systems.* Manufacturing systems are distinguished with warehouses for raw materials and finished items as well as work-in-process buffers. This includes all types of storage systems, from simple part bins to automated storage and retrieval systems.

4. *Material-handling systems.* Manufacturing systems are characterized by extensive manual or automatic material-handling systems, including forklifts, conveyors, automated guided vehicles, power and free systems, and others.

5. *Facility and physical structure.* The layout and physical structure of a manufacturing facility often represent the best configuration of plant resources that facilitates flow and increases effectiveness. Different types of layouts and flow methods can be utilized to this end. Also, meeting certain requirements and codes is essential when designing the manufacturing facility.

6. *Operating pattern.* The operating pattern in a manufacturing system defines the number shifts per day, the working hours per shift, and breaks the distribution.

Design Parameters in Manufacturing Systems   Model control factors include the set of parameters that can be changed by the manufacturing system designer in order to enhance the system performance. In 3S applications these factors are optimized with DOE to enhance process CTQs. Examples of manufacturing design parameters include:

1. Number of production resources
2. Capacity of production processes
3. Production batch size (lot size)
4. Schedule of input materials
5. Shipping schedule
6. Material flow and routing rules
7. Buffer capacity and location
8. Number of material-handling carriers
9. Speed of conveyance systems

Time-Based CTQs in Manufacturing Systems   This represents the set of measures that can be used to assess the performance of a manufacturing system as well as to compare the performance of several process designs. In 3S applications these measures can be estimated from available accumulated model statistics (or from a special code) to compute the measures' values. Examples of performance measures in a manufacturing system include:

1. Manufacturing lead time
2. Throughput per hour or per shift
3. Total inventory level or cost
4. Utilization of production resources
5. Number of late production orders
6. Equipment availability
7. Percent of units built to schedule

11.2.5 Simulation-Based Application to Service Systems

During the last two decades, the service industry has shown a remarkable growth in various aspects of national and international economies. Service companies, including banking, the food industry, health systems, telecommunication, transportation, and insurance, play a major role in today's market. As a result, many techniques, analytical methods, and software tools were developed to help design service systems, to solve operational problems, and to optimize their performance.

Both six-sigma and simulation modeling have been widely used in service systems applications. 3S applications can therefore include banks, fast-food restaurants, computer systems, clinics, traffic, airports, post offices, and many other service systems. This can include internal business operations such as product development processes, financial transactions, and information flow. 3S projects can be used to study system behavior, quantify the service provided, compare proposed alternatives, and set up and configure the system to provide the best performance possible.

Elements of Service Systems   Like manufacturing systems, service systems provide one or more services to entities (e.g., customers) through resources and operations. Entities are routed through a sequence of processing operations or stations at which system resources such as employees or automatic processing machines provide the service required. Basic elements that are modeled and tracked in any service system include:

1. *System entities.* Customers (humans) represent the entity in many service systems. Customers arrive at the service system, request the service, receive the service, and departure the service system. For example, customers arrive at a bank and request the kind of service they wish to do, such as making deposits, withdrawals, money transfers, and so on. Customers often wait for bank services in queues when bank tellers are not available. Once a customer gets to the server (bank teller), he or she receives the service (the transaction) and then departs the bank. Similar examples include patients arriving at a clinic, customers arriving at a fast-food restaurant, and customers arriving at the post office.

Other types of entities in addition to customers can also be part of a service system. Examples include paperwork in a governmental office, insurance claims in an insurance company, calls in a call center, and information bytes in a computer system. Such entities are processed to provide a certain service to customers, where value in such a context may refer to a percentage of completion or the benefit obtained from processing stations.

2. *Service providers.* Entities in a service system arrive at the service center, request the service, and wait in front of service providers. Service providers are the resources through which the service requested is provided. Examples include waitresses in a restaurant, window tellers in a fast-food restaurant, bank tellers in a bank, doctors and nurses in a clinic, customs officers at a border-crossing terminal, and receptionists.

The capacity of service providers determines the service time (time that a customer spends during service) and affects the waiting time (time that customers wait to get to the service providers). Thus, determining the best number of service providers is a key factor in designing service systems. The 3S approach can be utilized to model the service system and analyze the impact of varying the number of service providers on key system performance measures such as customer average waiting time and servers' utilization.

3. *Customer service.* Most service systems include a mean for customer service through which complaints and feedback from customers are received and analyzed. Free of charge phone numbers, centers of customer service, and a Web site for customer feedback are the major forms of customer service in service systems. In retail stores, customer service allows shoppers to return or replace merchandise, helps customers find merchandise, and reports direct feedback to store management.

4. *Staff and human resources.* Staff, business managers, and customer service associates are key building blocks that contribute greatly to the success or failure of a service system. Most service systems rely on people to provide services. For example, hospital doctors and nurses play the major role in providing medical services. Bank tellers, calling officers, and receptionists are other examples.

5. *Facility layout and physical structure.* The layout and physical structure of the service facility have a special importance in service systems. Designing the facility layout in an effective manner that assists both service providers and customers is often critical to the performance of the service system. Certain requirements and codes must be met when designing a service facility. Examples include a parking lot, handicapped parking, waiting-area capacity and features, location of reception and help desks, layout and structure of service areas or station, male and female rest room space and capacity, and facility environment, such as illumination, air conditioning, insulation, and heating.

Building codes and standards that are compliant to regulations of cities and provinces provide the specific requirements of service facilities. Such requirements vary based on the nature of the service. For example, what is required for a gas station and oil change facility is different from that required of banks and restaurants. The interior design of a facility, the layout and physical structure, is a combination of art and business needs.

6. *Operating policies.* Operating policies are also a key component in any service system. Operation pattern (open and close hours), routing customers, flow of each service, queue and service discipline, and departure rules are examples of operating policies.

Design Variables in Service Systems   Model design variables include the set of parameters that can be changed by the designer to enhance system performance. In general, the service system design is in control of the entities' acceptance or admission to the system, the entities' waiting and classification rules, the service-providing process, the logic of entity flow between servers, and the rules of system departure. Entities' arrival rate to the system is typically not within the control of system designers. Adjusting such processes may be translated into providing settings of key model control factors, such as:

1. Percent or rate of entities admitted to the service system.
2. Capacity of the waiting area or line.
3. Waiting discipline and rules of selecting customers to receive service. First-come-first-served is the most common waiting discipline in service systems. A preemptive method can also be used to expedite or select customers to service.
4. Number of servers in the service system and their configuration.
5. Service time at each server.
6. Percentages used to route entity flow among servers.
7. Rules of system departure (if applicable).

Time-Based CTQs in Service Systems   A set of measures can be used to assess the performance of a service system as well as to compare performance of several system designs. Quantified performance measures should be used to assess service system performance. Such measures can be estimated from model accumulated statistics, or a special code may be necessary to compute the measures' values. Examples of performance measures in a service system include:

1. Waiting time per customer
2. Number of customers left without receiving the service (in case there is no capacity or the waiting time was too long)
3. Time-in-system (the total time a customer spends in the service system; includes waiting time, transfer time, and service time)
4. Average and maximum size of the queue (length of the waiting line)

5. Server utilization (percent idle and percent busy)
6. Service system throughput (number of entities processed per time unit, such as the number of customers served per day)
7. Service level (number of customers who finished the service without waiting or with less than 5 minutes of waiting time)

11.3   INGREDIENTS FOR A SUCCESSFUL 3S PROGRAM

For an effective and successful introduction, development, and implementation of a 3S program in any organization, key ingredients will drive the synergy from integration of the six-sigma and simulation approach:

- Understanding the 3S methodology, tools, and metrics
- Strong leadership and top management commitment
- Effective organizational infrastructure
- Human resource cultural change and training
- Extending 3S project to customers and suppliers
- Proper project prioritization and selection

A key ingredient for a successful 3S application involves learning the theory and the principles behind six-sigma methodology and simulation. This was provided through Parts I and II of the book.

However, like any other deployment, 3S application will face major difficulty without leadership and sufficient management support. Good support from top management is imperative to allocate budget and technical resources to the project as well to motivate project team and employees. The leaders have to be strong advocates of both six-sigma and should communicate the synergy of the 3S approach.

In addition to top management support, 3S successful introduction and deployment require an effective organizational infrastructure to support the actions and measures taken. A part of creating an effective organizational infrastructure is to evolve the culture of the organization toward excellence and to change the decision making to be data driven. Resistance to change can be overcome through effective communication, customized training, and building an account of success stories. Many of 3S projects will require cross-functional teams to tackle the problem successfully and implement process improvements. Along with that, employees need to be motivated toward the introduction and development of 3S programs through various reward and recognition schemes.

Being a data-driven simulation-based approach, effective 3S implementation requires a decent IT system (computer hardware, networks, databases, and software). This is essential for effective data collection and analyses, information search, training, model building, six-sigma analyses, and communication.

Just like the six-sigma initiative, the 3S approach should begin and end with the customers. This requires linking the 3S project to customers and suppliers as well as to various business functions within the company. Selecting inappropriate 3S projects is also key. Thus, we need to manage the six-sigma pipeline using portfolio techniques to achieve better results. Organizations need a set of strategic criteria against which potential six-sigma projects can be evaluated, and projects that do not meet these should not be launched.

When selecting a 3S project, priority should be given to projects with maximum impact on customers. This will reflect on the company financial results. For many organizations, financial returns are the main criterion. Therefore, the projects should be selected in such a way that they are closely tied to the business objectives of the organization. Attention should be paid to conflicting priorities within these three business elements (customers, company, and suppliers). It is also imperative to keep projects small and focused so that they are meaningful and manageable. Projects that have been poorly selected and defined lead to delayed results and also a great deal of frustration and doubt to the validity of the 3S approach starts to emerge. The following three generic criteria may be used for the selection of a 3S projects:

1. Business benefits criteria
   - Impact on meeting external customer requirements
   - Impact on core competencies
   - Financial impact
   - Urgency
2. Feasibility criteria
   - Resources required
   - Complexity issues
   - Expertise available and required
   - Likelihood of success within a reasonable time frame
3. Organizational impact criteria
   - Learning benefits (new knowledge gained about business, customers, processes, etc.)
   - Cross-functional benefits

Several root cause factors may lead the 3S project to deliver results that are unacceptable and fall below the expectations. These include:

1. Inadequate information extracted from overwhelming data sources
2. Selecting the wrong 3S application
3. Targeting a noncritical problem
4. Failing to meet simulation requirements (data, logic, flow, etc.)
5. Failing to meet six-sigma requirements (data, measures, DMAIC, DFSS, etc.)

6. Failing to analyze the risks involved in the 3S project
7. Leaping to actions without analyzing root causes properly
8. Lack of project management and faulty implementation: illogical sequences, ineffective resource planning, weak evaluation, and so on
9. Failing to sustain momentum

11.4 FRAMEWORK FOR SUCCESSFUL 3S IMPLEMENTATION

The 3S approach provides a comprehensive and flexible system for achieving process improvement and business excellence through the effective use and integration of simulation and statistical methods. However, there are key project management techniques to consider for successful introduction and development of a 3S program in an organization. These techniques are essential for the successful implementation of a 3S program to any business process. The degree to which these techniques are applied represents the difference between a successful implementation of 3S program and a waste of resources.

Before discussing 3S project management techniques, it is imperative to keep in mind that the overall objective of a 3S project is to utilize simulation-based six-sigma as a powerful improvement business strategy that enables companies to achieve and sustain operational excellence. This operational excellence is evident in many leading companies that adopted six-sigma programs, including Motorola, GE, Honeywell, ABB, Sony, Texas Instruments, Rathyon, Textron, Bank of America, Ford, and Johnson Control Systems. These companies were able to use effective project management techniques to introduce and develop six-sigma project successfully. It is a good business practice to benchmark how these companies deploy six-sigma and strive to follow suit in the introduction and development of 3S projects.

In our analysis for 3S project management, we focus on the following aspects: (1) project selection, (2) project business case, (3) project charter, and (4) software tools.

11.4.1 3S Project Selection

Selecting the proper 3S project can have a tremendous effect on the production or service process targeted. As discussed in Section 11.2, a 3S application possesses certain characteristics through which the team can decide whether a potential improvement opportunity can be approached with 3S. If a 3S project is selected properly, the benefits of six-sigma and simulation integration can result in significant process improvement. Moreover, the project team will feel satisfied and appreciated for making business improvements, and ultimately, the shareholders will notice the benefit in financial terms. If the 3S project selection is done improperly, on the other hand, the team may not be able to complete the project or the project may not meet its goals, cost estimates, or time frame. Such negative results may also lead to losing management support and full business buy-in, and the project teams may feel ineffective. In this section we provide some guidelines for a proper 3S project selection.

The bottom line in a 3S project selection is to select projects that are in line with business priorities. Every business is different, and the 3S project team should ensure that business-specific priorities are taken into account when evaluating and prioritizing potential projects. Toward this end, two key considerations should be taken when assessing potential 3S projects:

1. *Project priority.* 3S projects should first be directed at the three greatest issues facing the business as seen from the eyes of customers. Selected issues are expected to have maximum benefit to the company tactically and strategically. A 3S project team should identify one or more measurable impacts on the business processes or on financial bottom lines. For example, reviewing a list of customer complaints may reveal an increased number of late or unfulfilled deliveries, an unacceptable waiting time of customers, noncompetitive prices of products or services, and so on. Focusing the effort on such issues will probably increase the level of project recognition and management support and will reflect project results on the business directly in terms of sales, profits, and possibly the market share. The project team will also be motivated when the potential expectations from the project are clearly defined and easily noticeable. Hoshin planning, coupled with graphical representation such as a Pareto chart, is often used to prioritize business objectives in a company. The well-known 80/20 rule can be applied to 3S project selection. For example, 3S projects are focused on that 20% of a company's processes that contribute 80% of the improvement possible. Organizations vary on what constitutes 20% of the processes. For example, mortgage and insurance companies often spend more time on the loan or insurance origination and underwriting cycle than on post-loan processes or claims processing.

2. *Project management.* It is essential to assess a potential 3S project from a project management and control perspective. For example, the team needs to check whether the potential project is manageable, whether the required project resources are realistic, and whether the project can be completed by the team within a reasonable time frame (typically, from 4 to 8 months). Addressing such concerns is essential for selecting the proper 3S project. Experience plays a key role in assessing 3S projects based on project management and control criteria.

Several methods can be utilized for selecting a 3S project from among a set of potential projects. For example, an analytical hierarchy process can be used to organize potential 3S projects by forming evaluation categories and ranking potential projects by assigning values and weightings to create a consistent selection process.

Table 11.1 presents ideas that can be used for identifying business issues at major functions in an organization. These ideas can then be evaluated and prioritized for 3S project selection.

TABLE 11.1   Ideas for 3S Projects by Business Function

Business Function	3S Project Ideas
Product or service development	Reduce product or service development time
	Increase product or service quality
	Reduce design changes
	Reduce product or service tests and reviews
	Reduce number of prototypes
Production	Increase productivity
	Reduce machine failures
	Reduce manufacturing lead time
	Reduce yearly inventory cost
	Reduce scrap and rework
Purchases	Reduce order cost
	Reduce order lead time
	Reduce transportation cost
	Improve supplies/logistics forecast
	Improve payment processing
Human resource	Reduce expenditures/time for recruitment
	Reduce the cost of health insurance
	Increase employee satisfaction
	Improve employee learning curve
	Improving workplace
Sales and marketing	Reduce time required to enter sales orders
	Reduce order fulfillment time
	Improve demand forecast
	Increase the number of returning customers
	Reduce cash-to-cash time

In summary, the following points are general guidelines for 3S project selection:

- 3S projects should possess the 3S application criteria (Section 11.2).
- 3S projects should have identifiable and measurable business benefits. Expected benefits from 3S projects typically include:
  - *Financial benefits:* reduced cost and increased sales, market share, revenues, and profit
  - *Operational benefits:* increased productivity, utilization, on-time deliveries, quality, and efficiency; reduced lead time, downtime, delivery tardiness and shortages, and number of defects
  - *Organizational benefits:* increased competitiveness, market reputation, and employee satisfaction
- To utilize the power of six-sigma, selected 3S projects should be approached from the perspective of understanding the variation in process inputs, controlling them, and eliminating waste and defects.

- To utilize the capability of simulation modeling, selected 3S projects should allow being approached from the perspective of modeling a production or business process and using the model as a flexible and inexpensive platform for six-sigma analyses.
- The project selected should be amenable to 3S-DFSS or a 3S-LSS.

11.4.2   3S Project Business Case

Six-sigma practitioners often view the project business case as a financial venue to justify the resources necessary to bring the 3S project effort to fruition. In fact, there are a number of valid reasons for writing a business case in any reengineering effort. Although all business cases should include financial justification, it should not be the only purpose of the endeavor. The 3S project business case should be viewed as an opportunity to *document* and link all relevant project facts into a cohesive story. This story summarizes the *why*, *what*, *how*, *where*, and *when* of the 3S project:

- Why is the 3S project needed (issues and opportunities)?
- What solution is recommended to address the issues or opportunities?
- How will the solution be deployed, and how much money, people, and time will be needed to deliver the solution and realize the benefits?
- Where will the solutions be deployed (location)?
- When will the solutions be deployed (time frame)?

Writing a business case to the 3S project forces the project team to diligently study, justify, plan, and document the project undertaken from concept to completion. This helps to summarize the knowledge developed by the team about the business in general and how the project fits into the company's overall short-term tactics and long-term company strategy. Therefore, the first key benefit of a 3S business case is to serve as an early wakeup call to the 3S team before committing to the project and allocating company resources.

The second key role of the 3S business case is to verify that the solution will substantiate or meet the business objectives. The 3S business case will simply state what would happen to the business if the 3S project is not undertaken. Financial justification usually helps identify holes or problems with the solution and provides a key measure for the project success. This forces the 3S project team to step back and review their facts and assumptions subjectively.

The third important role of the 3S project business case is in communicating the project to relevant business partners. From this perspective, the business case can serve as a high-level view of the entire 3S project, which enables all organizations affected by the project (suppliers, customers, management, sales, finance, etc.) to be cognizant and knowledgeable about the project.

The development of the business case will be a good opportunity to test the 3S project team competency and teamwork. A good business case is typically developed by a team that has an overall understanding of the entire project and can synthesize their understanding into one document and communicate it with all relevant parties. A best practice is to involve all members in discussing and approving the plan. In addition to creating team ownership, group participation provides multiple benefits through brainstorming and multidisciplinary interactions. If the team is not well organized or does not do a thorough job of substantiating facts throughout the project, the team probably will not be able to execute the project successfully.

Elements of a 3S Project Business Case at Launch   An effective 3S project business case states the issues, goals, and solution elements (the proposed changes) concisely, and clearly communicates the interrelationship. This business case is expected to involve the following main elements:

1. *Issues and opportunities (problem statement).* A thorough situation assessment in a production or business process often leads to identifying some issues and defining opportunities for improvement. These issues are often related to customer satisfaction and financial results. A strong situational assessment will assess the historic, current, and future issues related to operational performance, customers, employees, competitors, and market trends. A strong business case is expected to provide a thorough understanding of the issues facing the organization and the team conclusion about the opportunity for improvement.

2. *Project description.* In this section we describe the objective of the scope of the 3S project. Define the processes, systems, and organizations included within the scope of the 3S project. Summarize the activities performed by the 3S project team. Finally, include an overview of the stakeholders for whom this effort is being undertaken (customers, management, etc.).

3. *Project costs.* This section should include an estimate for every cost anticipated for the 3S project. This includes costs for the project team, development, quality assurance, testing, parallel operations during transition, and implementation. It should also include any ongoing maintenance or administrative costs. Calculate the impact on the operation due to finding a 3S solution and later, its implementation. This would include productivity losses or the need to hire temporary workers to cover for personnel while training. Project costs need to be revisited at project conclusion.

4. *Expected benefits.* The benefit section should quantify or qualify those benefits that were touched upon in the solution detail. Count benefits for any organization that will reap positive results from the solution. Benefits should be both qualitative and quantitative and include cost reductions, revenue increases, improved customer satisfaction, improved employee morale, and lower turnover. Categorize benefits into groups for ease of understanding. Some business cases link the benefits directly to the solution elements. However, this is not always possible. Other methods for categorization include:

- Organization affected (customers, development, parts, service, etc.)
- Type of benefit (cost reduction, increased revenue, etc.)
- Timing of benefit (immediate, first year, future, etc.)

5. *Implementation time line.* Depict each major step in implementation of the solution on a time line. Major steps should include development, testing, training, initial implementation, and rollout. Consider any impacts to the organization from a productivity or operational viewpoint. Most implementation plans get dictated by the systems development schedule because it is usually the least flexible and has the most dependencies. However, it is important that your team think about each solution element and define an implementation time line that will maximize benefits while having the fewest impacts on the organization. The implementation focus will enable you to prioritize the steps in the implementation. For example, if the focus is to streamline the operation, the priority would be to implement the automated functions quickly. To help you to focus your implementation, link your costs and benefits to the time line.

6. *Assumptions and risk assessment.* List all assumptions made by your 3S project team. Include assumptions about the current state of the business, the status quo of organizations, processes and systems that are outside the scope of the project, constants used in cost–benefit analysis, the approval of the business case, and so on. Your description should indicate the impact to the solution if the assumptions did not hold true. Discuss the risks of implementation. Discuss what will happen to the organization if the benefits from the reengineering effort are not realized. Include an assessment of the risks caused by implementation of the ongoing operation of the business. Discuss the steps that will be taken to minimize or mitigate each risk.

Elements of a 3S Project Business Case at Conclusion

7. *Solution overview.* In this section we define the desired end state for the 3S project. The end state provides the framework for a solution definition that includes a goal statement. In addition, the solution overview should provide a high-level description of the solution (it should paint a picture for the reader of what the end state will look like).

8. *Solution detail.* This section should walk the reader through all aspects of the solution implementation. This would include:

- Changes in the organization (people, culture, training, etc.)
- Changes in the processes
- Changes in the support systems

It is important that the solution be presented from the viewpoint of the organization receiving the benefit of the solution. For example, present the solution through the eyes of the customer if the goal of the effort is to improve customer service. The solution detail should point out clearly how issues presented earlier are being resolved by this solution.

9. *Solution alternatives.* Discuss the alternatives to the solution proposed. This must include discussion of the implications to the organization if this 3S project solution is not implemented (the do-nothing scenario). A discussion of solutions tried previously that failed may also be appropriate here if it helps to justify why the 3S project is required. This would typically be the case if past efforts were Band-Aid or incremental solutions instead of reengineered solutions.

10. *Project costs.* Project costs need to be revisited at project conclusion.

11. *Expected benefits.* Project benefits need to be revisited at project conclusion.

12. *Implementation time line.* The project time line needs to be revisited at project conclusion.

13. *SWOT analysis.* This analysis looks at the strengths, weaknesses, opportunities, and threats (SWOT) of the solution being proposed. Demonstrate how the organization will maximize strengths and minimize weaknesses of the solution. Include a discussion of the opportunities now possible because of the solution. Include a means to minimize and prevent threats to the organization caused by the solution. A FMEA is a good tool in this regard.

14. *Executive summary.* The executive summary provides management with a short (one to three pages) snapshot of the project findings and business case. It must be persuasive and provide sufficient language to paint an accurate picture. The focus of the executive summary should be on the bottom-line benefits to the organization. All of the other information in the business case should be summarized as supporting details. Conclusions should summarize the issues, costs, and benefits of the solution. Demonstrate that the financial benefits outweigh the costs by including a financial return-on-investment analysis. Convey a sense of urgency.

General Tips for Developing a 3S Project Business Case   Although initially, the primary goals of the 3S business case is to obtain funding, the chances of success will be greater when simplifying and organizing the business plan so that the financial justification is clear and any document generated by the team (e.g., project charter) is easy to read and understand. This can be attained by paying attention to the following:

- Focus on demonstrating the value the project brings to the organization's customers and financial bottom lines.
- Support the case with clear technical and financial illustrations.
- Conduct a team review for potential inconsistencies or weaknesses.
- Keep the format clear, interesting, and concise.
- Minimize jargon and conjecture.
- Present the case as a team product.
- Present all facts as part of the overall cohesive story.

11.5 3S PROJECT CHARTER

Impressive bottom-line results can be achieved from 3S projects when they meet certain fundamental requirements. A successful project should be defined properly (have clearly defined deliverables), be approved by management, not be unmanageable (so large in scope) or trivial, and relate directly to the organization's mission. The 3S project charter is prepared to address these requirements in the project definition. A project charter contains elements of its business case.

A project charter is a key first step in the 3S project to specify the scope and the necessary resources and boundaries that are essential to ensure project success and to create and maintain team focus, effectiveness, and motivation. The 3S project charter is a high-level document that is issued by the 3S project sponsor (owner) to authorize the 3S project and to provide a clear definition to the problem addressed by the project, the project mission, scope, deliverables, and a documentation of project resources, team, and time frame. This key part of the project plan consists of the following:

- Problem statement: a description of the business need to be addressed by the 3S project (i.e., the opportunity or issue addressed by the project and approved by the organization).
- The mission of the project team and the project link to the organization's mission, objectives, and metrics. A specific and informative mission statement that is focused on root causes rather than symptoms may require several revisions as the team gains more insights.
- A description of the project's objectives, scope, and major stakeholders.
- A work-breakdown structure with a detailed time frame.
- A summary of project deliverables (outcomes).
- The authorization to apply organizational resources to the project. A document that provides the 3S project manager with the authority to apply organizational resources to project activities.

Although developing a 3S project charter can be a fairly simple process, it represents an important step in the project. The process involves several steps defined at a high level and refined systematically as the project proceeds. These refinements will require frequent communication between the project team and the project sponsor. Generally speaking, there are six steps in the 3S project chartering process:

1. Obtain a problem statement.
2. Identify the principal stakeholders.
3. Create a macro flowchart of the process.
4. Select the team members.

5. Identify the training to be received by the team.
6. Select the team leader.

The content of a 3S project charter is very much similar to a six-sigma project charter. It includes a concise description of the elements above. Similarly, different formats can be used to structure a 3S project charter. A proposed template follows:

3S Project Charter

*   Project information
    1. Project title
    2. Sponsoring organization or division
    3. Project leader (black belt, green belt, master belt)
    4. Project mentor
    5. Project team
        *   Six-sigma black belt
        *   Simulation expert
        *   Subject matter experts
        *   Project management expert
*   Project description
    6. Problem statement
    7. Mission and objectives
        *   Mission statement
        *   Project objectives
    8. Work description
        *   Work focus
        *   Work elements
        *   Work plan
        *   Work resources
*   Project expectations
    9. Project deliverables
        *   Process measure 1
        *   Process measure 2
        *   Process measure 3
    10. Project time frame
        *   Start date
        *   Closing date
        *   Project milestones

11.6   3S SOFTWARE TOOLS

Many software tools can be utilized on 3S projects. Just as the popularity of six-sigma seems to continue its upward trend, so will the number of software companies vying for the business of providing six-sigma supporting software tools. Commercially available 3S software tools can be classified into three categories: six-sigma software tools, simulation software tools and packages, and simulation-six-sigma modules and software tools. We discuss each category next.

11.6.1   Six-Sigma Software

There is hardly a six-sigma project executed without utilizing one or more six-sigma software and supporting tools. Since the six-sigma initiative includes strict adherence to templates, methods, and tools, it makes sense that computer technology is one of the first resources to consider in six-sigma projects. Keeping in mind that six-sigma software tools were initially developed to focus the businesses on effective development of six-sigma projects, it is imperative to make sure that any software tool we select fits the specific needs of six-sigma projects in order to make our journey toward six-sigma quality as smooth as possible.

But selecting six-sigma software is not a trivial task. Six-sigma is a comprehensive, holistic approach. It is unlikely, however, to find any one software tool to meet all six-sigma needs. Six-sigma is synonymous with extensive data analysis. Consequently, a major sector of the six-sigma software market is statistical tools, which different software developers provide in different formats. For example, some six-sigma software tools provide a suite of statistical analysis that is based on Microsoft Excel spreadsheet applications. For example, MINITAB is popularly used as a stand-alone statistical and graphical analysis software application. Whereas the Excel-based applications are user-friendly with high popularity, the stand-alone MINITAB and similar products exceeds the limits of the base application.

Nevertheless, statistics are just one element of six-sigma software. As discussed in earlier chapters, applying six-sigma using the DMAIC method requires a wide spectrum of analyses, from basic analyses such as descriptive statistics, basic control charts, and hypothesis testing (e.g., t-test and F-test) to advanced tools such as DOE, Taguchi designs, and regression analysis with prediction and multiple-response optimization. In addition to statistical analyses tools (such as MINITAB), six-sigma practitioners also rely on many software tools, such as QFD, FMEA, MS Office tools (mainly PowerPoint, Excel, and Word), team collaboration tools (such as MS Outlook, virtual meeting tools, and a shared Access database), a good graphical tool (such as Visio), good statistical software, and of course, a simulation software. For DFSS application, teams may need some form of Monte Carlo simulation that can predict defects based on an engineering relationship or an empirical model, also known as a *transfer function*. Discrete event simulation can play the role of a transfer function, especially for transactional and service DFSS projects.

It is also important to consider the future needs of six-sigma projects when looking for the six-sigma software solution. Invest in packages that offer a room for potential growth in needs and applications. A few considerations when choosing six-sigma software include:

1. Consider Web-based software tools and solutions when people at multiple sites need access to project information.
2. Consider the training cost when selecting software with complex features, especially when project team computer literacy with the software is limited.
3. Trade off the software cost and software capabilities. For example, most data analysis software tools, including a suite of statistical tools, can cost up to $3000. Cost for good process mapping and simulation packages range from $1000 to 10,000. Commercial QFD, FMEA, and team tools are usually priced below $500 per license.

The following is a brief review of two popular six-sigma tools (i.e., MINITAB and NetSigma).

MINITAB MINITAB statistical software is a common package for six-sigma and other quality improvement projects. From statistical process control to design of experiments, MINITAB offers an array of methods that are essential to implement every phase in six-sigma projects, along with features that help in understanding and communicating the results.

MINITAB offers many features, such as a powerful graphics engine, that delivers engaging results that offer tremendous insight into data, an effortless method to create, edit, and update graphs, and the ability to customize menus and toolbars to conveniently access the most frequently used methods. Table 11.2 presents the main tools and features available in MINITAB.

NetSigma NextSigma is a comprehensive six-sigma software package that provides innovative software solutions that have been developed specifically

TABLE 11.2 Summary of MINITAB Tools for Six-Sigma

Data and File Management	Design of Experiment
Graphics	Reliability and survival analysis
Macros and command language	Power and sample size
Basic statistics	Multivariate analysis
Regression analysis	Time series and forecasting
Analysis of variance	Nonparametrics
Statistical process control	Tables
Measurement systems analysis	Simulation and distributions

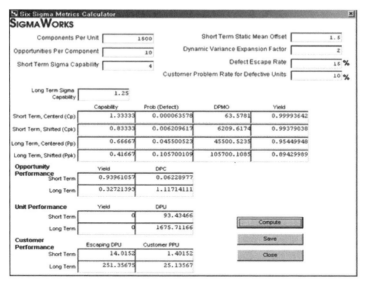

Figure 11.1   SigmaWorks six-sigma calculations.

for DMAIC six-sigma, DFSS, project management, quality improvement, simulation optimization, and test and measurement.

NextSigma has created several products specifically to fill gaps in the software tool sets that are currently available to six-sigma practitioners. These products include SigmaWorks and RiskWizard. SigmaWorks is a stand-alone Excel add-in that contains the following features: process mapping, project planning, Monte Carlo simulation, optimization, FMEA, cause-and-effect matrix, concept selection, quality function deployment, SIPOC, control plans, DMAIC road maps, statistical tables, queuing analysis, value stream analysis, cycle-time reduction, financial analysis routines, data analysis, and many more. Figure 11.1 shows a screen shot of SigmaWorks six-sigma calculations.

RiskWizard is a software tool that provides state-of-the-art test and measurement analysis using certain specification limits. The user may compute alpha and beta risks for actual test data based on specification limits. The user may also interact with a test and measurement model to determine the effect of different specification limits. The software supports random effects, fixed effects, and mixed models, as well as multiple correlated responses. The software does not require balanced data as do most standard statistical software programs. Table 11.3 summarizes the capabilities of RiskWizard and SigmaWorks NextSigma products.

The SigmaWorks pull-down menu provides access to various six-sigma tools, DMAIC planners, statistical calculators, simulation and optimization, data analysis, and project management tools. If you have selected multiple responses, you will be directed to the multivariate risk analysis page. This page displays the joint probabilities of true passes, false passes, false failures, and

TABLE 11.3 Summary of NextSigma Uses in Six-Sigma Projects

	SigmaWorks	RiskWizard
DMAIC champion		
DMAIC master black belt		×
DMAIC black belt		
Industrial	×	×
Service	×	
DMAIC green belt		
Industrial	×	×
Service	×	
DFSS champion		
DFSS master black belt		×
DFSS black belt		
Industrial	×	×
Service	×	
DFSS Green Belt		
Industrial	×	×
Service	×	
Project planner	×	
Quality specialist	×	×
Quality manager	×	
Test and measurement technologist		×

true failures based on the covariance structure of your data set as well as on the specifications you entered. Project selection software offers great functionality. The covariance matrices are displayed for both the factors and the errors. Figure 11.2 shows examples of process mapping and Monte Carlo simulation in SigmaWorks. A list of six-sigma software vendors is provided in Appendix F.

11.6.2 3S Simulation Software

In the corporate world, simulation is increasingly commonplace. Companies use simulation at all phases of product and process development throughout the entire life cycle of their production facilities. As the size and complexity of systems increase, simulation is no longer a luxury but a necessity for proper analysis to support good decisions.

Simulation software is a powerful tool for designing new systems, quantifying performance, and predicting behavior. The focus here is on products that run on desktop computers to perform discrete event simulation. These software tools are common in management science and operations research. Simulation products whose primary capability is continuous simulation (systems of differential equations observed in physical systems) or training (e.g., aircraft and vehicle simulators) are not the focus of this book.

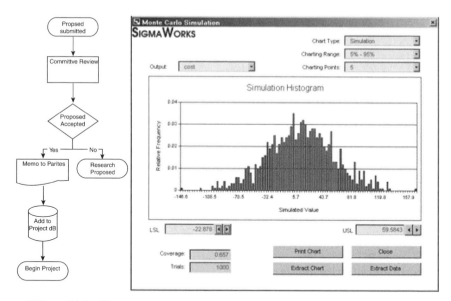

Figure 11.2 Process mapping and Monte Carlo simulation in SigmaWorks.

Whether the system is a production line, a distribution network, or a communications system, simulation can be used to study and compare alternative designs or to troubleshoot existing systems. Simulation software tools are also beneficial to model and analyze existing system, visualize and observe system behavior, and introduce and test changes and improvement actions. The ability to construct and execute models easily and to generate statistics and animations about results has always been one of the main attractions of simulation software.

Simulation software is built from entities and processes that mimic the objects and activities in the real system. The spectrum of simulation software currently available ranges from procedural languages and simulation languages to special "simulators." The vast amount of simulation software available can be overwhelming for new users. The following are only a sample of common software in the market today: Arena, AutoMod, GPSS, Taylor II, Simul8, Quest, and WITNESS.

The important aspects to look for in a simulation package depend first on the specific applications. Several factors make an ideal simulation package: ease-of-use, comprehensiveness, optimization functionality, cost of package and cost of training, strong animation, statistical analysis, summary reports, modeling capabilities, and support, among others. Selecting a simulation software package with specific properties depends primarily on the application. Hence, asking which package is best should be addressed after studying the portfolio of applications that need it. Is it for business modeling, consultations, making decisions, six-sigma projects, education, teaching, student projects, or research?

In 3S, building a simulation model is rarely an end in itself; instead, the goal is usually to run simulation analyses and make a decision regarding a certain system. To assist in analysis, simulation software tools are increasingly able to exchange information with other software tools in an integrated manner, and simulation tools are being positioned as modeling tools within a general decision-making framework. For instance, simulation may be integrated with presentation software and graphical tools to document and report on findings. Examples include simulation integration with process mapping tools, the Visio graphical tool, and MS Office tools. They can also be integrated with spreadsheet or statistical software to facilitate analyses of model outputs. A common example is simulation integration with MINITAB in WITNESS and with spreadsheets in Simul8. Similarly, simulation models may be driven by solutions obtained from experimental design, scheduling, or optimization software and then run to provide insight into the proposed solution or to aid in the evaluation among alternative solutions. Examples include simulation integration with experimental design in AutoMod AutoStat and with optimization search engines such as simulated annealing in Witness Optimizer and the genetic algorithm in ProModel SimRunner. Such integration speeds the process of design evaluation and optimization, which makes simulation a valuable component within the decision-making process.

Many simulation products now also include tools for data analysis for both input and output. For instance, products include distribution fitting software or convenient access to tools such as Expertfit, Bestfit, or Statfit distribution-fitting software. Another example is input modeling and fitting probability distributions in ARENA simulation package and the capability of WITNESS and AutoMod to import AutoCAD files. Support for run control and statistical analysis of results simplifies simulation analysis and makes simulation packages more common. Examples include allowing the replication of simulation runs and the analyses of run controls, running complicated experimental designs, and supporting the production of summary graphics and reports for the storage of production runs.

Simulation is only one modeling methodology. Since simulation models operate by the construction of an explicit trajectory or realization of a process, each replication provides only a single sample of the possible behavior of a system. The accuracy of inference can be improved through repetition, so that the analysis provides a statistical estimate of any particular system parameter. This is in contrast to many analytical models that provide exact results over all possible results.

Another key feature of simulation software tools is the flexibility of simulation in coding and logical design and the direct representation of entities within the simulation using a window-driven approach. Since simulation directly represents a realization of the system behavior dynamically, simulation packages are often joined to animated outputs and used for validation, analysis, and training. The interplay between simulation and visualization is a powerful combination. Visualization adds impact to simulation output and

Figure 11.3   Manufacturing facilities three-dimensional layout and workstation design software.

often enhances the credibility of the model. Almost all simulation software has some level of animation, from the basic to products that make you feel that you are there. Developers are already taking the next step, combining simulation with virtual reality to provide a simulation-based background for exploration, much as might be done in a training simulator. Movies and video clips are just another type of visualization available in simulation packages. There is an immediate difference in visualizations based on simulations and those used for entertainment. Figure 11.3 shows an example of three-dimensional simulation animation in AutoMod.

In addition to user groups, there are a number of organizations and conferences devoted to the application and methodology of simulation. The INFORMS publications *Management Science, Operations Research* (MS/OR), and *Interfaces* publish articles on simulation. The INFORMS College on Simulation sponsors sessions on simulation at the national meeting and makes awards for both the best simulation publication and service awards in the area, including the Lifetime Achievement Award for service to the area of simulation. Further information about the College on Simulation can be obtained from the Web site www.informs-cs.org. This site now provides the complete contents of the proceedings of the conference for the last three years and also contains links to many vendors of simulation products and sources of information about simulation, simulation education, and references about simulation. The Winter Simulation Conference (www.wintersim.org) also provides an intensive yearly update on simulation software, research, and application.

In addition to the changes in products in the survey, the corporate world of simulation has undergone significant changes in the last few years, as many of the well-known names of simulation have been acquired by larger concerns. For instance, Pritsker Corporation has been a division of Symix (itself newly renamed Frontstep) for several years, Systems Modeling is now a division of Rockwell Software, and Autosimulations is now a part of Brooks Automation. These acquisitions often provide a stronger financial base for the divisions to promote and develop their products. Meanwhile, CACI has reorganized its software sales, selling several of its products, and is emphasizing its large governmental consulting services.

The simulation software market appears to be fairly well established and international. Within the U.S. market, it appears that simulation software is now a standard software category, much as a spreadsheet program or a word processor is in the general business software environment. As with spreadsheets and word-processing software, simulation software needs to work well with the other programs within the suite of software used in analysis. Beyond working with business software, changes to simulation products suggest that simulation tools will increasingly be linked to other decision software, such as production scheduling software.

A recent simulation software survey appeared in the August 2003 issue of *OR/MS Today* (www.lionhrtpub.com). The information in the survey was provided by vendors in response to a questionnaire developed by James Swain. The survey should not be considered comprehensive, but rather, as a representation of available simulation packages. Questionnaires were sent to simulation vendors drawn from previous survey participants, the OR/MS Today database, and other sources. It includes the products of those vendors who responded by the deadline.

This survey is not intended as a review of the products or a comparison of their relative merits. Such target can be tackled taking into account the general qualities of the software as well as the fitness of specific products for particular classes of problems, the client requirements, and the special features of the problem. Like all software, the relation between the vendor and the user is ongoing. Products evolve over time, and new versions of the software become available periodically. In addition, the vendor is a source of information about both its products and their application. Most vendors of simulation software now maintain contact with their users through mailings, newsletters, their own Web sites, and annual user group conferences. These conferences showcase the application and usefulness of products, nurture contact with users, and provide a way for users to learn from each other.

The survey summarizes important product features of the software, including price. The range and variety of these products continue to grow, reflecting the robustness of the products and the increasing sophistication of users. The information elicited in the survey is intended to give a general gauge of the product's capability, special features, and usage.

The survey is divided into seven separate pages. Following is an index of the pages and the information they contain:

- Page 1
  - Vendors
  - Typical applications of the software
  - Primary markets for which the software is applied
  - System requirements: RAM, operating systems
- Page 2
  - Model building: graphical model construction (icon or drag-and-drop)
  - Model building using programming or access to programmed modules
  - Run-time debug
- Page 3
  - Model building (continued): input distribution fitting (specify), output analysis support (specify), batch run, or experimental design (specify)
- Page 4
  - Model building (continued): optimization (specify), code reuse (e.g., objects, templates)
  - Model packaging (e.g., can completed model be shared with others who might lack the software to develop their own model?)
  - Tools to support packaging (specify) (Does this feature cost extra?)
- Page 5
  - Animation: animation, real-time viewing, export animation (e.g., MPEG version that can run independent of simulation for presentation), compatible animation software
  - Price: standard, student version
- Page 6
  - Major new features (since February 1999)
  - Vendor comments
- Vendor list
  - Software vendor contact information (Appendix F)

11.6.3 Simulation-Based Six-Sigma Software

There is no specific simulation-based six-sigma software tool. In this section we introduce a framework for such software and discuss some commercial software tools that started to include some six-sigma aspects.

For six-sigma software, simulation is the process of building and using a time-based visual model, which emulates every significant step that occurs in a process and every significant interaction between resources in a process so as to gain insight about the impact of potential decisions on that process. The model shows you visually what will happen in the process if you make changes to it and it records performance measures of your system under different scenarios. Simulation lets you explore an electronic model of the project you manage: whether the project is a process, a portion of your process, a hospital, an administrative center, or whatever. The type of model it provides is time-based, and takes into account all the resources and constraints involved and the way that all these things interact with each other as time passes.

WITNESS Six-Sigma Module WITNESS (Lanner Group) has a built-in module for six-sigma calculations. Enhancements to WITNESS 2002 edition included the ability to run six-sigma tools. The software module provides an analysis for six-sigma metrics such as scrap and rework and links the output directly to the MINITAB package. The WITNESS OPTIMIZER module also uses six-sigma in optimization to identify best areas and options available for specific process or machine improvements. WITNESS MINER, Lanner's pattern-matching and rule-finding engine, completes the six-sigma tool kit by helping identify key patterns in project data. Figure 11.4 shows a screen shot of the WITNESS six-sigma module.

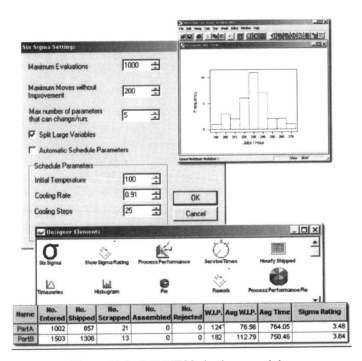

Figure 11.4 WITNESS six-sigma module.

The fundamental objective of a six-sigma methodology in WITNESS is to implement a measurement-based strategy focused on process improvement and reducing internal or external variables in model response. Because six-sigma success is driven by metrics and process methodology, long-term success requires agile and robust technology and solutions that deliver the metrics and measures, an integrated view of the business, simulation of "what-if" scenarios, and optimization of stated measures. WITNESS supports continuous improvement programs using six-sigma, a data-driven methodology for eliminating defects in processes. This represents dramatic improvements and key enhancements to the WITNESS 2002 release. WITNESS provides in-depth support for six-sigma projects including:

- Direct linkage to MINITAB statistical software
- Unique WITNESS six-sigma optimization algorithm
- Direct reports on sigma ratings
- Function calculation for multidefect sigma ratings
- WITNESS designer element palette for six-sigma

These enhancements in WITNESS build on the powerful and flexible nature of WITNESS models, which allows for the flexible rules that govern process operation, including aspects such as resource allocation, distributional timings, complex routing, and work schedules and rates. WITNESS easily handles concepts such as scrap and rework and also allows easy use and export of data to and from databases and Microsoft Excel. Some of the new WITNESS screens offer rating statistics and a view of the results of the unique WITNESS six-sigma optimizer algorithm that applies intelligent search methods but limits the overall levels of change to fit in with the process.

@Risk Software   An available software for six-sigma and Monte Carlo simulation is @Risk. It works as a companion to any six-sigma or quality professional to allow for a quick analysis of the effect of variation within processes and designs. Furthermore, @RISK enhances any Microsoft Excel spreadsheet model by adding the power of Monte Carlo simulation. Monte Carlo simulation is a technique that examines the range of possible outcomes in a given situation and tells you how likely different scenarios are to occur. Six-sigma is defined as identifying, and minimizing, the percentage occurrence of defects and errors in a process. Figure 11.5 shows a framework for @Risk software with six-sigma calculations.

The various @RISK functions can be used to customize underlying models with the performance measures that are suited for the six-sigma project. @Risk can be combined with RISKOptimizer (a genetic algorithm-based optimization tool that finds the best solution to complex nonlinear real-life problems). Applications of @RISK and RISKOptimizer in six-sigma analyses include the following:

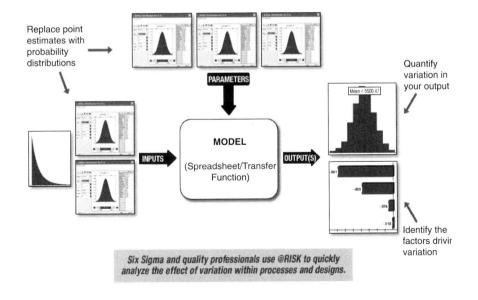

Figure 11.5   Six-Sigma calculations in @Risk software.

* Easy, accurate definition of variation using 38 probability distribution functions
* Sensitivity analysis to identify key factors that drive variation and uncertainty
* Distribution fitting to allow quick definition of a data set
* Scenario analysis to determine which combinations of input variables led to different outcomes
* Use of the fastest Monte Carlo simulation engine on the market to save valuable time
* Ability to use multiple CPUs in a single machine to speed up large simulations
* Correlation of uncertain inputs to reflect real-life dependencies between elements

- Risk analysis to determine the extent of quality issues and identify the key drivers
- Optimization to generate a workable solution and meet project goals
- Seamless integration of risk analysis and optimization to perform multiple analyses on the same models

A six-sigma analysis with @RISK consists of the following basic steps:

1. Define the model in Excel by outlining the structure of the problem in a spreadsheet format.
2. Replace uncertain factors in the model with probability distribution functions using historical data or expert judgment.
3. Identify the bottom lines (performance measures) by tracking the values in certain model cells (e.g., the number of service complaints in a year, a target dimension, the number of defective parts per batch).
4. Run the Monte Carlo spreadsheet simulation. @RISK will recalculate the spreadsheet model thousands of times while choosing values at random from the input distributions defined and record the resulting outcome. The results summarize all possible outcomes and their probabilities of occurring. These results can be used by six-sigma analysts to improve the quality of products and services.
5. Graph and analyze the results and make decisions regarding the model. @RISK provides a wide variety of graphs and statistical reports representing the data from the simulation. @RISK also offers sensitivity and scenario analysis that identifies which factors or combinations of related factors contribute to the results obtained.
6. Practice simulation optimization. For example, RISKOptimizer can be used to optimize process settings to maximize yield or minimize cost, optimize tolerance allocation to maximize quality levels, and optimize staffing schedules to maximize service levels.

Crystal Ball   Crystal Ball (www.crystalball.com) is an easy-to-use simulation program that helps you analyze variation and uncertainties associated with Microsoft Excel spreadsheet models. Crystal Ball is a risk-analysis tool. The software provides a spreadsheet environment to better calculate, communicate, and diminish business risks based on Monte Carlo simulation and global optimization. With Crystall Ball, professionals make critical business decisions using innovative charts, reports, and simulation technology. Practical uses of the software include six-sigma, reserve analysis, financial valuation, environmental and waste assessment, market evaluation, portfolio planning, and sales forecasting.

Key Crystal Ball features of interest to six-sigma practitioners include sensitivity analysis, correlation, and historical data fitting. The sensitivity analysis helps you to understand which of the variables are most critical and drive the uncertainty in your process. Correlation lets you link uncertain variables and account for their positive or negative dependencies. If historical data do exist, the data-fitting feature will compare the data to the distribution algorithms and calculate the best possible distribution fit for your data.

Crystal Ball is used to predict the quality and variability of processes and products prior to production. The models are derived from the transfer functions, and it is critical to add uncertainty and risk to these models. The ability to quantify this uncertainty easily using Crystal Ball is critical for six-sigma analysis. As a tool for risk analysis, therefore, Crystal Ball is common in six-sigma projects, training, and education at leading companies and prominent six-sigma consulting firms. Six-sigma black belts and green belts, quality professionals, design engineers, supply chain management or material control professionals, operations, project managers, and service organizations rely on Crystal Ball to improve the quality of their decision to improve processes. With Crystal Ball, six-sigma practitioners attempt to understand the effects of uncertainty within their current process and products and to obtain early visibility into the performance and variability of new processes and product designs. The Professional Edition of Crystal Ball includes a suite of additional analysis tools for optimization and time-series forecasting. Figure 11.6 shows a screen shot of Crystal Ball software.

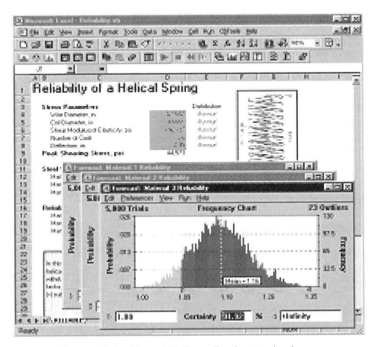

Figure 11.6   Crystal Ball application to six-sigma.

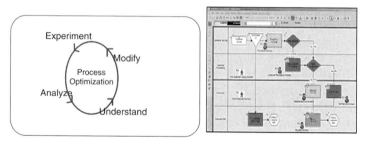

Figure 11.7   Simulation with the SigmaFlow modeler.

Crystal Ball can be also used in DFSS to:

- Identify key parameters driving variation
- Understand and reduce variation
- Obtain early visibility into design performance
- Create robust designs
- Optimize parameters and tolerances

SigmaFlow   SigmaFlow seamlessly integrates the complex data and relationships that make up the organization, and provides an integrated one-stop-shop tool kit for managing lean and six-sigma projects. The SigmaFlow product family includes the following:

- Modeler (discrete event simulation)
- Workbench (six-sigma, compliance, and statistics)
- Mapper – lean (value process mapping)

For the simulation side of a six-sigma project, the SigmaFlow modeler provides an integrated simulation engine for discrete event simulation. As shown in Figure 11.7, with the Modeler you can experiment, analyze, understand, and modify your processes as follows to produce optimized processes as new information or data are realized:

- Experiment and analyze results in a risk-free environment.
- Improve throughput by studying bottlenecks and cycle times.
- Understand the impact of process changes and resource adjustments.
- Perform financial analysis.

For the six-sigma side, the SigmaFlow WorkBench provides an integrated tool kit for six-sigma analyses. It provides three-dimensional Vital Cause holistic transparency between the six-sigma decision tools to demystify six-sigma

TABLE 11.4 Features of SigmaFlow Products

Tool/Feature/Report	Mapper	Workbench	Modeler
Drawing productivity features: 50% fewer clicks	×	×	×
Data panel: Always visible, provides easy data-entry mechanism	×	✓	✓
Display/surface process data on map: Always reconciled	×	✓	✓
Automatic reports: Integrated with your process map	×	✓	✓
Value-added report: Analyze as-is vs. to-be, value-added vs. non-value-added	×	✓	✓
Vital path report: Automatic rolled throughput yield calculation	×	✓	✓
Project prioritization: Contrast project effort to value		✓	✓
CT-tree (hierarchy view of customer requirements)		✓	✓
Fishbone: N levels with an output report		✓	✓
Detailed process map: Display six-sigma inputs and outputs		✓	✓
SIPOC: Scope your project . . . easier than Excel		✓	✓
Cause-and-effect matrix: Automatic charting and 3D vital cause transparency		✓	✓
FMEA: Built-in rankings and charts allow risk reduction		✓	✓
Control plan: Tightly integrated with other decision tools		✓	✓
Project charter		✓	✓
Project action plan		✓	✓
SigmaLink: Connectivity hub		✓	✓
Audit plan		✓	✓
Funnel report: Vital cause dashboard and report-out		✓	✓
Data collection plan: Tightly integrated with other tools		✓	✓
Six-sigma input/output summary report		✓	✓
Variability modeling: Over 20 standard built-in statistical distributions			✓
Resources: Flexible resources and resource pools			✓
Trials: Quickly run trials with different random number seeds			✓
Priority modeling: Assign work, resources, workstation priorities			✓
Workflow management: Advanced routing-in and routing-out control			✓
Scenario planning: What-if analysis, compare results side by side			✓
Queuing time, service level, utilization, bottlenecks (constraints) analysis			✓
Arrivals modeling: Time-based/independent			✓
Simulation reporting: Profit and loss statement, transactional cycle-time report, custom reports			✓

to save time and improve results sustainability. On the lean side, the SigmaFlow mapper provides a tool for value process mapping, waste eliminating, and process improvement. Table 11.4 summarizes the features of SigmaFlow products. A list of simulation vendors is provided in Appendix F.

APPENDIX A

BASIC STATISTICS

A working knowledge of statistics is necessary to an understanding of simulation-based six-sigma, design for six-sigma, and lean six-sigma. In this appendix we provide a basic review of appropriate terms and statistical methods that are encountered in this book. Note that this appendix barely touches the surface, and we encourage the reader to consult other resources for further reference.

Statistics is the science of data. It involves collecting, classifying, summarizing, organizing, analyzing, and interpreting data. The purpose is to extract information to aid decision making. Statistical methods can be categorized as descriptive or inferential. Descriptive statistics involves collecting, presenting, and characterizing data. The purpose is to describe the data graphically and numerically. Inferential statistics involves estimation and hypothesis testing in order to make decisions about population characteristics. The statistical analysis presented here is applicable to all analytical data that involve multiple measurements.

COMMON PROBABILITY DISTRIBUTIONS USED IN SIMULATION STUDIES

Statistical methods of input modeling, such as descriptive statistics, removing outliers, fitting data distributions, and others play an important role in

Simulation-Based Lean Six-Sigma and Design for Six-Sigma, by Basem El-Haik and Raid Al-Aomar
Copyright © 2006 John Wiley & Sons, Inc.

TABLE A.1 Probability Distributions Commonly Used in Simulation

Density Function	Graph

Bernoulli distribution:

$$p(x) = \begin{cases} 1-p & \text{if } x = 0 \\ p & \text{if } x = 0 \\ 0 & \text{otherwise} \end{cases}$$

Generalized random
   experiment of two
   outcomes

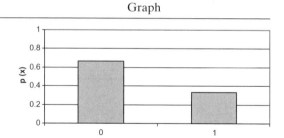

Binomial distribution:

$$p(x) = \binom{n}{x} p^x (1-p)^{n-x}$$

Number of successes in n
   experiments (number of
   defective items in a batch)

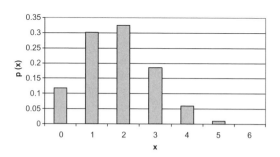

Poisson distribution:

$$p(x) = \frac{e^{-\lambda}\lambda^x}{x!}, \quad x = 0, 1, \ldots$$

Stochastic arrival processes;
   λ: average number of
   arrivals per time unit

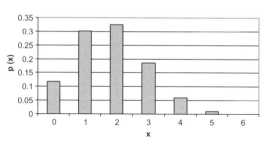

Geometric distribution:

$$p(x) = p(1-p)^x$$

Number of failures before
   a success in a series of
   independent Bernoulli trials

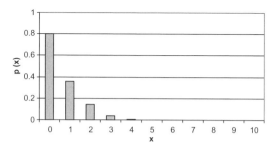

TABLE A.1 *Continued*

Density Function	Graph

Uniform distribution:

$$f_U(x) = \frac{1}{b-a}, \quad a \le x \le b$$

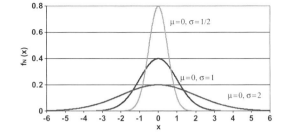

Random number
  generation

Normal distribution:

$$f_N(x) = \frac{1}{\sqrt{2\pi}\sigma}\exp\left[-\frac{(x-\mu)^2}{2\sigma^2}\right]$$

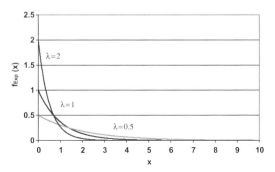

Natural phenomena of large
  population size

Exponential distribution:

$$f_{Exp}(x) = \lambda e^{-\lambda x}$$

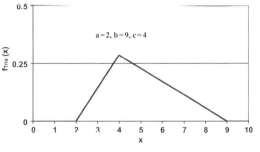

Reliability models:
  Lifetime of a component
  Service time
  Time between arrivals

Triangular distribution:

$$f_{tria}(x) = \begin{cases} \dfrac{2(x-a)}{(b-a)(c-a)}, & if \ a \le x \le c \\ \dfrac{2(b-x)}{(b-a)(b-c)}, & if \ c < x \le b \end{cases}$$

TABLE A.1 *Continued*

Density Function	Graph
Gamma distribution: $$f_{\text{gamma}}(x) = \frac{\lambda}{\Gamma(\lambda)}\lambda x^{k-1}e^{-\lambda x}$$ Failure due to repetitive disturbances Duration of a multiphase task	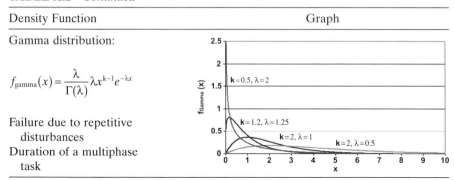

analyzing historical data. The largest value-added result of simulation modeling is achieved from analyzing simulation outputs to draw statistical inferences
and from optimizing the model design parameters through experimental
design and optimization. DES models provide a flexible and cost-effective
platform for running experimental design, what-if analysis, and optimization
methods. Using the results obtained from the simulation analysis, decision
makers can draw better inferences about model behavior, compare multiple
design alternatives, and optimize the model performance. The following are
some prerequisites for running simulation experiments and optimization
studies:

1. Understanding the type of simulation (terminating or steady state)
2. Selecting the system performance measure(s), such as throughput, lead
   time, cost, and utilization
3. Selecting model control and random factors that affect model performance and representing them statistically in the model (examples include
   arrival rate, cycle times, failure rate, available resources, speeds)
4. Determining the simulation run controls (warm-up period, run length,
   number of simulation replications)
5. Understanding the stochastic nature of the simulation model and including the variability in simulation outputs in the analysis
6. Familiarity with various statistical tools, experimental design, and optimization and search methods that can be utilized in output analysis

Along with statistical and analytical methods, graphical methods, animation, a practical sense of the underlying system, and effective modeling techniques can greatly assist the analysis of simulation outputs. Combined, such
statistical and modeling techniques often lead to arriving at accurate analysis
and clear conclusions. Several statistical methods and modeling skills are
coupled together at each main modeling activity to facilitate the analysis of
simulation output. Table A.2 summarizes the statistical methods and the mod-

TABLE A.2   Modeling and Statistical Methods in Simulation Analysis

Modeling Activity	Statistical Methods	Modeling Skills
Input modeling	Sampling techniques	Data collection
	Probability models	Random generation
	Histograms	Data classification
	Theoretical distributions	Fitting distributions
	Parameter estimation	Modeling variability
	Goodness of fit	Conformance test
	Empirical distributions	Using actual data
Model Running	Model type	Steady state/terminal
	Transient time	Warm-up period
	Data collection	Run length
	Sample size	Replications
	Performance estimation	Recording response
Output Analysis	Graphical tools	Output representation
	Descriptive statistics	Results summary
	Inferential statistics	Drawing inferences
	Experimental design	Design alternatives
	Optimization search	Best design

eling skills that are essential at each of the three major modeling activities: input modeling, model running, and output analysis.

Simulation output analysis focuses on measuring and analyzing certain model time-based critical-to-satisfaction (CTQ) or a functional requirement. A CTQ response is any measured variable that differs from one subject to another or from one time unit to another. For example, the manufacturing lead time varies from one unit produced to another unit, and the throughput of a production system varies from one hour to another.

Variables can be quantitative or qualitative. *Quantitative variables* are measured numerically in a discrete or continuous manner, whereas *qualitative variables* are measured in a descriptive manner. For example, system through-put is a quantitative variable and product color is a qualitative variable. Variables are also dependent and independent. Simulation input variables such as number of machines, entities arrival rates, and cycle times are *independent variables*, whereas model outcomes such as throughput and lead time are *dependent variables*. Finally, variables are either continuous or discrete. A *continuous variable* is one for which any value is possible within the limits the variable ranges. For example, the time between arrivals of customers at a bank is continuous since it can take real values, such as 3.00 minutes, 4.73 minutes, and so on. The variable "number of customers who arrive each hour at the bank" is a *discrete variable* since it can only take countable integer values such as 1, 2, 3, 4, and so on. It is clear that statistics computed from continuous variables have many more possible values than those of the discrete variables themselves.

TABLE A.3   Examples of Parameters and Statistics

Measure	Parameter	Statistic
Mean	μ	$\bar{X}$
Standard deviation	σ	s
Proportion	π	p
Correlation	ρ	r

The word *statistics* is used in several different senses. In the broadest sense, *statistics* refers to a range of techniques and procedures for analyzing data, interpreting data, displaying data, and making decisions based on data. The term also refers to the numerical quantity calculated from a sample of size *n*. Such statistics are used for parameter estimation.

In analyzing simulation outputs, it is also essential to distinguish between statistics and parameters. Whereas statistics are measured from data samples of limited size (*n*), a *parameter* is a numerical quantity that measures some aspect of the data population. *Population* consists of an entire set of objects, observations, or scores that have something in common. The distribution of a population can be described by several parameters, such as the mean and the standard deviation. Estimates of these parameters taken from a sample are called *statistics*. A *sample* is, therefore, a subset of a population. Since it is usually impractical to test every member of a population, a sample from the population is typically the best approach available. For example, the average of system throughput in 100 hours of run time is a statistic, whereas the generic throughput mean over the plant history is a parameter. Population parameters are rarely known and are usually estimated by statistics computed using samples. Certain statistical requirements are, however, necessary in order to estimate the population parameters using computed statistics. Table A.3 shows examples of selected parameters and statistics.

Descriptive Statistics

One important use of statistics is to summarize a collection of data in a clear and understandable way. Data can be summarized numerically and graphically. In a numerical approach a set of descriptive statistics are computed using a set of formulas. These statistics convey information about the data's central tendency measures (mean, median, and mode) and dispersion measures (range, inter-quartiles, variance, and standard deviation). Using a graphical approach, data central and dispersion tendencies are represented graphically (such as dot plots, histograms, probability density functions, stem and leaf plots, and box plots).

For example, recording the throughput in terms of units produced per hour (UPH) for 100 consecutive hours results in the data set shown in Table A.4. Throughput observations are collected after the model response passes the

TABLE A.4 Hourly Throughput for a 100-Hour Run Time

65	62	59	56	53	50	47	44	41	38
49	46	43	40	37	34	31	28	25	22
55	52	55	52	50	55	52	49	55	52
48	45	42	39	36	48	45	48	48	45
64	61	64	61	64	64	61	64	64	61
63	60	63	58	63	63	60	66	63	63
60	57	54	51	60	44	41	60	63	50
65	62	65	62	65	65	62	65	66	65
46	43	46	43	46	46	43	46	63	46
56	53	56	53	56	56	53	56	60	66

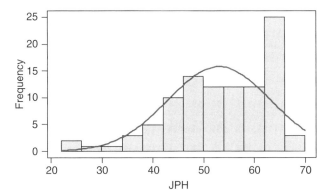

Figure A.1 Histogram and normal curve of throughput distribution.

warm-up period. The changing hourly throughput reflects the variability of simulation outcomes that is typically caused by elements of randomness in model inputs and processes.

Graphical representations of simulation outputs help us understand the distribution and behavior of model outcomes. For example, a histogram representation can be established by drawing the cells or intervals of data points versus each cell's frequency of occurrence. A probability density function curve can be constructed and added to the graph by connecting the centers of data cells. As discussed in Chapter 6, histograms help in selecting the proper distribution that represents simulation data. Figure A.1 shows the histogram and normal curve of the data in Table A.4.

Several other types of graphical representation can be used to summarize and represent the distribution of a certain simulation output. For example, Figures A.2 and A.3 show another two types of graphical representation of simulation outputs using box and dot plots, respectively.

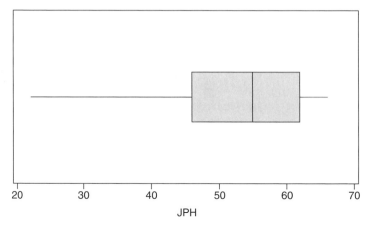

Figure A.2   Box plot of throughput data.

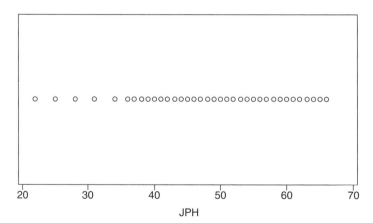

Figure A.3   Dot plot of throughput data.

Measures of Central Tendency   Measures of central tendency are measures of the location of the middle or the center of a distribution. The mean is the most commonly used measure of central tendency. The *arithmetic mean* is what is commonly called the *average*, the sum of all the observations divided by the number of observations in a sample or in a population. The mean of a population is expressed mathematically as

$$\mu = \frac{\sum_{i=1}^{N} x_i}{N}$$

where N is the number of population observations. The average of a sample j is expressed mathematically as

$$\overline{X}_j = \frac{\sum_{i=1}^{n} x_i}{n}$$

where n is the sample size.

The mean is a good measure of central tendency for roughly symmetric distributions but can be misleading in skewed distributions since it can be greatly influenced by extreme observations. Therefore, other statistics, such as the median and mode, may be more informative for skewed distributions. The mean, median, and mode are equal in symmetric distributions. The mean is higher than the median in positively skewed distributions and lower than the median in negatively skewed distributions.

The *median* is the middle of a distribution, where half the scores are above the median and half are below the median. The median is less sensitive than the mean to extreme scores, and this makes it a better measure than the mean for highly skewed distributions.

The *mode* is the most frequently occurring score in a distribution. The advantage of the mode as a measure of central tendency is that it has an obvious meaning. Further, it is the only measure of central tendency that can be used with nominal data (it is not computed). The mode is sensitive to sample fluctuation and is therefore not recommended to be used as the only measure of central tendency. A further disadvantage of the mode is that many distributions have more than one mode. These distributions are called *multimodal*.

Measures of Dispersion   A variable's *dispersion* is the degree to which scores on the variable differ from each other. *Variability* and *spread* are synonyms for *dispersion*. There are many measures of spread. The *range* is the simplest measure of dispersion. It is equal to the difference between the largest and the smallest values. The range can be a useful measure of spread because it is so easily understood. However, it is very sensitive to extreme scores since it is based on only two values. The range should almost never be used as the only measure of spread but can be informative if used as a supplement to other measures of spread, such as the standard deviation and interquartile range. For example, the range is determined for the following set ot numbers:

$$[10, 12, 4, 6, 13, 15, 19, 16] \qquad R = 19 - 4 = 15$$

The range is a useful statistic to know, but it cannot stand alone as a measure of spread since it takes into account only two scores.

The *variance* is a measure of how spread out a distribution is. It is computed as the average squared deviation of each number from its mean. Formulas for the variance are as follows. For a population:

$$\sigma^2 = \frac{\sum_{i=1}^{N} (x_i - \mu)^2}{N}$$

where N is the number of population observations. For a sample j:

$$s_j^2 = \frac{\sum_{i=1}^{n}\left(x_i - \overline{X}_j\right)^2}{n-1}$$

where n is the sample size.

The *standard deviation* is the most commonly used measure of dispersion. The formula for the standard deviation is the square root of the variance. An important attribute of the standard deviation is that if the mean and standard deviation of a normal distribution are known, it is possible to compute the percentile rank associated with any given observation. For example, the empirical rule states that in a normal distribution, about 68.27% of the scores are within one standard deviation of the mean, about 95.45% of the scores are within two standard deviations of the mean, and about 99.73% of the scores are within three standard deviations of the mean.

The standard deviation is not often considered a good measure of spread in highly skewed distributions and should be supplemented in those cases by the *interquartile range* ($IQ_3 - IQ_1$). The interquartile range is rarely used as a measure of spread because it is not very mathematically tractable. However, it is less sensitive than the standard deviation to extreme scores, it is less subject to sampling fluctuations in highly skewed distributions, and it has a good intuitive meaning.

For the data set shown in Table A.4, a set of descriptive statistics, shown in Table A.5, is computed using an Excel sheet to summarize the behavior of system throughput.

TABLE A.5 Descriptive Statistics Summary for the Data in Table A.4

Mean	53.06
Standard error	1.01
Median	55
Mode	63
Standard deviation	10.11
Sample variance	102.24
Range	44
Minimum	22
Maximum	66
First quartile (IQ_1)	46
Third quartile (IQ_3)	62
Interquartile range	16
Count	100
Sum	5306

Inferential Statistics

Inferential statistics are used to draw inferences about a population from a sample of n observations. Inferential statistics generally require that sampling be both random and representative. Observations are selected by randomly choosing the sample that resembles the population's most important characteristics. This can be obtained through the following:

1. A sample is random if the method for obtaining the sample meets the criterion of randomness (each item or element of the population having an equal chance of being chosen). Hence random numbers are typically generated from a uniform distribution $U(a, b)$ (see Appendix B).
2. Samples are drawn independently with no sequence, correlation, or auto-correlation between consecutive observations.
3. The sample size is large enough to be representative, usually $n \geq 30$.

The two main methods used in inferential statistics are: parameter estimation and hypothesis testing.

Parameter Estimation In estimation, a sample is used to estimate a parameter and to construct a confidence interval around the estimated parameter. Point estimates are used to estimate the parameter of interest. The mean (μ) and standard deviation (σ) are the most commonly used point estimates. As discussed earlier, the population mean (μ) and standard deviation (σ) are estimated using sample average ($\overline{X}_j$) and standard deviation (s_j), respectively.

A point estimate, by itself, does not provide enough information regarding variability encompassed in the simulation response (output measure). This variability represents the differences between the point estimates and the population parameters. Hence, an interval estimate in terms of a confidence interval is constructed using the estimated average ($\overline{X}_j$) and standard deviation (s_j). A *confidence interval* is a range of values that has a high probability of containing the parameter being estimated. For example, the 95% confidence interval is constructed in such a way that the probability that the estimated parameter is contained with the lower and upper limits of the interval is 95%. Similarly, 99% is the probability that the 99% confidence interval contains the parameter.

The confidence interval is symmetric about the sample mean $\overline{X}_j$. If the parameter being estimated is μ, for example, the 95% confidence interval constructed around an average of $\overline{X}_1 = 28.0$ is expressed as follows:

$$25.5 \leq \mu \leq 30.5$$

This means that we can be 95% confident that the unknown performance mean (μ) falls within the interval (25.5, 30.5).

As discussed earlier, three statistical assumptions must be met in a sample of observations to be used in constructing the confidence interval. The observations should be normally, independently, and identically distributed:

1. Observations are identically distributed through the entire duration of the process (i.e., they are stationary or time invariant).
2. Observations are independent, so that no interdependency or correlation exists between consecutive observations.
3. Observations are normally distributed (i.e., have a symmetric bell-shaped probability density function).

It is worth mentioning that individual observations $(x_1, x_2, \ldots, x_n)$ are non-stationary, since they vary by time based on model randomness and dynamic interactions. Also, consecutive individual observations from the model are often autocorrelated and dependent on each other's values. Values of the average of individual observations $(\overline{X})$ are, however, stationary and independent. Hence multiple simulation replications (n replications) are often run to obtain several values of the average (i.e., $\overline{X}_1, \overline{X}_2, \ldots, \overline{X}_n$). Dependency is avoided using different random streams in the model random number generator. Also, at similar run conditions, $\overline{X}$ values collected at the same point in time are typically stationary. For example, the values of hourly throughput of a production system within the simulation run or replication are nonstationary since they vary from one hour to another and they do not follow the same identical distribution throughout the simulation run time. Throughput at one hour may also depend on the throughput of the preceding hour. On the other hand, the average run throughput that is measured from independent simulation replications at a certain hour results in an identical distribution after reaching steady state.

The following formula is typically used to compute the confidence interval for a given significance level (α):

$$\overline{X} - \frac{t_{\alpha/2, n-1} s}{\sqrt{n}} \leq \mu \leq \overline{X} + \frac{t_{\alpha/2, n-1} s}{\sqrt{n}}$$

where $\overline{X}$ is the average of multiple performance means, $\overline{X}_1, \overline{X}_2, \ldots \overline{X}_m$ (determined from m simulation replications), $t_{n-1;\alpha/2}$ is a value from Student's t-distribution for an α level of significance.

For example, using the data in Table A.4, Figure A.4 summarizes the graphical and descriptive statistics along with the 95% CI computed for the mean, median, and standard deviation. The graph is created with MINITAB statistical software.

The normality assumption can be met by increasing the sample size (n) so that the central limit theorem (CLT) is applied. Each average performance, $\overline{X}_j$ (e.g., average throughput), is determined by summing individual perform-

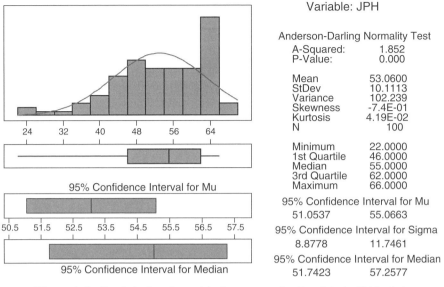

Anderson-Darling Normality Test
A-Squared: 1.852
P-Value: 0.000

Mean   53.0600
StDev   10.1113
Variance   102.239
Skewness   -7.4E-01
Kurtosis   4.19E-02
N   100

Minimum   22.0000
1st Quartile   46.0000
Median   55.0000
3rd Quartile   62.0000
Maximum   66.0000

95% Confidence Interval for Mu
51.0537   55.0663
95% Confidence Interval for Sigma
8.8778   11.7461
95% Confidence Interval for Median
51.7423   57.2577

95% Confidence Interval for Mu

95% Confidence Interval for Median

Figure A.4 Statistical and graphical summary for the data in Table A.4.

ance values $(x_1, x_2, \ldots, x_n)$ and dividing them by n. The CLT states that the variable representing the sum of several independent and identically distributed random values tends to be normally distributed. Since $(x_1, x_2, \ldots, x_n)$ are not independent and identically distributed, the CLT for correlated data suggests that the average performance $(\overline{X}_j)$ will be approximately normal if the sample size (n) used to compute $\overline{X}_j$ is large, $n \geq 30$. The $100\%(1 - \alpha)$ confidence interval on the true population mean is expressed as follows:

$$\overline{X} - \frac{Z_{\alpha/2}\sigma}{\sqrt{n}} \leq \mu \leq \overline{X} + \frac{Z_{\alpha/2}\sigma}{\sqrt{n}}$$

Hypothesis Testing   *Hypothesis testing* is a method of inferential statistics that is aimed at testing the viability of a *null hypothesis* about a certain population parameter based on some experimental data. It is common to put forward the null hypothesis and to determine whether the available data are strong enough to reject it. The null hypothesis is rejected when the sample data are very different from what would be expected under the assumption that the null hypothesis is true. It should be noticed, however, that failure to reject the null hypothesis is not the same thing as accepting the null hypothesis.

In simulation, hypothesis testing is used primarily for comparing systems. Two or more simulated design alternatives can be compared with the goal of identifying the superior design alternative relative to some performance measure. In testing a hypothesis with simulation modeling, the null hypothesis is often defined to be the reverse of what the simulation analyst actually believes about model performance. Thus, the simulation data collected are used to contradict the null hypothesis, which may result in its rejection. For example, if the simulation analyst has proposed a new design alternative to a certain production process, he or she would be interested in testing experimentally whether the design proposed works better than the existing production process. To this end, the analyst would design an experiment comparing the two methods of production. The hourly throughput of the two processes could be collected and used as data for testing the viability of the null hypothesis. The null hypothesis would be, for example, that there is no difference between the two methods (i.e., the throughput population means of the two production processes μ_1 and μ_2 are identical). In such a case, the analyst would be hoping to reject the null hypothesis and conclude that the proposed method that he or she developed is a better approach.

The symbol H_0 is used to indicate the null hypothesis, where *null* refers to the hypothesis of no difference. This is expressed as follows:

$$H_0: \mu_1 - \mu_2 = 0 \quad \text{or} \quad H_0: \mu_1 = \mu_2$$

The alternative hypothesis (H_1 or H_a) is simply set to state that the mean throughput of the method proposed (μ_1) is higher than that of the current production process (μ_2). That is,

$$H_a: \mu_1 - \mu_2 > 0 \quad \text{or} \quad H_a: \mu_1 > \mu_2$$

Although H_0 is called the null hypothesis, there are occasions when the parameter of interest is not hypothesized to be zero. For instance, it is possible for the null hypothesis to be that the difference (d) between population means is of a particular value ($H_0: \mu_1 - \mu_2 = d$); or the null hypothesis could be that the population mean is of a certain value ($H_0: \mu = \mu_0$).

The test statistics used in hypothesis testing depends on the hypothesized parameter and the available process information. In practical simulation studies, most tests involve comparisons of a mean performance with a certain value or with another process mean. When the process variance (σ^2) is known, which is rarely the case in real-world applications, Z_0 is used as a test statistic for the null hypothesis $H_0: \mu = \mu_0$, assuming that the population observed is normal or the sample size is large enough so that the CLT applies. Z_0 is computed as follows:

$$Z_0 = \frac{\overline{X} - \mu_0}{\sigma/\sqrt{n}}$$

The null hypothesis $H_0: \mu = \mu_0$ would be rejected if $|Z_0| > Z_{\alpha/2}$ when $H_a: \mu \neq \mu_0$, $Z_0 < -Z_\alpha$ when $H_a: \mu < \mu_0$, and $Z_0 > Z_\alpha$ when $H_a: \mu > \mu_0$.

Depending on the test situation, several test statistics, distributions, and comparison methods can also be used at several hypothesis tests. The following are some examples: For the null hypothesis H_0: $\mu_1 = \mu_2$, Z_0 is computed as follows:

$$Z_0 = \frac{\overline{X}_1 - \overline{X}_2}{\sqrt{\sigma_1^2/n_1 + \sigma_2^2/n_2}}$$

The null hypothesis H_0: $\mu_1 = \mu_2$ would be rejected if $|Z_0| > Z_{\alpha/2}$ when H_a: $\mu_1 \neq \mu_2$, $Z_0 < -Z_\alpha$ when H_a: $\mu_1 < \mu_2$, and $Z_0 > Z_\alpha$ when H_a: $\mu_1 > \mu_2$.

When the process variance (σ^2) is unknown, which is typically the case in real-world applications, t_0 is used as a test statistic for the null hypothesis H_0: $\mu = \mu_0$, t_0 is computed as follows:

$$t_0 = \frac{\overline{X} - \mu_0}{s/\sqrt{n}}$$

The null hypothesis H_0: $\mu = \mu_0$ would be rejected if $|t_0| > t_{\alpha/2,n-1}$ when H_a: $\mu \neq \mu_0$, $t_0 < -t_{\alpha,n-1}$ when H_a: $\mu < \mu_0$, and $t_0 > t_{\alpha,n-1}$ when H_a: $\mu > \mu_0$. For the null hypothesis H_0: $\mu_1 = \mu_2$, t_0 is computed as

$$t_0 = \frac{\overline{X}_1 - \overline{X}_2}{\sqrt{s_1^2/n_1 + s_2^2/n_2}}$$

Similarly, the null hypothesis H_0: $\mu_1 = \mu_2$ would be rejected if $|t_0| > t_{\alpha/2,v}$ when H_a: $\mu_1 \neq \mu_2$, $t_0 < -t_{\alpha,v}$ when H_a: $\mu_1 < \mu_2$, and $t_0 > t_{\alpha,v}$ when H_a: $\mu_1 > \mu_2$, where $v = n_1 + n_2 - 2$.

The examples of null hypotheses discussed involved the testing of hypotheses about one or more population means. Null hypotheses can also involve other parameters, such as an experiment investigating the variance (σ^2) of two populations, the proportion, and the correlation (ρ) between two variables. For example, the correlation between job satisfaction and performance on the job would test the null hypothesis that the population correlation (ρ) is 0. Symbolically, H_0: $\rho = 0$.

Sometimes it is necessary for the simulation analyst to compare more than two simulated alternatives for a system design or an improvement plan with respect to a given performance measure. Most practical simulation studies tackle this challenge by conducting multiple paired comparisons using several paired-t confidence intervals, as discussed above. Bonferroni's approach is another statistical approach for comparing more than two alternative systems. This approach is also based on computing confidence intervals to determine if the true mean performance of one system (μ_i) is significantly different from the true mean performance of another system (μ_i'). Analysis of variance (ANOVA) is another advanced statistical method that is often utilized for comparing multiple alternative systems. ANOVA's multiple comparison tests are widely used in experimental designs.

To draw the inference that the hypothesized value of the parameter is not the true value, a significance test is performed to determine if an observed value of a statistic is sufficiently different from a hypothesized value of a parameter (null hypothesis). The significance test consists of calculating the probability of obtaining a statistic that is different or more different from the null hypothesis (given that the null hypothesis is correct) than the statistic obtained in the sample. This probability is referred to as *p-value*. If this probability is sufficiently low, the difference between the parameter and the statistic is considered to be *statistically significant.* The probability of a type I error (α), called the *significance level*, is set by the experimenter. The significance level (α) is commonly set to 0.05 and 0.01. The significance level is used in hypothesis testing as follows:

1. Determine the difference between the results of the statistical experiment and the null hypothesis.
2. Assume that the null hypothesis is true.
3. Compute the probability (*p*-value) of the difference between the statistic of the experimental results and the null hypothesis.
4. Compare the *p*-value to the significance level (α). If the probability is less than or equal to the significance level, the null hypothesis is rejected and the outcome is said to be statistically significant.
5. The lower the significance level, therefore, the more the data must diverge from the null hypothesis to be significant. Therefore, the 0.01 level is more conservative since it requires stronger evidence to reject the null hypothesis then that of the 0.05 level.

There are two types of errors that can be made in significance testing: *type I error* (α), where a true null hypothesis can be incorrectly rejected, and *type II error* (β), where a false null hypothesis can be accepted incorrectly. A type II error is an error only in the sense that an opportunity to reject the null hypothesis correctly was lost. It is not an error in the sense that an incorrect conclusion was drawn since no conclusion is drawn when the null hypothesis is accepted. Table A.6 summarized the two types of test errors.

Type I error is generally considered more serious than type II error since it results in drawing a conclusion that the null hypothesis is false when, in fact, it is true. The experimenter often makes a trade-off between type I and type II errors. An experimenter protects himself or herself against type I errors by choosing a low significance level. This, however, increases the chance of type II error. Requiring very strong evidence to reject the null hypothesis makes it very unlikely that a true null hypothesis will be rejected. However, it increases the chance that a false null hypothesis will be accepted, thus lowering the test power.

Test power is the probability of correctly rejecting a false null hypothesis. Power is therefore defined as $1 - \beta$, where β is the type II error probability. If

**TABLE A.6 Summary of the Two Types of Test
Errors**

Statistical Decision	True State of Null Hypothesis (H_0)	
	H_0 Is True	H_0 Is False
Reject H_0	Type I error (α)	Correct
Accept H_0	Correct	Type II error (β)

the power of an experiment is low, there is a good chance that the experiment
will be inconclusive. There are several methods for estimating the test power
of an experiment. For example, to increase the test power, the experiment can
be redesigned by changing one of the factors that determines the power, such
as the sample size, the process standard deviation (σ), and the size of difference between the means of the tested processes.

EXPERIMENTAL DESIGN

In practical simulation studies, experimental design is usually a main objective
for building a simulation model. Simulation models are built with an extensive effort spent on data collection, verification, and validation to provide
a flexible platform for experimentation rather than for process animation.
Experimenting with simulation models is a typical practice for estimating performance under various running conditions, conducting what-if analysis,
testing hypothesis, comparing alternatives, factorial design, and system optimization. The results of such experiments and methods of analysis provide the
analyst and decision maker with insight, data, and necessary information for
making decisions, allocating resources, and setting strategies.

An *experimental design* is a plan that is based on a systematic and efficient
application of certain treatments to an experimental unit or subject. Being a
flexible and efficient experimenting platform, the simulation model represents
the subject of experimentation at which different treatments (simulation scenarios) are applied systematically and efficiently. The planned simulation scenarios (treatments) may include both structural and parametric changes
applied to the simulated system. Structural changes include altering the type
and configuration of simulation elements, the logic and flow of simulation entities, and the structure of the system layout. Examples include adding a new
machine, changing the sequence of operation, changing the direction of flow,
and so on. Parametric changes, on the other hand, include making adjustments
to model parameters or variables such as changes made to cycle times, buffer
sizes, conveyors speeds, and so on.

Parameter design is, however, more common than structural design in simulation-based experimental design, In practical simulation applications,
designers often adopt a certain model (system) structure and then use the

simulation to enhance performance by designing system parameters so that the best attainable performance is achieved. Hence, in most simulation experiments, design parameters are defined in the model as decision variables, and the model is set to receive and run at different levels of these decision variables in order to study their impact on a certain model response (performance measure). Partial or full factorial design is used for two purposes:

- Finding those decision variables of greatest significance on the system performance.
- Determining the levels of parameter settings at which the *best* performance level is obtained. Best performance can be maximizing, minimizing, or meeting a preset target of a certain performance measure.

The success of experimental design techniques is highly dependent on providing an efficient experiment setup. This includes the appropriate selection of decision variables, performance measure, levels of decision variables, and number of experiments required. To avoid conducting a large number of experiments, especially when the run length is long, certain experimental design techniques can be used. Using screening runs to eliminate insignificant design parameters and using Taguchi's fractional factorial designs instead of full factorial designs are approaches that can be used to reduce the size of the experimental task.

In experimental design, decision variables are referred to as *factors*, and the output measures are referred to as *response*. Factors are often classified into control and noise factors. *Control factors* are within the control of the designer, whereas *noise factors* are imposed by operating conditions and other internal or external uncontrollable factors. The objective of simulation experiments is usually to determine settings to system control factors so that system response is optimized and system random (noise) factors have the least impact on system response. Figure A.5 depicts the structure of parameter design using a simulation model.

The general representation of the number of experiments is l^k, where l is the number of levels of decision variables and k is the number of factors. Using only two levels of each control factor (low, high) often results in 2^k factorial designs. The easiest testing method is to change one design factor while other factors are kept fixed. In reality, however, control factors are often interrelated, which often results in masking the factors' main effects. Hence, the interactions among experiment factors should also be analyzed. Factorial design therefore looks at the combined effect of multiple factors on system performance. Using three levels (low, medium, high) of model control factors is also common in simulation studies.

In simulation applications, several model control factors can be set to be of two levels (low, high). For example, assume that we have three control factors in a certain production system: buffer size, conveyor speed, and number of forklifts. An experimental design is required to study the effect of the three

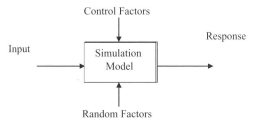

Figure A.5   Parameter design with simulation.

TABLE A.7   DOE for a 2^3 Factorial Design

Experiment Number	Model Control Factor			Response, R
	Buffer Size	Conveyor Speed	Forklift Number	Throughput
1	5 (−)	12.5 (−)	3 (−)	R_1
2	10 (+)	12.5 (−)	3 (−)	R_2
3	5 (−)	15.0 (+)	3 (−)	R_3
4	10 (+)	15.0 (+)	3 (−)	R_4
5	5 (−)	12.5 (−)	4 (+)	R_5
6	10 (+)	12.5 (−)	4 (+)	R_6
7	5 (−)	15.0 (+)	4 (+)	R_7
8	10 (+)	15.0 (+)	4 (+)	R_8

control factors on the throughput of the production system. Two levels of each control factor were defined: buffer size (5, 10) units, conveyor speed (12.5, 15.0) feet per minute, and number of forklifts (3, 4). In design of experiment it is common to use a minus sign to indicate the low factor level and a plus sign to indicate the high factor level. Thus, 2^3 factorial designs are required in this experiment. The model response (R), which is the throughput in this case, is evaluated at the designed experiments. Table A.7 summarizes the design of experiments of the eight factorial designs.

Multiple simulation replications are run at each of the 2^3 factor-level combinations. The average throughput is estimated from the simulation runs to represent the model response (R) at each factor-level combination. For example, R_5 represents the response resulting from running the simulation with low buffer size (5), low conveyor speed (12.5), and a high number of forklifts (4). The results in Table A.7 can then be used to calculate each factor effect on the model response R along with the degree of interactions among factors. The main effect of a factor i is the average change in the response due to changing factor i from "−" to "+" while holding all other factors fixed. For example, the main effect (e) of the buffer size on the model throughput is defined as follows:

$$e_1 = \frac{(R_2 - R_1) + (R_4 - R_3) + (R_6 - R_5) + (R_8 - R_7)}{4}$$

The interaction effect between any two factors occurs when the effect of a factor depends on the level of some other factor. This interaction is measured by the half of the difference between the average effect of a factor at the "–" and "+" levels of the other factor while all other factors are held constant. For example, the interaction between buffer size and conveyor speed can be defined as

$$e_1 = \frac{1}{2}\left[\frac{(R_4 - R_3) + (R_8 - R_7)}{4} - \frac{(R_2 - R_1) + (R_6 - R_5)}{4} \right]$$

With multiple simulation replications at each factorial design, the $100(1 - \alpha)$ confidence interval can be estimated for the expected effect. Also, a common random number can be used for the 2^k factor-level combinations to reduce the variance and width of the confidence interval and to increase the precision the effects of comparisons.

As the number and levels of control factors increase, a full factorial design may lead to running a huge number of experiments, especially when running multiple simulation replications. For example, experimenting with 10 control factors at each of two levels requires $2^{10} = 1024$ experiments. Running only five replications at each factorial design results in 5120 simulation runs. This increases the run time and analysis effort significantly. Hence, strategies of factor screening and developing fractional factorial design are often used to cope with a large number of control factors. With simulation, the cost and time of experimentation is marginal compared to physical experimentation.

Fractional factorial design is used to select a subset of combinations to test. The subset is selected by focusing on the main effects of control factors and some interactions. The subset size represents 2^{k-p} of the 2^k possible factorial designs. Determining the p-value, which can be 1, 2, 3, and so on, depends on the idea of confounding 2^{k-p} fractional factorial designs, where one effect is confounded with another if the two effects are determined using the same formula. Factor screening provides a smaller experiment subset by screening out factors with little or no impact on system performance using methods such as Plackett–Burman designs, supersaturated design, group-screening designs, and frequency-domain methods.

APPENDIX B

RANDOM NUMBERS FROM UNIFORM DISTRIBUTION U(0,1)

0.241672	0.707607	0.149551	0.244088	0.023288	0.605842
0.130395	0.412656	0.782611	0.163801	0.681001	0.805216
0.125521	0.289602	0.796590	0.818269	0.021933	0.724844
0.581766	0.061480	0.541614	0.136515	0.258806	0.095035
0.362065	0.185742	0.803744	0.726605	0.069880	0.619215
0.707612	0.081286	0.909875	0.630796	0.553827	0.008884
0.086201	0.970514	0.386577	0.784513	0.961142	0.973885
0.229614	0.774073	0.321826	0.662116	0.814679	0.647288
0.994100	0.890597	0.222232	0.006551	0.066242	0.328966
0.776734	0.099722	0.983406	0.608468	0.454539	0.168190
0.330102	0.042425	0.912484	0.889895	0.732549	0.426840
0.176164	0.043708	0.252851	0.004765	0.246006	0.639727
0.586119	0.252616	0.860748	0.224356	0.996291	0.265973
0.901310	0.269835	0.776467	0.046893	0.620921	0.974530
0.308797	0.159000	0.576900	0.413772	0.970986	0.060331
0.485659	0.789567	0.350096	0.260982	0.213692	0.234481
0.706428	0.700755	0.796194	0.366989	0.692362	0.074966
0.293048	0.775144	0.275982	0.779875	0.170110	0.793882
0.992831	0.876263	0.170610	0.440961	0.670306	0.828897
0.388389	0.972619	0.796626	0.723891	0.275735	0.173476
0.075031	0.581027	0.013501	0.577752	0.291566	0.655352
0.826204	0.399335	0.476663	0.772933	0.794377	0.608949
0.484815	0.656158	0.420684	0.007693	0.128676	0.911585

Simulation-Based Lean Six-Sigma and Design for Six-Sigma, by Basem El-Haik and Raid Al-Aomar
Copyright © 2006 John Wiley & Sons, Inc.

0.009105	0.192471	0.529841	0.997775	0.039371	0.830326
0.138261	0.428193	0.856544	0.424917	0.956624	0.184543
0.336794	0.369537	0.492827	0.444803	0.918587	0.866013
0.429948	0.674859	0.449968	0.226274	0.672120	0.522587
0.026548	0.055979	0.077013	0.832308	0.737676	0.811486
0.060978	0.009979	0.419899	0.621385	0.386700	0.968712
0.835398	0.056052	0.119755	0.630656	0.145757	0.808184
0.165149	0.975078	0.749635	0.710942	0.946238	0.171463
0.644001	0.193590	0.580653	0.827341	0.777910	0.781526
0.977190	0.228938	0.947335	0.503840	0.219351	0.691770
0.612895	0.124858	0.484572	0.674955	0.690415	0.248279
0.145467	0.513736	0.054678	0.519341	0.384625	0.877074
0.930567	0.107505	0.589492	0.171577	0.006004	0.002633
0.863939	0.165742	0.081077	0.249367	0.292524	0.202207
0.761715	0.866625	0.928880	0.848111	0.968593	0.732202
0.076297	0.973984	0.749542	0.768398	0.061307	0.249665
0.240833	0.539028	0.833231	0.264718	0.358686	0.246055
0.303655	0.091629	0.270470	0.306339	0.833383	0.425771
0.417309	0.539226	0.698344	0.199290	0.774944	0.386841
0.469692	0.174354	0.257337	0.466115	0.191161	0.194037
0.101630	0.525791	0.999856	0.061957	0.366740	0.646317
0.943076	0.808649	0.162622	0.005758	0.411454	0.741447
0.030540	0.275032	0.858510	0.891611	0.243260	0.633339
0.011997	0.097299	0.536898	0.021660	0.604345	0.240126
0.464901	0.694229	0.352010	0.659914	0.750387	0.839731
0.673109	0.866755	0.862720	0.242346	0.888004	0.759723
0.365594	0.490212	0.686469	0.329679	0.043072	0.556681
0.346779	0.690360	0.023762	0.242668	0.037572	0.802095
0.683103	0.961705	0.959999	0.899816	0.752486	0.733685
0.253665	0.721084	0.221340	0.083459	0.919328	0.393684
0.738435	0.099173	0.551225	0.847279	0.559015	0.346928
0.908274	0.206417	0.982413	0.022704	0.422405	0.786648
0.145980	0.888231	0.546014	0.847230	0.280300	0.959627
0.202452	0.486459	0.017531	0.584267	0.253479	0.102189
0.090778	0.930960	0.928910	0.316658	0.796783	0.511968
0.770998	0.758533	0.012261	0.497374	0.818692	0.032710
0.128388	0.133861	0.597927	0.784951	0.367985	0.042289
0.417637	0.736534	0.207259	0.323696	0.859506	0.305170
0.074911	0.700896	0.404523	0.873332	0.007001	0.421198
0.630289	0.202250	0.212295	0.061042	0.511563	0.181749
0.494760	0.916657	0.692431	0.245803	0.725729	0.388206
0.248164	0.516881	0.279786	0.479146	0.113645	0.457275
0.365842	0.591782	0.587998	0.733199	0.108353	0.613953
0.909322	0.534984	0.723551	0.771499	0.160171	0.213261
0.614181	0.913437	0.646528	0.083388	0.975430	0.468150
0.554715	0.331901	0.180196	0.931803	0.794580	0.363353
0.221737	0.575398	0.094961	0.277677	0.827929	0.231800
0.531803	0.099274	0.414635	0.513009	0.681185	0.189719
0.118799	0.756729	0.486304	0.941948	0.008911	0.036873

0.848188	0.053202	0.713632	0.155574	0.729220	0.110161
0.097187	0.139497	0.550116	0.285555	0.014146	0.559444
0.051598	0.279954	0.480825	0.603676	0.809057	0.292454
0.921528	0.562801	0.306958	0.238190	0.469674	0.071950
0.742770	0.233234	0.823157	0.208410	0.135838	0.584728
0.137195	0.099826	0.630742	0.250818	0.011010	0.050181
0.013453	0.231198	0.640598	0.584914	0.017724	0.692732
0.216712	0.041705	0.953708	0.564366	0.578372	0.866246
0.210460	0.879081	0.653108	0.180763	0.624162	0.282409
0.344797	0.493295	0.487461	0.119369	0.207022	0.346422
0.467670	0.089673	0.985351	0.264070	0.287505	0.120937
0.847443	0.698103	0.193953	0.290000	0.235702	0.667696
0.621303	0.342916	0.431170	0.482074	0.758102	0.682821
0.077937	0.203815	0.815869	0.576006	0.219599	0.328205
0.038442	0.879841	0.088095	0.500207	0.662453	0.259159
0.860394	0.092990	0.391612	0.843762	0.520961	0.118902
0.138558	0.381449	0.407744	0.417618	0.874152	0.832447
0.122147	0.472264	0.166580	0.086747	0.880771	0.700122

APPENDIX C

AXIOMATIC DESIGN

Axiomatic design (see Suh, 1990, 2001; El-Haik, 2005) is a prescriptive engineering design method.[1] Systematic research in engineering design began in Germany during the 1850s. The recent contributions in the field of engineering design include axiomatic design (Suh, 1984–2001), product design and development (Ulrich and Eppinger, 1995), the mechanical design process (Ulman, 1992), Pugh's total design (Pugh, 1991, 1996), and TRIZ (Altshuller, 1988, 1990; Rantanen, 1988; Arciszewsky, 1988). These contributions demonstrate that research in engineering design is an active field that has spread from Germany to most industrialized nations around the world. To date, most research in engineering design theory has focused on design methods. As a result, a number of design methods are now being taught and practiced in both industry and academia. However, most of these methods overlook the need to integrate quality methods in the concept stage. Therefore, the assurance that only healthy concepts are conceived, optimized, and validated with no (or minimal) vulnerabilities cannot be guaranteed.

Axiomatic design is a design theory that constitutes basic and fundamental design element knowledge. In this context, a *scientific theory* is defined as a

---

[1] Prescriptive design describes how design should be processed. Axiomatic design is an example of prescriptive design methodologies. Descriptive design methods such as design for assembly are descriptive of best practices and algorithmic in nature.

Simulation-Based Lean Six-Sigma and Design for Six-Sigma, by Basem El-Haik and Raid Al-Aomar
Copyright © 2006 John Wiley & Sons, Inc.

theory comprising fundamental knowledge areas in the form of perceptions and understandings of different entities, and the relationship between these fundamental areas. These perceptions and relations are combined by the theorist to produce consequences that can be, but are not necessarily, predictions of observations. Fundamental knowledge areas include mathematical expressions, categorizations of phenomena or objects, models, and so on, and are more abstract than observations of real-world data. Such knowledge and relations between knowledge elements constitute a theoretical system. A theoretical system may be one of two types: axioms or hypotheses, depending on how the fundamental knowledge areas are treated. Fundamental knowledge that is generally accepted as true, yet cannot be tested, is treated as an *axiom*. If the fundamental knowledge areas are being tested, they are treated as *hypotheses* (Nordlund, 1996). In this regard, axiomatic design is a scientific design method but with the premise of a theoretical system based on two axioms.

Motivated by the absence of scientific design principles, Suh (1984–2001) proposed the use of axioms as the scientific foundations of design. The following are two axioms that a design needs to satisfy:

- *Axiom 1: Independence axiom.* Maintain the independence of the functional requirements.
- *Axiom 2: Information axiom.* Minimize the information content in a design.

The term *quality* in our context can be defined as the degree to which the design vulnerabilities do *not* adversely affect product performance. In the context of the *axiomatic quality* process (El-Haik, 2005), the major design vulnerabilities are be categorized as one of the following:

- *Conceptual vulnerabilities:* established due to the violation of design principles.
- *Operational vulnerabilities:* created as a result of factors beyond the control of the designer, called *noise factors*. These factors, in general, are responsible for causing a product's functional characteristic or process to deviate from target values (Appendix D). Controlling noise factors is very costly or difficult, if not impossible. Operational vulnerability is usually addressed by robust design (Taguchi, 1989).

Conceptual vulnerabilities will always result in operational vulnerabilities. However, the reverse is not true. That is, it is possible for a healthy concept that is in full obedience to design principles to be operationally vulnerable.

Conceptual vulnerabilities are usually overlooked during product development, due to the lack of understanding of the principles of design, the absence of a compatible systemic approach to find ideal solutions, the pressure of deadlines, and budget constraints. These vulnerabilities are usually addressed by traditional quality methods. These methods can be characterized as after-the-fact practices since they use lagging information relative to developmental activities such as bench tests and field data. Unfortunately, these practices drive development toward endless design–test–fix–retest cycles, creating what is broadly known in the manufacturing industry as the firefighting operation mode. Companies that follow these practices usually suffer from high development costs, longer time to market, lower quality levels, and a marginal competitive edge. In addition, firefighting actions to improve the conceptual vulnerabilities are not only costly but also difficult to implement, as pressure to achieve design milestones builds during the development cycle. Therefore, it should be a goal to implement high-quality thinking in the conceptual stages of the development cycle. This goal can be achieved when systematic design theories are integrated with quality concepts and methods up front. El-Haik (2005) developed an integration framework, a process for quality in design, by borrowing from quality engineering (Taguchi, 1986) and the axiomatic design principles of Suh (1990). This framework is referred to as *axiomatic quality*. The objective axiomatic quality process is to address design vulnerabilities, both conceptual and operational, by providing tools, processes, and formulations for their quantification, then elimination or reduction.

Operational vulnerabilities necessitate the pursuit of variability reduction as an objective. Variability reduction has been a popular field of study. The method of robust design advanced by Taguchi (1986), Taguchi and Wu (1986), and Taguchi et al. (1989), time-domain control theory (Dorf, 2000), tolerance design, and tolerancing technique[2] are just some of the works in this area.

The independence axiom is used to address the conceptual vulnerabilities, while the information axiom will be tasked with the operational type of the design vulnerabilities. Operational vulnerability is usually minimized but cannot be totally eliminated. Reducing the variability of the design functional requirements and adjusting their mean performance to desired targets are two steps toward achieving such minimization. Such activities will also result in reducing design information content, a measure of design complexity according to the information axiom. Information content is related to the probability of successful manufacture of the design as intended by the customer.

The design process involves three mappings between four domains (Figure C.1). The first mapping involves the mapping between customer attributes (CAs) and the functional requirements (FRs). This mapping is very important, as it yields the definition of the high-level minimum set of functional requirements needed to accomplish the design intent. This definition can be accomplished by the application of quality function deployment (QFD). Once the minimum set of FRs is defined, the *physical mapping* may be started. This mapping involves the FRs domain and the design parameter codomain (DPs).

---

[2] Some developments in this arena, including M-space theory, offset-solids theory, and virtual boundaries, were discussed by Srinivasan and Wood (1992), Vasseur et al. (1993) and Wood et al. (1993), Zhang and Huq (1995), who linked process variation with inspection methods.

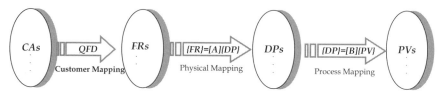

Figure C.1   Design mapping process.

It represents the product development activities and can be depicted by design matrices; hence, the term *mapping* is used. This mapping is conducted over design hierarchy as the high-level set of FRs, defined earlier, is cascaded down to the lowest hierarchical level. Design matrices reveal coupling, a conceptual vulnerability discussed in El-Haik (2005), and provide a means to track the chain of effects of design changes as they propagate across the design structure.

Process mapping is the last mapping of axiomatic design and involves the DP domain and the process variable (PV) codomain. This mapping can be represented formally by matrices as well and provides the process elements needed to translate DPs to PVs in manufacturing and production domains. A conceptual design structure called the *physical structure* is generally used as a graphical representation of the design mappings.

The mapping equation FR = f(DP) or, in matrix notation, $\{FR\}_{m\times1} = [A]_{m\times p} \cdot \{DP\}_{p\times1}$, is used to reflect the relationship between the domain, array $\{FR\}$, and the codomain, array $\{DP\}$, in the physical mapping where the array $\{FR\}_{m\times1}$ is a vector with m requirements, $\{DP\}_{p\times1}$ is the vector of design parameters with p characteristics, and **A** is the design matrix. According to the independence axiom, the ideal case is to have one-to-one mapping so that a specific DP can be adjusted to satisfy its corresponding FR without affecting the other requirements. However, perfect deployment of the design axioms may be infeasible due to technological and cost limitations. Under these circumstances, different degrees of conceptual vulnerabilities are established in the measures (criteria) related to the unsatisfied axiom. For example, a degree of *coupling* may be created because of axiom 1 violation, and this design may function adequately for some time in the use environment; however, a conceptually weak system may have a limited opportunity for continuous success, even with the aggressive implementation of an operational vulnerability improvement phase.

When matrix **A** is a square diagonal matrix, the design is called *uncoupled*, (i.e., each FR can be adjusted or changed independent of the other FRs). An uncoupled design is a one-to-one mapping. Another design that obeys axiom 1, although with a known design sequence, is called *decoupled*. In a decoupled design, matrix **A** is a *lower* or *upper triangular* matrix. The decoupled design may be treated as an uncoupled design when the DPs are adjusted in some sequence conveyed by the matrix. Uncoupled and decoupled design entities

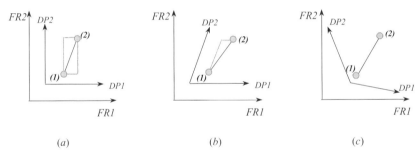

Figure C.2 Design categories according to Axiom 1: (*a*) uncoupled design, path independence; (*b*) decoupled design, path dependence; (*c*) coupled design, path dependence.

possess conceptual robustness (i.e., the DPs can be changed to affect specific requirements without affecting other FRs unintentionally). A coupled design definitely results in a design matrix with a number of requirements, m, greater than the number of DPs, p. Square design matrices ($m = p$) may be classified as a coupled design when the off-diagonal matrix elements are nonzeros. Graphically, the three design classifications are depicted in Figure C.2 for the 2×2 design matrix case. Notice that we denote the nonzero mapping relationship in the respective design matrices by X. On the other hand, 0 denotes the absence of such a relationship.

Consider the uncoupled design in Figure C.2*a*. The uncoupled design possesses the *path independence* property; that is, the design team could set the design to level 1 as a start point and move to setting 2 by changing DP1 first (moving east to the right of the page or parallel to DP1) and then changing DP2 (moving toward the top of the page or parallel to DP2). Due to the path independence property of the uncoupled design, the team could start from setting 1 to setting 2 by changing DP2 first (moving toward the top of the page or parallel to DP2) and then changing DP1 second (moving east or parallel to DP1). Both paths are equivalent; that is, they accomplish the same result. Notice also that the FR independence is depicted as orthogonal coordinates as well as perpendicular DP axes that parallel the respective FR in the diagonal matrix.

Path independence is characterized mathematically by a diagonal design matrix (uncoupled design). Path independence is a very desirable property of an uncoupled design and implies full control by the design team and ultimately, the customer (user), over the design. It also implies a high level of design quality and reliability since interaction effects between the FRs are minimized. In addition, a failure in one (FR, DP) combination of the uncoupled design matrix is not reflected in the other mappings within the same design hierarchical level of interest.

For the decoupled design, the path independence property is somehow fractured. As depicted in Figure C.2*b*, decoupled design matrices have a design setting sequence that need to be followed for the functional requirements to maintain their independence. This sequence is revealed by the matrix as follows: First, we need to set FR2 using DP2, and fix DP2, and second, set FR1 by leveraging DP1. Starting from setting 1, we need to set FR2 at setting 2 by changing DP2, and then change DP1 to the desired level of FR1.

The discussion above is a testimony to the fact that uncoupled and decoupled designs have conceptual robustness; that is, coupling can be resolved with proper selection of the DPs, path sequence application and employment of design theorems.

The coupled design matrix in Figure C.2c indicates the loss of path independence, due to the off-diagonal design matrix entries (on both sides), and the design team has no easy way to improve the controllability, reliability, and quality of their design. The design team is left with compromise practices (e.g., optimization) among the FRs as the only option since a component of the individual DPs can be projected on all orthogonal directions of the FRs. The uncoupling or decoupling step of a coupled design is a conceptual activity that follows the design mapping and will be explored later.

An example of design coupling is presented in Figure C.3, where two possible arrangements of the generic water faucet are displayed. There are two functional requirements: water flow and water temperature. The faucet shown in Figure C.3a has two design parameters, the water valves (knobs), one for each water line. When the hot water valve is turned, both flow and temperature are affected. The same happens when the cold water valve is turned. That

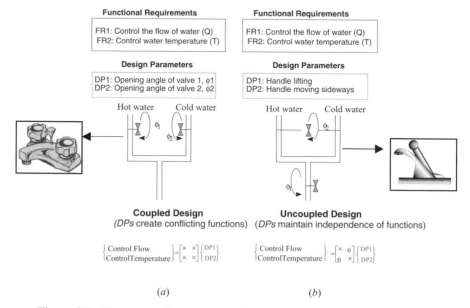

Figure C.3   Faucet coupling example. (From Swenson and Nordlund, 1996.)

is, the functional requirements are not independent, and a coupled design matrix below the schematic reflects this fact. From a consumer perspective, optimization of the temperature will require reoptimization of the flow rate until a satisfactory compromise among the FRs, as a function of the DP settings, is obtained over several iterations.

Figure C.3*b* exhibits an alternative design with one handle system delivering the FRs but with a new set of design parameters. In this design, flow is adjusted by lifting the handle, while moving the handle sideways will adjust the temperature. In this alternative, adjusting the flow does not affect the temperature, and vice versa. This design is better since the functional requirements maintain their independence according to axiom 1. The uncoupled design will give the customer path independence to set either requirement without affecting the other. Note also that in the uncoupled design case, design changes to improve an FR can be done independently as well, a valuable design attribute.

The importance of design mapping has many perspectives. Chief among them is the identification of coupling among the functional requirements, due to the physical mapping process with the design parameters, in the codomain. Knowledge of coupling is important because it provides the design team with clues from which to find solutions, make adjustments or design changes in proper sequence, and maintain their effects over the long term with minimal negative consequences.

The design matrices are obtained in a hierarchy and result from employment of the *zigzagging method of mapping*, as depicted in Figure C.4. The zigzagging process requires a solution-neutral environment, where the DPs are chosen after the FRs are defined, and not vice versa. When the FRs are defined, we have to *zig* to the physical domain, and after proper DP selection, we have to *zag* back to the functional domain for further decomposition or

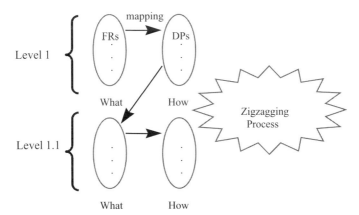

Figure C.4   Zigzagging process. (From Suh, 1990.)

cascading, although at a lower hierarchical level. This process is in contrast to the traditional cascading processes, which utilize only one domain at a time, treating the design as the sum of functions or the sum of parts.

At lower levels of hierarchy, entries of design matrices can be obtained mathematically from basic physical and engineering quantities, enabling the definition and detailing of transfer functions, an operational vulnerability treatment vehicle. In some cases, these relationships are not readily available, and some effort needs to be paid to obtaining them empirically or via modeling (e.g., CAE). Lower levels represent the roots of the hierarchical structure where robust design and six-sigma concepts can be applied with some degree of ease.

APPENDIX D

TAGUCHI'S QUALITY ENGINEERING

In the context of this book, the terms *quality* and *robustness* are used interchangeably. *Robustness* is defined as reducing the variation of the functional requirements of the system and having them on target as defined by the customer (Taguchi, 1986; Taguchi and Wu, 1986; Phadke, 1989; Taguchi et al., 1989, 1999).

Variability reduction has been the subject of robust design (Taguchi, 1986) through methods such as parameter design and tolerance design. The principal idea of *robust design* is that statistical testing of a product should be carried out at the design stage, also referred to as the *off-line stage*. To make the product *robust* against the effects of variation sources in the manufacturing and use environments, the design problem is viewed from the point of view of quality and cost (Taguchi, 1986; Taguchi and Wu, 1986; Taguchi et al., 1989, 1999; Nair, 1992).

Quality is measured through quantifying statistical variability through measures such as standard deviation or mean square error. The main performance criterion is to achieve the functional requirements (FR) target on average, while minimizing variability around this target. Robustness means that a system performs its intended functions under all operating conditions (different causes of variations) throughout its intended life. The undesirable and uncontrollable factors that cause the FR under consideration to deviate from target value are called *noise factors*. Noise factors affect quality

Simulation-Based Lean Six-Sigma and Design for Six-Sigma, by Basem El-Haik and
Raid Al-Aomar
Copyright © 2006 John Wiley & Sons, Inc.

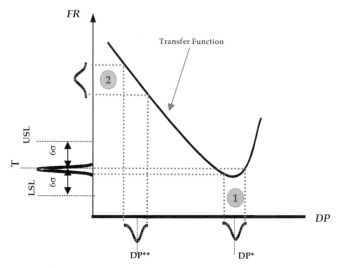

Figure D.1 Robustness optimization definition.

adversely, and ignoring them will result in a system not optimized for conditions of use. Eliminating noise factors may be expensive. Instead, we seek to reduce the effect of the noise factors on FR performance.

Robust design is a disciplined engineering process that seeks to find the best expression of a system design. *Best* is carefully defined to mean that the design is the lowest-cost solution to the specification, which itself is based on the identified customer needs. Taguchi has included design quality as one more dimension of cost. High-quality systems minimize these costs by performing consistently at target levels specified by the customer.

Taguchi's philosophy of robust design is aimed at reducing the loss due to variation of performance from the target value based on a portfolio of concepts and measures such as *quality loss function* (QLF), *signal-to-noise ratio*, optimization, and experimental design. *Quality loss* is the loss experienced by customers and society and is a function of how far performance deviates from target. The QLF relates quality to cost and is considered a better evaluation system than the traditional binary treatment of quality (i.e., within/outside specifications). The QLF of an FR has two components: mean (μ_{FR}) deviation from targeted performance value (T) and variability (σ_{FR}^2). It can be approximated by a quadratic polynomial of the functional requirement.

Consider two settings or means of a design parameter: setting 1 (DP*) and setting 2 (DP**), each with the same variance and probability density function (statistical distribution), as depicted in Figure D.1. Consider also the given curve of a hypothetical transfer function,[1] which is in this case a nonlinear function in the design parameter, DP. It is obvious that setting 1 produces less

---

[1] A mathematical form of design mapping.

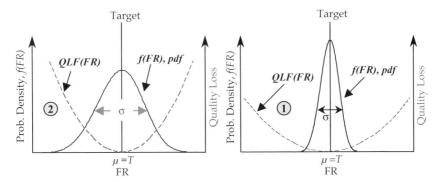

Figure D.2   Quality loss function scenarios of Figure D.1.

variation in the functional requirement (FR) than setting 2 by capitalizing on nonlinearity.[2] This also implies lower information content and thus a lower degree of complexity based on the information axiom (Appendix C). Setting 1 (DP*) will also produce a lower quality loss value, similar to the scenario on the right of Figure D.2. In other words, the design produced by setting 1 (DP*) is more robust than that produced by setting 2. Setting 1 (DP*) robustness is evident in the amount of variation transferred through the transfer function to the FR response of Figure D.1 and the flatter quadratic QLF in Figure D.2. When the distance between the specification limits is six times the standard deviation ($6\sigma_{FR}$), a six-sigma level optimized FR is achieved. When all design FRs are released at this level, a six-sigma design is obtained.

The important contribution of robust design is the systematic inclusion into experimental design of noise variables, the variables over which the designer has no or little control. A distinction is also made between internal noise, such as component wear and material variability, and environmental noise, which the designer cannot control (e.g., humidity, temperature). Robust design's objective is to suppress, as far as possible, the effect of noise by exploring the levels of the factors to determine their potential for making the system insensitive to these sources of variation.

QUALITY LOSS FUNCTION

Inspection schemes represent the heart of online quality control. Inspection schemes depend on the binary characterization of design parameters (i.e., being within or outside the specification limits). An entity is called *confirming* if all DPs inspected are within their respective specification limits; otherwise, it is *nonconfirming*. This binary representation of the acceptance criteria based

---

[2] In addition to nonlinearity, leveraging interactions between the noise factors and the design parameters is another popular empirical parameter design approach.

on a DP is not realistic since it characterizes equally entities that are marginally off specification limits and those marginally within the limits. In addition, this characterization does not make justice for parts marginally off with those that are off significantly. Taguchi proposed a continuous and better representation, the *quality loss function* (Taguchi and Wu, 1980), which provides a better estimate of the monetary loss incurred by manufacturing and customers as an FR deviate from its targeted performance value, T. The determination of the target T implies the nominal-the-best and dynamic classifications.

A quality loss function can be interpreted as a means to translate variation and target adjustment to a monetary value. It allows the design teams to perform a detailed optimization of cost by relating engineering terminology to economical measures. In its quadratic form, a quality loss is determined by first finding the functional limits, $T \pm \Delta$FR, of the relevant FR. The functional limits are the points at which the solution entity would fail (i.e., produces unacceptable performance in approximately half of the customer applications). In a sense, these represent performance levels that are equivalent to average customer tolerance. Kapur (1988) continued with this path of thinking and illustrated the derivation of specification limits using Taguchi's quality loss function. A quality loss is incurred due to the deviation in the FRs, as caused by the noise factors (NFs), from their intended targeted performance, T. Let L denote the QLF that takes the numerical value of the FR and the targeted value as arguments. By Taylor series expansion[3] at FR = T, we have

$$L(\text{FR},T) = L(T,T) + \left.\frac{\partial L}{\partial \text{FR}}\right|_{\text{FR}=T}(\text{FR}-T) + \frac{1}{2!}\left.\frac{\partial^2 L}{\partial \text{FR}^2}\right|_{\text{FR}=T}$$

$$(\text{FR}-T)^2 + \frac{1}{3!}\left.\frac{\partial^3 L}{\partial \text{FR}^3}\right|_{\text{FR}=T}(\text{FR}-T)^3 + \cdots \tag{D.1}$$

The target T is defined such that L is minimum at FR = $T \Rightarrow \partial L/\partial \text{FR}_{\text{FR}=T} = 0$. In addition, the robustness theme implies that most entities should be delivering the FR at target (T) on a continuous basis or at least in the very near neighborhood of FR = T to minimize the quality loss. In the latter case, the expansion quadratic term is the most significant term. This condition results in the following approximation:

$$L(\text{FR},T) \cong K(\text{FR}-T_{\text{FR}})^2 \tag{D.2}$$

where

$$K = \frac{1}{2!}\left.\frac{\partial^2 L}{\partial \text{FR}^2}\right|_{\text{FR}=T}$$

---

[3] The assumption here is that L is a higher-order continuous function such that derivatives exist and is symmetrical around FR = T.

Let $FR \in (T_{FR} - \Delta FR, T_{FR} + \Delta FR)$, where T_{FR} is the target value and ΔFR is the functional deviation from the target. Let A_Δ be the quality loss incurred due to the symmetrical deviation, ΔFR; then

$$K = \frac{A_\Delta}{(\Delta FR)^2} \tag{D.3}$$

In Taguchi's tolerance design method, the quality loss coefficient K can be determined on the basis of losses in monetary terms by falling outside the *customer tolerance limits (design range* or *manufacturer's limits)* instead of the specification limits generally used in process capability studies, for example. The specification limits are most often associated with the DPs. Customer tolerance limits are used to estimate the loss from customer perspective or the quality loss to society as proposed by Taguchi. Usually, customer tolerance is wider than manufacturer tolerance.

Let $f(FR)$ be the probability density function of the FR; then via the expectation operator, E, we have

$$E[L(FR,T)] = \int_{-\infty}^{\infty} K(FR - T_{FR})^2 f(FR)\, dFR$$

$$= K\left[\sigma_{FR}^2 + (\mu_{FR} - T_{FR})^2\right] \tag{D.4}$$

Quality loss has two ingredients: loss incurred due to variability, σ_{FR}^2, and loss incurred due to mean deviation from target, $(\mu_{FR} - T_{FR})^2$. Usually, the second term is achieved by adjustment of the mean of a few critical DPs, a typical engineering problem. Other forms of loss function are described below.

Larger-the-Better Loss Function

For functions such as "increase strength" (FR = strength) or "multiply torque" (FR = torque), we would like a very large target: ideally, $T_{FR} \rightarrow \infty$. The FR is bounded by the lower functional specifications limit, FR_l. The loss function is then given by

$$L(FR, T_{FR}) = \frac{K}{FR^2} \qquad \text{where} \quad FR \geq FR_l \tag{D.5}$$

Let μ_{FR} be the average FR numerical value of the system range, the average around which performance delivery is expected. Then by Taylor series expansion around $FR = \mu_{FR}$, we have

$$L(FR, T_{FR}) = K\left[\frac{1}{\mu_{FR}^2} - \frac{2(FR - \mu_{FR})}{\mu_{FR}^3} + \frac{3(FR - \mu_{FR})^2}{\mu_{FR}^4} - \cdots\right]_{FR=\mu_{FR}} \tag{D.6}$$

If the higher-order terms are negligibly small, we have

$$E[L(\text{FR},T_{\text{FR}})] = K\left(\frac{1}{\mu_{\text{FR}}^2} + \frac{3}{\mu_{\text{FR}}^4}\sigma_{\text{FR}}^2\right) \tag{D.7}$$

Smaller-the-Better Loss Function

Functions such as "reduce noise" or "reduce wear" would like to have zero as their target value. The loss function in this category and its expected values are given by

$$L(\text{FR},T) = K \cdot \text{FR}^2 \tag{D.8}$$

$$E[L(\text{FR},T)] = K(\sigma_{\text{FR}}^2 + \mu_{\text{FR}}^2) \tag{D.9}$$

In the development above, the average loss can be estimated from a parameter design or even a tolerance design experiment by substituting the experiment variance S^2 and average $\overline{\text{FR}}$ as estimates for σ_{FR}^2 and μ_{FR} as discussed in Appendix A. Equations (D.4), (D.7), and (D.9) were used in the Chapter 10 case study.

APPENDIX E

PROCESS MAPPING

Like most six-sigma tools, process mapping requires a cross-functional team effort with involvement from process owners, process members, customers, and suppliers. Brainstorming, operation manuals, specifications, operator experience, and process walk are very critical inputs to the mapping activity. A detailed process map provides input to other tools, such as FMEA, P-diagrams, DOEs, capability studies, and control plans.

Process maps can be created for many different levels, zooming in and out of the targeted process delivering the service under consideration in a 3S project. Depending on the detail the team requires, the map should be created at that level. If more details are required, a more detailed map of the sub-process should be completed. The team should objectively demand verification with hands-on exposure to local activities, identify rework loops and redundancies, and seek insight into bottlenecks, cycle times, and inventory. Rework loops and redundancies are non-value-added, costing time, effort, money, and other resources, and are often referred to in six-sigma literature as the hidden factory.

To make the most improvements in any existing process, it is necessary to understand the actual way that the process occurs. Within this context, it is easy to understand the reasons for the flow and then address the causes. In process mapping, there need to be symbols for a process step (Table E.1), a measurement, a queue, a storage, a transportation (movement), and a decision.

Simulation-Based Lean Six-Sigma and Design for Six-Sigma, by Basem El-Haik and Raid Al-Aomar
Copyright © 2006 John Wiley & Sons, Inc.

TABLE E.1   Process Mapping Standard Symbols

☐	Process step or operation (white)
◗	Delay (red)
◯	Quality check, inspection, or measurement (yellow)
▽	Storage (yellow)
◆	Decision (blue)
⇨	Transport or movement of material or transmission of information (yellow)

What You *Think* It Is...   What It *Actually* Is...   What You Would *Like* It To Be...

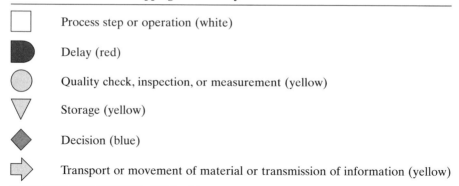
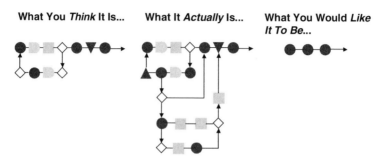

Figure E.1   Three versions of a process.

There are three versions of a process map (Figure E.1). There is "what it was designed to be," which is usually a clean flow. There is the "as-is" process map, with all the gritty variety that occurs because of varying suppliers, customers, operators, and conditions. Take an ATM machine, for example. Does everyone follow the same flow for getting money out? Some may check their balance first and then withdraw, while others may type in the wrong PIN and have to retype it. The third version is "what we would like it to be," with only value-added steps; it is clean, intuitive, and works correctly every time.

A process map is a pictorial representation showing all of the steps in a process. As a first step, team members should familiarize themselves with the mapping symbols, then walk the process by asking such questions as: "What really happens next in the process?", "Does a decision need to be made before the next step?", and "What approvals are required before moving on to the next task?" The team then draws the process using the symbols on a flipchart or overhead transparency. Every process will have a start and an end (elongated circles). All processes will have tasks, most will have decision points (a diamond), and upon completion the team should analyze the map for such items as non-value-added steps, rework loops, duplication, and cycle time. A typical high-level process map is shown in Figure E.2.

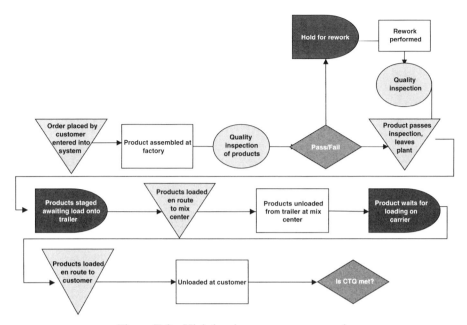

Figure E.2 High-level process map example.

A good process map should identify all process steps or operations as well as visible measurement, inspection, rework loops, and decision points. In addition, *swim lanes* are often used when mapping information flow for service-type transactional and business processes. We believe that swim lanes, which segregate steps by who does them or where they are done and makes hand-offs visual, are appropriate for all types of 3S projects. The map is arranged on a table where the rows indicate "who" owns or performs the process step (process owner), and the process flows that change "lanes" indicate hand-offs. Hand-off points are where lack of coordination and communication can cause process problems. An example is depicted in Figure E.3.

Clear distinctions can be made between the warehousing and those process steps where customer interactions occur. The swim lane process map example shows a high-level portion of the order receiving process. Another level of detail in each of the process steps will require further mapping if they are within the scope of the project.

Process mapping is a methodology composed of the following steps and actions:

- *Step 1:* Define the process.
  - Define the objective or scope of the project.
  - Revisit the objective to select and prioritize processes to be mapped.
  - Focus on times, inputs, functions, hand-offs, and authority and responsibility boundaries and outputs.

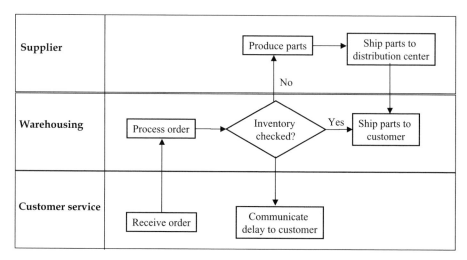

Figure E.3 High-level swim lanes process map.

- Discuss the appropriate level of detail.
- Introduce linkages to other analyses.
- List high-level process steps.
- Document the overall flow of the process between the start and stop boundaries.
- *Step 2:* Collect information.
  - Use sampling and collection techniques (focus groups, interviews, observations) to obtain data.
  - Acquire the necessary resources in this mapping step and the mapping exercise in general.
  - Process the walk schedule based on the shifts, days, and weeks where unique work occurs.
  - Determine when to zoom in to map every process step.
  - Assure continuous flow.
  - Process stops (when flow is disrupted or disconnected).
  - Determine whether process inputs and output variables have been identified. Remember that we are after a transfer function that fully describes what is happening or should be happening, to confirm that the solution is really optimal.
  - Consider merging of multiple steps.
  - Add operating specifications.
  - Be clear about terminology and nomenclature so that everyone is speaking the same language. Avoid biased, nontechnical lingo. Be consistent.

- Accumulate a list of quick-hit items and develop an action plan for their implementation.
- *Step 3:* Validate and verify.
  - Ensure the use of common terms and nomenclature.
  - Consolidate, compile, and reconcile information.
  - Summarize the findings
  - Validate the information.
  - Confirm controversial data.
- *Step 4:* Build the process maps.
  - Follow a sequence of activities and work steps based on the earlier steps.
  - Establish swim lanes (if necessary).
  - Determine the beginning and end of the process.
  - Assign work steps to participants.
  - Parcel out times for each work step.
  - Conduct an initial assessment of the control plan.
  - Start an initial assessment after the process map is complete.
  - Add a measurement technique.
  - Define the operating specifications.
  - Determine the targets.
  - Know the context of the process mapping within the project.
  - Decide whether the process mapping fits with previous, parallel, and ongoing analyses.
  - Conduct benchmarking.
  - Decide on the baseline for current performance.
  - Document all comments.
- *Step 5:* Analyze and draw conclusions.
  - Identfy the characteristics of the process steps.
  - Define hypotheses about how process step input links to output variables relative to targets or historical means and variances.
  - Plan follow-up work, if any (e.g., new measurement systems, SPC charting).
  - View improvement or redesign opportunities.
  - Apply "if-then" simulation scenarios for any layout changes.
- *Step 6:* Communicate recommendations, findings, and conclusions.
  - Customize the presentation to the audience with insights into change implications.
  - Update specifications, control plans, procedures, training, and so on.

APPENDIX F

VENDORS

SIX-SIGMA SOFTWARE VENDORS AND CONSULTANTS

Six Sigma Professionals, Inc.
39505 Dorchester Circle
Canton, MI 48188
Phone: 877-900-3210 or 734-765-5229
Fax: 734-728-8507
www.Six-SigmaPro.com

Achiever Business Solutions
355 E. Campus View Blvd., Ste. 285
Columbus, OH 43235
Phone: 888-645-4933
Fax: 614-410-9004
www.goalachiever.com

Advanced Surface Microsopy Inc.
6009 Knyghton Road
Indianapolis, IN 46220
Phone: 800-374-8557
Fax: 317-254-8690
www.asmicro.com

Advanced Systems Consultants
P.O. Box 1176
Scottsdale, AZ 85252
Phone: 480-423-0081
Fax: 480-423-8217
www.mpcps.com

Advanced Systems & Designs Inc.
3925 Industrial Drive
Rochester Hills, MI 48309
Phone: 866-772-0012
Fax: 248-844-2199
www.spcanywhere.com

Air Academy Associates
1650 Telstar Drive, Ste. 110
Colorado Springs, CO 80920
Phone: 800-748-1277
Fax: 719-531-0778
www.airacad.com

Simulation-Based Lean Six-Sigma and Design for Six-Sigma, by Basem El-Haik and Raid Al-Aomar

Alamo Learning Systems
3160 Crow Canyon Road, Ste. 280
San Ramon, CA 94583
Phone: 800-829-8081
Fax: 925-277-1919
www.alamols.com

Breakthrough Management Group (BMG)
2101 Ken Pratt Blvd.
Longmont, CO 80501
Phone: 800-467-4462
Fax: 303-827-0011
www.bmgi.com

Corel Corp. – iGrafx
1600 Carling Ave.
Ottawa, ON K1Z 8R7 Canada
Phone: 800-772-6735
www.corel.com/iGrafx

DataNet Quality Systems
24567 Northwestern Highway
Southfield, MI 48075
Phone: 248-357-2200
Fax: 248-357-4933
www.qualtrend.com

Deltek Systems Inc.
8280 Greensboro Drive
McLean, VA 22102
Phone: 800-456-2009
Fax: 703-734-1146
www.deltek.com

EGsoft
257 S.W. Madison Ave.
Corvallis, OR 97333
Phone: 800-478-2892
Fax: 541-758-4666
www.statware.com

General Engineering Methods Inc.
12459 White Tail Court
Plymouth, MI 48170
Phone: 734-455-8980
Fax: 734-455-8980

George Group Inc.
13355 Noel Road, Ste. 1100
Dallas, TX 75240
Phone: 972-789-3211
Fax: 972-458-9229
www.georgegroup.com

H.J. Steudel & Associates Inc.
6417 Normandy Lane, Ste. 205
Madison, WI 53719
Phone: 866-271-3121
Fax: 608-271-4755
www.hjsteudel.com

The Harrington Group Inc.
11501 Lake Underhill Road
Orlando, FL 32825
Phone: 800-476-9000
Fax: 407-382-6141
www.hginet.com

Helm Instrument Co. Inc.
361 W. Dussel Drive
Maumee, OH 43537
Phone: 419-893-4356
Fax: 419-893-1371
www.helminstrument.com

Hertzler Systems Inc.
2312 Eisenhower Drive N.
Goshen, IN 46526
Phone: 219-533-0571
Fax: 219-533-3885
www.hertzler.com

InfinityQS International
7998 Donegan Drive
Manassas, VA 20109
Phone: 800-772-7978
Fax: 703-393-2211
www.infinityqs.com

Intelex Technologies Inc.
165 Spadina Ave., Ste. 300
Toronto, ON M5T 2C3 Canada
Phone: 416-599-6009
Fax: 416-599-6867
www.intelex.com

IQS Inc.
19706 Center Ridge Road
Cleveland, OH 44116
Phone: 800-635-5901
Fax: 440-333-3752
www.iqs.com

ISO9
1433 Webster St., Ste. 200
Oakland, CA 94612
Phone: 866-321-4769
Fax: 509-357-8673
www.iso9.com

JMP, A Business Unit of SAS
SAS Campus Drive
Cary, NC 27513
Phone: 877-594-6567
Fax: 919-677-4444
www.jmpdiscovery.com

Juran Institute Inc.
115 Old Ridgefield Road
Wilton, CT 06897
Phone: 800-338-7726
Fax: 203-834-9891
www.juran.com

Lighthouse Systems Inc.
6780 Pittsford-Palmyre Road
Building 3, Ste. C
Fairport, NY 14450
Phone: 716-223-0600
Fax: 716-223-0620
www.lighthouse-usa.com

Lilly Software Associates Inc.
500 Lafayette Road
Hampton, NH 03842
Phone: 603-926-9696
Fax: 603-929-3975
www.lillysoftware.com

Minitab Inc.
3081 Enterprise Drive
State College, PA 16801
Phone: 800-488-3555
Fax: 814-238-1702
www.minitab.com

New Paradigms Inc.
5630 Mahoney Ave.
Minnetonka, MN 55345
Phone: 612-388-2944
Fax: 952-934-5509
www.newparadigms.com

Nutek Inc.
3829 Quarton Road
Bloomfield Hills, MI 48302
Phone: 248-540-4827
Fax: 248-540-4827
www.rkroy.com

Omnex
3025 Boardwalk, Ste. 190
Ann Arbor, MI 48106
Phone: 734-761-4940
Fax: 734-761-4966
www.omnex.com

Palisade Corp.
31 Decker Road
Newfield, NY 14867
Phone: 800-432-7475
Fax: 607-277-8001
www.palisade.com

Pilgrim Software Inc.
2807 W. Busch Blvd., Ste. 200
Tampa, FL 33618
Phone: 813-915-1663
Fax: 813-915-1948
www.pilgrimsoftware.com

Pister Group Inc.
550 Eglinton Ave. W.
Toronto, ON M5N 3A8 Canada
Phone: 905-886-9470
Fax: 905-764-6405
www.pister.com

PQ Systems Inc.
10468 Miamisburg–Springboro Road
Miamisburg, OH 45342
Phone: 800-777-3020
Fax: 937-885-2252
www.pqsystems.com

Prism eSolutions LLC
512 Township Line Road
3 Valley Square, Ste. 362
Blue Bell, PA 19422
Phone: 888-386-2330
Fax: 267-468-0199
www.prismesolutions.com

ProcessModel Inc.
32 W. Center St., Ste. 301
Provo, UT 84601
Phone: 801-356-7165
Fax: 801-356-7175
www.processmodel.com

Proxima Technology
1350 17th St., Ste. 200
Denver, CO 80202
Phone: 720-946-7200
Fax: 720-932-9499
www.proxima-tech.com

QSE/Quality Systems Engineering
2724 Kathleen Drive
Brighton, MI 48114
Phone: 810-229-7329
Fax: 810-494-5001
www.qseprocess.com

Qualitron Systems Inc.
1673 Star Batt Drive
Rochester Hills, MI 48309
Phone: 889-569-7094
Fax: 248-299-0000
www.qualitronsystems.com

Quality America Inc.
7650 E. Broadway, Ste. 208
Tucson, AZ 85710
Phone: 800-722-6154
Fax: 520-722-6705
www.qualityamerica.com

The Quality Group
6059 Boylston Drive, Ste. 250
Atlanta, GA 30328
Phone: 800-772-3071
Fax: 404-252-4475
www.thequalitygroup.net

Quality Systems Integrators
P.O. Box 91
Eagle, PA 19480
Phone: 800-458-0539
Fax: 610-458-7555
www.qsi-inc.com

Resource Engineering Inc.
P.O. Box 219
Tolland, CT 06084
Phone: 800-810-8326
Fax: 860-872-2488
www.reseng.com

Roth-Donleigh
29002 La Carreterra
Laguna Niguel, CA 92677
Phone: 949-364-5907
Fax: 949-364-5907

S-Matrix Corp.
835 Third St.
Eureka, CA 95501
Phone: 800-336-8428
Fax: 707-441-0410
www.s-matrix-corp.com

SigmaPro Inc.
3131 S. College Ave., Ste. 2C
Fort Collins, CO 80525
Phone: 970-207-0077
Fax: 970-207-0078
www.sigmapro.com

Sigmetrix
105 W. Virginia St.
McKinney, TX 75069
Phone: 972-542-7517
Fax: 972-542-7520
www.sigmetrix.com

Six Sigma Academy
8876 Pinnacle Peak Road, Ste. 100
Scottsdale, AZ 85255
Phone: 800-726-2030
Fax: 480-515-9507
www.6-sigma.com

Six Sigma Qualtec
1295 W. Washington St., Ste. 215
Tempe, AZ 85281
Phone: 800-247-9871
Fax: 480-586-2599
www.sixsigmaqualtec.com

Skymark Corp.
7300 Penn Ave.
Pittsburgh, PA 15208
Phone: 800-826-7284
Fax: 412-371-0681
www.skymark.com

Stat-Ease Inc.
2021 E. Hennepin Ave., Ste. 480
Minneapolis, MN 55413
Phone: 612-378-9449
Fax: 612-378-2152
www.statease.com

Statgraphics, Manugistics Inc.
2115 E. Jefferson St.
Rockville, MD 20852
Phone: 800-592-0050
Fax: 301-255-8406
www.statgraphics.com

StatPoint LLC
P.O. Box 1124
Englewood Cliffs, NJ 07632
Phone: 800-232-7828
Fax: 201-585-8589
www.statpoint.com

StatSoft Inc.
2300 E. 14th St.
Tulsa, OK 74104
Phone: 918-749-1119
Fax: 918-749-2217
www.statsoft.com

Technicomp
1111 Chester Ave.
Cleveland, OH 44114
Phone: 800-735-4440
Fax: 216-687-1168
www.technicomp.com

SIMULATION SOFTWARE VENDORS

Aptech Systems, Inc.
23804 SE Kent-Kangley Rd.
Maple Valley, WA 98038
Phone: 425-432-7855
Fax: 425-432-7832
www.aptech.com

Averill M. Law & Associates
P.O. Box 40996
Tucson, AZ 85717
Phone: 520-795-6265
Fax: 520-795-6302
www.averill-law.com

Brooks Automation
5245 Yeager Rd.
Salt Lake City, UT 84116
Phone: 801-736-3201
Fax: 801-736-3443
www.automod.com

CACI Products Company
1011 Camino Del Rio South, Ste.
  230
San Diego, CA 92108
Phone: 619-542-5228
Fax: 619-692-1013
www.simprocess.com

CMS Research Inc.
1610 S. Main Street
Oshkosh, WI 54902
Phone: 920-235-3356
Fax: 920-235-3816
www.cmsres.com

CreateASoft, Inc.
1212 S. Naper Blvd., Ste 119
Naperville, IL 60540
Phone: 630-428-2850
Fax: 630-357-2590
www.createasoft.com

Custom Simulations
1178 Laurel Street
Berkeley, CA 94708
Phone: 510-898-0692
Fax: 510-898-0692
www.customsimulations.com

Decisioneering, Inc.
1515 Arapahoe St., Ste. 1311
Denver, CO 80202
Phone: 800-289-2550 or 303-534-1515
Fax: 303-534-1515
www.crystalball.com

Flexsim Software Products, Inc.
University Office Park
1366 S. 740 E.
Orem, UT 84097
Phone: 801-224-6914
Fax: 801-224-6984
www.flexsim.com

FLUX Software Engineering
Anders Carlssons Gata 14
SE-417 55 Göteborg, Sweden
Phone: +46 31 230 727
Fax: +46 31 230 728
www.webgpss.com

Global Strategy Dynamics Ltd.
P.O. Box 314
Princes Risborough
Bucks, HP27 0JS UK
Phone: +44 1844 275518
Fax: +44 1844 275507
www.strategydynamics.com

GoldSim Technology Group
18300 NE Union Hill Rd., Ste. 200
Redmond, WA 98052
Phone: 425-883-0777
Fax: 425-882-5498
www.goldsim.com

Highpoint Software Systems, LLC
S42 W27451 Oak Grove Lane
Phone: 262-893-5400
Fax: 262-650-0781
www.highpointsoftware.com

Imagine That, Inc.
6830 Via Del Oro, Ste. 230
San Jose, CA 95119
Phone: 408-365-0305
Fax: 408-629-1251
www.imaginethatinc.com

Incontrol Enterprise Dynamics
Planetenbaan 21
3606 AK Maarssen, The
   Netherlands
Phone: +31-346-552500 or
   313-441-4460 Ext. 1131
Fax: +31-346-552451 or 313-441-6098
www.enterprisedynamics.com

Interfacing Technologies
3333 Côte Vertu., Ste. 400
Saint-Laurent, Quebec, Canada H4R
   2N1
Phone: 514-737-7333
Fax: 514-737-0856
www.interfacing.com

Lumina Decision Systems, Inc.
26010 Highland Way
Los Gatos, CA 95033
Phone: 650-212-1212
Fax: 650-240-2230
www.lumina.com

Micro Analysis & Design, Inc.
4949 Pearl East Circle, Ste. 300
Boulder, CO 80301
Phone: 303-442-6947
Fax: 303-442-8274
www.maad.com

Minuteman Software
P.O. Box 131
Holly Springs, NC 27540
www.minutemansoftware.com

MJC2 Limited
33 Wellington Business Park
Crowthorne, RG45 6LS UK
Phone: +44 1344 760000
Fax: +44 1344 760017
www.mjc2.com

Numerical Algorithms Group
1431 Opus Place, Ste. 220
Downers Grove, IL 60515
Phone: 630-971-2337
Fax: 630-971-2706
www.nag.com

Orca Computer, Inc.
Virginia Tech Corporate Research
  Center
1800 Kraft Drive, Ste. 111
Blacksburg, VA 24060
Phone: 540-961-6722
Fax: 540-961-4162
www.OrcaComputer.com

ProcessModel, Inc.
32 West Center, Ste. 301
Provo, UT 84601
Phone: 801-356-7165
Fax: 801-356-7175
www.processmodel.com

Proforma Corporation
26261 Evergreen Rd., Ste. 200
Southfield, MI 48076
Phone: 248-356-9775
Fax: 248-356-9025
www.Proformacorp.com

ProModel Solutions
3400 Bath Pike, Ste. 200
Bethlehem, PA 18017
Phone: 888-900-3090
Fax: 610-867-8240
www.promodel.com

Proplanner
2321 North Loop Drive
Ames, IA 50010
Phone: 515-296-9914
Fax: 515-296-3229
www.proplanner.com

ProtoDesign, Inc.
2 Sandalwood Court
Bolingbrook, IL 60440
Phone: 630-759-9930
Fax: 630-759-4759
www.protodesign-inc.com

Rockwell Software
504 Beaver Street
Sewickley, PA 15143
Phone: 412-741-3727
Fax: 412-741-5635
www.arenasimulation.com

SAS Institute Inc.
SAS Campus Drive
Cary, NC 27513
Phone: 919-677-8000
Fax: 919-677-4444
www.sas.com

SIMUL8 Corporation
2214 Rock Hill Road, Ste. 501
Herndon, VA 20170
Phone: 800-547-6024
Fax: 800-547-6389
www.SIMUL8.com

Simulation Dynamics
416 High Street
Maryville, TN 37804
Phone: 865-982-7046
Fax: 865-982-2813
www.simulationdynamics.com

Stanislaw Raczynski
P.O. Box 22-783
14000 Mexico City, Mexico
Phone: +55 56 55 44 67
Fax: +55 56 55 44 67
www.raczynski.com/pn/pn.htm

Tecnomatix Technologies Inc.
21500 Haggerty Rd., Ste. 300
Northville, MI 48167
Phone: 248-699-2500
Fax: 248-699-2595
www.tecnomatix.com

User Solutions, Inc.
11009 Tillson Drive
South Lyon, MI 48178
Phone: 248-486-1934
Fax: 248-486-6376
www.usersolutions.com

Vanguard Software Corporation
1100 Crescent Green
Cary, NC 27511
Phone: 919-859-4101
Fax: 919-851-9457
www.vanguardsw.com

Visual Solutions
487 Groton Road
Westford, MA 01886
Phone: 978-392-0100
Fax: 978-692-3102
www.vissim.com

Webb Systems Limited
Warrington Business Park, Long
   Lane
Warrington, WA2 8TX UK
Phone: +44 8700 110 823
Fax: +44 8700 110 824
www.showflow.co.uk

Wright Williams & Kelly
2020 SW 4th Ave., 3rd Floor
Portland, OR 97201
Phone: 925-399-6246
Fax: 925-399-6001
www.wwk.com

XJ Technologies
21 Polytechnicheskaya Street
St. Petersburg, 194021 Russia
Phone: +7 812 2471674
Fax: +7 812 2471639
www.xjtek.com

REFERENCES AND
FURTHER READING

Akao, 1997. QFD: past, present, and future, *Proceedings of the 3rd Annual International QFD Conference*, Linköping, Sweden, October 1997, www.qfdi.org/qfd_history.pdf.

Al-Aomar, R. 2000. Model-mix analyses with DES, *Proceedings of the 2000 Winter Simulation Conference*, pp. 1385–1392.

Al-Aomar, R. 2003. A methodology for determining system and process-level manufacturing performance metrics, *SAE Transactions Journal of Materials and Manufacturing*, Vol. 111, No. 5, pp. 1043–1050.

Al-Aomar, R. 2004. Achieving six-sigma rating in a system simulation model, *Proceedings of the 2nd International IE Conference*, IIEC'04, CD-ROM.

Al-Aomar, R. 2006. Incorporating robustness into genetic algorithm search of stochastic simulation outputs, *Simulation Practice and Theory*, Vol. 14, No. 3, pp. 201–223.

Al-Aomar, R., and D. Cook. 1998. Modeling at the machine-control level with DES, *Proceedings of the 1998 Winter Simulation Conference*, pp. 927–933.

Al-Aomar, R., and M. Philips. 2002. Determining FPS measurables in vehicle's final assembly operations with DES, *Ford Technical Journal*, Vol. 5, No. 1.

Altshuler, G. S. 1990. On the theory of solving inventive problems, *Design Methods and Theories*, Vol. 24, No. 2, pp. 1216–1222.

Altshuler, G. S. 1988. *Creativity as Exact Science*, Gordon & Breach, New York.

Antony, J., and M. Kaye. 2000. *Experimental Quality: A Strategic Approach to Achieve and Improve Quality*, Kluwer Academic Publishers, Boston.

Automotive Industry Action Group. 2001. *Potential Failure Mode and Effects Analysis (FMEA) Reference Manual*, 3rd ed., AIAG, Southfield, MI.

Azadivar, F. 1999. Simulation optimization methodologies, *Proceedings of the 1999 Winter Simulation Conference*, pp. 93–100.

Banks, J. (Ed.). 1998. *Handbook of Simulation: Principles, Methodology, Advances, Applications, and Practice*, Wiley, New York.

Banks, J., J. S. Carson II, B. L. Nelson, and D. M. Nicol. 2001. *Discrete Event System Simulation*, 3rd ed. Prentice Hall, Upper Saddle River, NJ.

Benedetto, R. 2003. Adapting manufacturing-based six sigma methodology to the service environment of a radiology film library, *Journal of Healthcare Management*, Vol. 48, No. 4.

Benjamin, P. C., M. Erraguntla, and R. J. Mayer. 1995. Using simulation for robust design, *Simulation*, Vol. 65, No. 2, pp. 116–127.

Box, G. E. P., W. G. Hunter, and J. S. Hunter. 1978. *Statistics for Experimenters*, Wiley, New York.

Breyfogle, F. W. 1999. *Implementing Six Sigma: Smarter Solutions Using Statistical Methods*, Wiley, New York.

Bulent, D. M., O. Kulak, and S. Tufekci. 2002. An implementation methodology for transition from traditional manufacturing to cellular manufacturing using axiomatic design, *Proceedings of the International Conference on Axiomatic Design*, Cambridge, MA.

Burchill, G., and C. H. Brodie. 1997. *Voices into Choices: Acting on the Voice of the Customer*, Joiner Associates, Inc., Madison, WI.

Cohen, L. 1995. *Quality Function Deployment: How to Make QFD Work for You*, Addison-Wesley, Reading, MA.

Ealey, L. A. 1994. *Quality by Design: Taguchi Methods and U.S. Industry*, American Supplier Institute, Dearborn, MI.

El-Haik, B. 1996. Vulnerability reduction techniques in engineering design, Ph.D. dissertation, Wayne State University, Detroit, MI.

El-Haik, B. 2005. *Axiomatic Quality and Reliability*, Wiley, New York.

El-Haik, B., and D. M. Roy. 2005. *Service Design for Six-Sigma: A Roadmap for Excellence*, Wiley, New York.

El-Haik, B., and K. Yang. 2000a. An integer programming formulation for the concept selection problem with an axiomatic perspective, Part I: Crisp formulation, *Proceedings of the First International Conference on Axiomatic Design*, MIT, Cambridge, MA, pp. 56–61.

El-Haik, B., and K. Yang. 2000b. An integer programming formulation for the concept selection problem with an axiomatic perspective, Part II: Fuzzy formulation, *Proceedings of the First International Conference on Axiomatic Design*, MIT, Cambridge, MA, pp. 62–69.

Feld, W. M. 2000. *Lean Manufacturing: Tools, Techniques, and How to Use Them*, CRC Press, Boca Raton, FL.

Ferng, J., and A. D. F. Price. 2005. An exploration of the synergies between six sigma, total quality management, lean construction and sustainable construction, *International Journal of Six Sigma and Competitive Advantage*, Vol. 1, No. 2, pp. 167–187.

Ferrin, D. M., D. Muthler, and M. J. Miller. 2002. Six sigma and simulation: so what's the correlation, *Proceedings of the 2002 Winter Simulation Conference*, pp. 1439–1433.

Fowlkes, W. Y., and C. M. Creveling. 1995. *Engineering Methods for Robust Product Design*, Addison-Wesley, Reading, MA.

General Electric Company. 2002. What is six sigma: the roadmap to customer impact, http://www.ge.com/sixsigma/.

George, M. L. 2002. *Lean Six Sigma: Combining Six Sigma Quality with Lean Speed*, McGraw-Hill, New York.

George, M. L. 2003. *Lean Six Sigma for Service: How to Use Lean Speed and Six Sigma Quality to Improve Services and Transactions*, McGraw-Hill, New York.

George, M. L., D. Rowland, and B. Kastle. 2004. *What Is Lean Six Sigma*, McGraw-Hill, New York.

Goel, P., P. Gupta, and R. Jain. 2005. *Six Sigma for Transactions and Service*, McGraw-Hill, New York.

Goldratt, E. 1990. *Theory of Constraints*, North River Press, New York.

Goldratt, E. 1998. Computerized shop floor scheduling, *International Journal of Production Research*, Vol. 26, pp. 429–442.

Goldratt, E., and J. Cox. 1992. *The Goal: A Process of Ongoing Improvement*, 2nd ed., North River Press, New York.

Groover, M. P. 2002. *Fundamentals of Modern Manufacturing: Materials, Processes, and Systems*, 2nd ed., Wiley, New York.

Harry, M. J., 1994. *The Vision of 6-Sigma: A Roadmap for Breakthrough*, Sigma Publishing Company, Phoenix, AZ.

Harry, M. J. 1998. Six sigma: a breakthrough strategy for profitability, *Quality Progress*, May, pp. 60–64.

Harry, M., and R. Schroeder. 2000. *Six Sigma: The Breakthrough Management Strategy Revolutionizing the World's Top Corporations*, Doubleday, New York.

Hauser, J. R., and D. Clausing. 1988. The house of quality, *Harvard Business Review*, Vol. 66, No. 3, pp. 63–73.

Heizer, J., and B. Render. 2001. *Principles of Operations Management*, 4th ed., Prentice Hall, Upper Saddle River, NJ.

iSixSigma LLC. 2004. http://www.isixsigma.com/.

Kelton, D. 1988. Designing computer simulation experiments, *Proceedings of the 1988 Winter Simulation Conference*, pp. 15–18.

Kacker, R. N. 1986. Taguchi's quality philosophy: analysis and comment, *Quality Progress*, Vol. 19, No. 12, pp. 21–29.

Kapur, K. C. 1988. An approach for the development of specifications for quality improvement, *Quality Engineering*, Vol. 1, No. 1, pp. 63–77.

Kapur, K. C. 1991. Quality engineering and tolerance design, *Concurrent Engineering: Automation, Tools and Techniques*, pp. 287–306.

Kapur, K. C., and L. R. Lamberson. 1977. *Reliability in Engineering Design*, Wiley, New York.

Krajewski, L. J., and L. P. Ritzman. 2002. *Operations Management: Strategy and Analysis*, 6th ed., Prentice Hall, Upper Saddle River, NJ.

Law, A., and D. Kelton. 1991. *Simulation Modeling and Analysis*, 3rd ed., McGraw-Hill, New York.

Law, A., and M. G. McComas. 2001. How to build valid and credible simulation models, *Proceedings of the 2001 Winter Simulation Conference*, pp. 22–28.

McClave, J., P. G. Benson, and T. Sincich. 1998. *Statistics for Business and Economics*, 7th ed., Prentice Hall, Upper Saddle River, NJ.

Moeeni, F., S. M. Sanchez, and A. J. Vakharia. 1997. A robust design methodology for Kanban system design, *International Journal of Production Research*, Vol. 35, No. 10, pp. 2821–2838.

Montgomery, D. C. 2003. *Design and Analysis of Experiments*, 5th ed., Wiley, New York.

Montgomery, D. C., and G. C. Runger. 2003. *Applied Statistics and Probability for Engineers*, 3rd ed., Wiley, New York.

Motorola Company. 2002. Motorola six sigma services, http://mu.motorola.com/sigmasplash.htm/.

Nair, V. N. 1992. Taguchi's parameter design: a panel discussion, *Technometrics*, Vol. 34, No. 2, pp. 127–161.

Nakui, S. 1991. Comprehensive QFD system, *Transactions of the 3rd International Symposium on QFD*, Novi, MI, pp. 137–152.

Niebel, B., and A. Freivalds. 2003. *Methods, Standards, and Work Design*, 7th ed., McGraw-Hill, New York.

Nikoukaran, J., and R. J. Paul. 1999. Software selection for simulation in manufacturing: a review, *Simulation Practice and Theory*, Vol. 7, pp. 1–14.

Nordlund, M. 1996. An information framework for engineering design based on axiomatic design, doctoral dissertation, Royal Institute of Technology (KTH), Department of Manufacturing Systems, Stockholm, Sweden.

Nordlund, M., D. Tate, and N. P. Suh. 1996. Growth of axiomatic design through industrial practice, *Proceedings of the 3rd CIRP Workshop on Design and Implementation of Intelligent Manufacturing Systems*, Tokyo, June 19–21, pp. 77–84.

Ohno, T. 1998. *Toyota Production System: Beyond Large-Scale Production*, Productivity Press, New York.

Phadke, S. M. 1989. *Quality Engineering Using Robust Design*, Prentice Hall, Englewood Cliffs, NJ.

Pugh, S. 1991. *Total Design: Integrated Methods for Successful Product Engineering*, Addison-Wesley, Reading, MA.

Pugh, S. 1996. In *Creating Innovative Products Using Total Design*, Clausing and Andrade (Eds.), Addison-Wesley, Reading, MA.

Ross, P. J. 1996. *Taguchi Techniques for Quality Engineering*, 2nd ed., McGraw-Hill, New York.

Sanchez, S. M., P. J., Sanchez, J. S. Ramberg, and F. Moeeni. 1996. Effective engineering design through simulation, *International Transactions in Operational Research*, Vol. 3, No. 2, pp. 169–185.

Sharad, K. M., V. Misra, and N. K. Mehta. 1991. A new approach to process parameter selection: integration of quality, *Computers and Industrial Engineering*, Vol. 21, No. 1, pp. 57–61.

Smith, J. S. 2003. Survey on the use of simulation for manufacturing system design and operation, *Journal of Manufacturing Systems*, Vol. 22, No. 2, pp. 157–171.

Snee, R. D. 2004. Six-sigma: the evolution of 100 years of business improvement methodology, *International Journal of Six Sigma and Competitive Advantage*, Vol. 1, No. 1, pp. 4–20.

Snee, R. D., and R. W. Hoerl. 2005. *Six Sigma Beyond the Factory Floor*, Prentice Hall, Upper Saddle River, NJ.

Suh, N. P. 1984. Development of the science base for the manufacturing field through the axiomatic approach, *Robotics and Computer Integrated Manufacturing*, Vol. 1, No. 3–4.

Suh, N. P. 1990. *The Principles of Design*, Oxford University Press, New York.

Suh, N. P. 1995. Design and operation of large systems, *Journal of Manufacturing Systems*, Vol. 14, No. 3.

Suh, N. P. 1996. Impact of axiomatic design, *Proceedings of the 3rd CIRPWorkshop on Design and the Implementation of Intelligent Manufacturing Systems*, Tokyo, June 19–22, pp. 8–17.

Suh, N. P. 1997. Design of systems, *Annals of CIRP*, Vol. 46, No. 1, pp. 75–80.

Suh, N. P. 2001. *Axiomatic Design: Advances and Applications*, Oxford University Press, New York.

Swisher, J., P. Hyden, S. Jacobson, and L. Schruben. 2000. A survey of simulation optimization techniques and procedures, *Proceedings of the 2000 Winter Simulation Conference*, pp. 119–128.

Taguchi, G. 1986. *Introduction to quality engineering: designing quality into products and processes, Quality Resources*.

Taguchi, G., and Y. Wu. 1980. *Introduction to Off-line Quality Control*, Central Japan Quality Association, Nagoya, Japan.

Tsai, C. S. 2002. Evaluation and optimization of integrated manufacturing system operations using Taguchi's experimental design in computer simulation, *Computers and Industrial Engineering*, Vol. 43, No. 3, pp. 591–604.

Tsui, K. L. 1992. An overview of Taguchi method and newly developed statistical methods for robust design, *IIE Transactions*, Vol. 24, No. 5, pp. 44–57.

Wild, R. H., and J. J. Pignatiello. 1991. An experimental design strategy for designing robust systems using discrete-event simulation, *Simulation*, Vol. 57, No. 6, pp. 358–368.

Womack, J. P., and D. T. Jones. 1996. *Lean Thinking*, Simon & Schuster, New York.

Womack, J. P., D. T. Jones, and D. Roos. 1991. *The Machine That Changed the World: The Story of Lean Production*, HarperCollin, New York.

Yang, K., and B. El-Haik. 2003. *Design for Six Sigma: A Roadmap for Product Development*, McGraw-Hill, New York.

Zairi, M., and M. A. Youssef. 1995. Quality function deployment: a main pillar for successful total quality management and product development, *International Journal of Quality and Reliability Management*, Vol. 12, No. 6, pp. 9–23.

Zeigler, B. P., K. Praehofer, and T. G. Kim. 2000. *Theory of Modeling and Simulation*, 2nd ed., Academic Press, San Diego, CA.

INDEX

Simulation-Based Lean Six-Sigma and Design for Six-Sigma, by Basem El-Haik and Raid Al-Aomar
Copyright © 2006 John Wiley & Sons, Inc.